T0176698

Robust Correlation

Robust Correlation

Theory and Applications

Georgy L. Shevlyakov

Peter the Great Saint-Petersburg Polytechnic University, Russia

Hannu Oja

University of Turku, Finland

Library of Congress Cataloging-in-Publication Data

Names: Shevlyakov, Georgy L. | Oja, Hannu.
Title: Robust correlation : theory and applications / Georgy L. Shevlyakov,
  Peter The Great Saint-Petersburg Polytechnic University, Russia, Hannu
  Oja, University Of Turku, Finland.
Description: Chichester, West Sussex : John Wiley & Sons, Inc., 2016. |
  Series: Wiley series in probability and statistics | Includes
  bibliographical references and index.
Identifiers: LCCN 2016017308 (print) | LCCN 2016020693 (ebook) | ISBN
  9781118493458 (cloth) | ISBN 9781119264538 (pdf) | ISBN 9781119264491
  (epub)
Subjects: LCSH: Correlation (Statistics) | Mathematical statistics.
Classification: LCC QA278.2 .S4975 2016 (print) | LCC QA278.2 (ebook) | DDC
  519.5/37–dc23
LC record available at https://lccn.loc.gov/2016017308

A catalogue record for this book is available from the British Library.

Set in 10/12pt, Times by SPi Global, Chennai, India.

To our families

Contents

Preface

Robust statistics as a branch of mathematical statistics appeared due to the seminal works of John W. Tukey (1960), Peter J. Huber (1964), and Frank R. Hampel (1968). It has been intensively developed since the sixties of the last century and is definitely formed by the present. The term "robust" (Latin: strong, sturdy, tough, vigorous) as applied to statistical procedures was proposed by George E.P. Box (1953).

The principal reason for research in this field of statistics is of a general mathematical nature. Optimality (accuracy) and stability (reliability) are the mutually complementary characteristics of many mathematical procedures. It is well-known that the performance of optimal procedures is, as a rule, rather sensitive to "small" perturbations of prior assumptions. In mathematical statistics, the classical example of such an unstable optimal procedure is given by the least squares method: its performance may become disastrously poor under small deviations from normality.

Roughly speaking, robustness means stability of statistical inference under the departures from the accepted distribution models. Since the term "stability" is generally overloaded in mathematics, the term "robustness" may be regarded as its synonym.

Peter J. Huber and Frank R. Hampel contributed much to robust statistics: they proposed and developed two principal approaches to robustness, namely, the minimax approach and the approach based on influence functions, which were applied to almost all areas of statistics: robust estimation of location, scale, regression, and multivariate model parameters, as well as to robust hypothesis testing. It is remarkable that although robust statistics involves mathematically highly refined asymptotic tools, nevertheless robust methods show a satisfactory performance in small samples, being quite useful in applications.

The main topic of our book is robust correlation. Correlation analysis is widely used in multivariate statistics and data analysis: computing correlation and covariance matrices is both an initial and a basic step in most procedures of multivariate statistics, for example, in principal component analysis, factor and discriminant analysis, detection of multivariate outliers, etc.

Our work represents new results generally related to robust correlation and data analysis technologies, with definite accents both on theoretical aspects and practical needs of data processing: we have written the book to be accessible both to the users of statistical methods as well as to professional statisticians. However, the mathematical background requires the basics of calculus, linear algebra, and mathematical statistics.

Chapter 1 is an introduction into the book, providing historical aspects of the origin and development of the notion "correlation" in science as well as ontological remarks on the subject of statistics and data processing. Chapter 2 delivers a survey of the classical measures of correlation aimed most at estimating linear dependencies.

Chapter 3 represents Huber's and Hampel's principal approaches to robustness in mathematical statistics, with novel additions to them, namely, a stable estimation approach and an essay on robustness versus Gaussianity, the latter of which could be helpful for students and their teachers. Except for a few paragraphs on the application of Huber's minimax approach to distribution classes of a non-neighborhood nature, Chapters 1 to 3 are accessible to a wide reader audience.

Chapters 4 to 8 comprise the core of the book, which contains most of the new theoretical and experimental (Monte Carlo) results. Chapter 4 treats the problems of robust estimation of a scale parameter, and the obtained results are used in Chapter 5 for the design of highly robust and efficient estimates of a correlation coefficient including robust minimax (in the Huber sense) estimates. Chapter 6 provides an overview of classical multivariate correlation measures and inference tools based on the covariance and correlation matrix. Chapter 7 deals with robust correlation measures and inference tools that are based on various robust covariance matrix functionals and estimates; in particular, robust versions of principal component and canonical correlation analysis are given. Chapter 8 comprises correlation measures and inference tools based on various concepts of univariate and multivariate signs and ranks.

Chapters 9 to 11 are devoted to the applications of the aforementioned robust estimates of correlation, as well as of location and scale, to different problems of statistical data and signal analysis, with a few examples of real-life data and signal processing. Chapter 9 is confined to the applications to exploratory data analysis and its technologies, mostly treating an important problem of detection of outliers in the data. Chapter 10 outlines a few novel approaches to robust estimation of time series power spectra: although the obtained results are preliminary, they are profitable, deserving a further thorough study. In Chapter 11, various problems of robust signal detection are posed and treated, in the solution of which the Huber's minimax and stable approaches to robust detection are successfully exploited.

Chapter 12 outlines several open problems in robust multivariate analysis and its applications.

From the aforementioned it follows that there are two main blocks of the book: Chapters 1 to 3 and 9 to 11 aim at the applied statistician and statistics user audience, while Chapters 4 to 8 focus on the theoretical aspects of robust correlation.

Most of the contents of the book, namely Chapters 1 to 5 and 9 to 11, have been written by the first author. The second author contributed Chapters 6 to 8 on general multivariate analysis.

Acknowledgements

John W. Tukey, Peter J. Huber, Frank R. Hampel, Elvezio M. Ronchetti, and Peter J. Rousseeuw have essentially influenced, directly or indirectly, our views on robustness and data analysis.

The first author is deeply grateful to his teachers and colleagues for their helpful and constructive discussions, namely, to Igor B. Chelpanov, Peter Filzmoser, Eugene P. Gilbo, Jana Jureckova, Abram M. Kagan, Vladimir Ya. Katkovnik, Yuriy S. Kharin, Kiseon Kim, Lev B. Klebanov, Stephan Morgenthaler, Yakov Yu. Nikitin, Boris T. Polyak, Alexander M. Shurygin, and Nikita O. Vilchevski.

Some results presented in Chapters 4, 5, and 9 to 11 by the first author are based on the Ph.D. and M.Sc. dissertations of his former students, including Kliton Andrea, JinTae Park, Pavel Smirnov, Galina Lavrentyeva, Nickolay Lyubomishchenko, and Nikita Vassilevskiy—we would like to thank them.

Research on multivariate analysis reported by the second author is to some degree based on the thesis works of his several ex-students, including Jyrki Möttönen, Samuli Visuri, Esa Ollila, Sara Taskinen, Seija Sirkiä, and Klaus Nordhausen. We wish to thank them all. The second author is naturally also indebted to many colleagues and coauthors for valuable discussions and express his sincere thanks for discussions and cooperation in this specific research area with Christopher Croux, Tom Hettmansperger, Annaliisa Kankainen, Visa Koivunen, Ron Randles, Bob Serfling, and Dave Tyler.

We are also grateful to Igor Bezdvornyh and Maksim Sovetnikov for their technical help in the preparation of the manuscript.

Finally, we wish to thank our wives, Elena and Ritva, for their patience, support, and understanding.

About the companion website

Don't forget to visit the companion website for this book:

www.wiley.com/go/Shevlyakov/Robust

 There you will find valuable material designed to enhance your learning, including:

- Datasets

- R codes

Scan this QR code to visit the companion website.

1

Introduction

This book is most about correlation, association and partially about regression, i.e., about those areas of science where the dependencies between random variables that mathematically describe the relations between observed phenomena and associated with them features are studied. Evidently, these concepts and terms firstly appeared in applied sciences, not in mathematics. Below we briefly overview the historical aspects of the considered concepts.

1.1 Historical Remarks

The word *"correlation"* is of late Latin origin meaning *"association"*, *"connection"*, *"correspondence"*, *"interdependence"*, *"relationship"*, but relationship not in the conventional for that time deterministic functional form.

The term *"correlation"* was introduced into science by a French naturalist Georges Cuvier (1769–1832), one of the major figures in natural sciences in the early 19th century, who had founded paleontology and comparative anatomy. Cuvier discovered and studied the relationships between the parts of animals, between the structure of animals and their mode of existence, between the species of animals and plants, and many others. This experience made him establish the general principles of *"the correlation of parts"* and of *"the functional correlation"* (Rudwick 1997):

Today comparative anatomy has reached such a point of perfection that, after inspecting a single bone, one can often determine the class, and sometimes even the genus of the animal to which it belonged, above all if that bone belonged to the head or the limbs. ... This is because the number, direction, and shape of the bones that compose each part of an animal's body are always in a necessary relation to all the

Robust Correlation: Theory and Applications, First Edition. Georgy L. Shevlyakov and Hannu Oja.
© 2016 John Wiley & Sons, Ltd. Published 2016 by John Wiley & Sons, Ltd.
Companion Website: www.wiley.com/go/Shevlyakov/Robust

other parts, in such a way that – up to a point – one can infer the whole from any one of them and vice versa.

From Cuvier to Galton, correlation had been understood as a qualitatively described relationship, not deterministic but of a statistical nature, however observed at that time within a rather narrow area of phenomena.

The notion of *regression* is connected with the great names of Laplace, Legendre, Gauss, and Galton (1885), who coined this term. Laplace (1799) was the first to propose a method for processing the astronomical data, namely, the least absolute values method. Legendre (1805) and Gauss (1809) independently of each other introduced the least squares method.

Francis Galton (1822–1911), a British anthropologist, biologist, psychologist, and meteorologist, understood that correlation is the interrelationship in average between any random variables (Galton 1888):

Two variable organs are said to be co-related when the variation of the one is accompanied on the average by more or less variation of the other, and in the same direction. ... It is easy to see that co-relation must be the consequence of the variations of the two organs being partly due to common cause. ... If they were in no respect due to common causes, the co-relation would be nil.

Correlation analysis (this term also was coined by Galton) deals with estimation of the value of correlation by number indexes or coefficients.

Similarly to Cuvier, Galton introduced regression dependence observing live nature, in particular, processing the heredity and sweet peas data (Galton 1894). Regression characterizes the correlation dependence between random variables functionally in average. Studying the sizes of sweet peas beans, he noticed that the offspring seeds did not reveal the tendency to reproduce the size of their parents being closer to the population mean than them. Namely, the seeds were smaller than their parents in the case of large parent sizes, and vice versa. Galton called this dependence regression, for the reverse changes had been observed: firstly, he used the term *"the law of reversion"*. Further studies showed that on average the offspring regression to the population mean was proportional to the parent deviations from it – this allowed the observed dependence to be described using the linear function. The similar linear regression is described by Galton as a result of processing the heights of 930 adult children and their 205 parents (Galton 1894).

The term *"regression"* became popular, and now it is used in the case of functional dependencies in average between any random variables. Using modern terminology, we may say that Galton considered the slope r of the simple linear regression line as a measure of correlation (Galton 1888):

Let y = the deviation of the subject [in units of the probably error, Q], whichever of the two variables may be taken in that capacity; and let x_1, x_2, x_3, & c., be the corresponding deviations of the relative, and let the mean of these be X. Then we find: (1) that y = r X for all values of y; (2) that r is the same, whichever of the two variables is taken for the subject; (3) that r is always less than 1; (4) that r measures the closeness of co-relation.

Now we briefly comment on the above-mentioned properties (1)–(4): the first is just the simple linear regression equation between the standardized variables X and y; the second means that *the co-relation r* is symmetric with regard to the variables X and y; the third and fourth show that Galton had not yet recognized the idea of negative correlation: stating that r could not be greater than 1, he evidently understood r as a positive measure of *"co-relation"*. Originally r stood for the regression slope, and that is really so for the standardized variables; Galton perceived the correlation coefficient as a scale invariant regression slope.

Galton contributed much to science studying the problems of heredity of qualitative and quantitative features. They were numerically examined by Galton on the basis of the concept of correlation. Until the present, the data on demography, heredity, and sociology collected by Galton with the corresponding numerical examples of correlations computed are used.

Karl Pearson (1857–1936), a British mathematician, statistician, biologist, and philosopher, had written out the explicit formulas for the population product-moment correlation coefficient (Pearson 1895)

$$\rho = \rho(X, Y) = \frac{cov(X, Y)}{[var(X) \, var(Y)]^{1/2}} \tag{1.1}$$

and its sample version

$$r = \frac{\sum_{i=1}^{n}(x_i - \bar{x})(y_i - \bar{y})}{\left[\sum_{i=1}^{n}(x_i - \bar{x})^2 \sum_{i=1}^{n}(y_i - \bar{y})^2\right]^{1/2}} \tag{1.2}$$

(here $\bar{x}$ and $\bar{y}$ are the sample means of the observations $\{x_i\}$ and $\{y_i\}$ of random variables X and Y). However, Pearson did not definitely distinguish the population and sample versions of the correlation coefficient, as it is commonly done at present.

Thus, on the one hand, the sample correlation coefficient r is a statistical counterpart of the correlation coefficient ρ of a bivariate distribution, where $var(X)$, $var(Y)$, and $cov(X, Y)$ are the variances and the covariance of the random variables X and Y, respectively.

On the other hand, it is an efficient maximum likelihood estimate of the correlation coefficient ρ of the bivariate normal distribution (Kendall and Stuart 1963) with density

$$N(x, y; \mu_X, \mu_Y, \sigma_X, \sigma_Y, \rho) = \frac{1}{2\pi\sigma_X\sigma_Y\sqrt{1-\rho^2}} \exp\left\{-\frac{1}{2(1-\rho^2)}\right.$$

$$\left. \times \left[\frac{(x-\mu_X)^2}{\sigma_X^2} - 2\rho\frac{(x-\mu_X)(y-\mu_Y)}{\sigma_X\sigma_Y} + \frac{(y-\mu_Y)^2}{\sigma_Y^2}\right]\right\}, \tag{1.3}$$

where $\mu_X = E(X)$, $\mu_Y = E(Y)$, $\sigma_X^2 = var(X)$, $\sigma_Y^2 = var(Y)$.

Galton (1888) derived the bivariate normal distribution (1.3), and he was the first who used it to scatter the frequencies of children's stature and parents' stature. Pearson noted that "in 1888 Galton had completed the theory of bivariate normal correlation" (Pearson 1920).

Like Galton, Auguste Bravais (1846), a French naval officer and astronomer, came very near to the definition (1.1) when he called one parameter of the bivariate normal distribution "une correlation", but he did not recognize it as a measure of the interrelationship between variables. However, "his work in Pearson's hands proved useful in framing formal approaches in those areas" (Stigler 1986).

Pearson's formulas (1.1) and (1.2) proved to be fruitful for studying dependencies: correlation analysis and most of multivariate statistical analysis tools are based on the pair-wise Pearson correlations; we may also add the correlation and spectral theories of stochastic processes, etc.

Since the time Pearson introduced the sample correlation coefficient (1.2), many other measures of correlation have been used aiming at estimation of the closeness of interrelationship (the coefficients of association, determination, contingency, etc.). Some of them were proposed by Karl Pearson (1920).

It would not be out of place to note the contributions to correlation analysis of the other British statisticians.

Ronald Fisher (1890–1962) is one of the creators of mathematical statistics. In particular, he is the originator of the analysis of variance and together with Karl Pearson he stands at the beginning of the theory of hypothesis testing. He introduced the notion of a sufficient statistic and proposed the maximum likelihood method (Fisher 1922). Fisher also payed much attention to correlation analysis: his tools for verifying the significance of correlation under the normal law are used until now.

George Yule (1871–1951) is a prominent statistician of the first half of the 20th century. He contributed much to the statistical theories of regression, correlation (Yule's coefficient of contingency between random events), and spectral analysis.

Maurice Kendall (1907–1983) is one of the creators of nonparametric statistics, in particular, of the nonparametric correlation analysis (the Kendall τ-rank correlation) (Kendall 1938). It is noteworthy that he is the coauthor of the classical course in mathematical statistics (Kendall and Stuart 1962, 1963, 1968).

In what follows, we represent their contributions to correlation analysis in more detail.

1.2   Ontological Remarks

Our personal research experience in applied statistics and real-life data analysis is relatively broad and long. It is concerned with the problems of data processing in medicine (cardiology and ophthalmology), biology (genetics), economics and finances (financial mathematics), industry (mechanical engineering, energetics, and material science), and analysis of semantic data and informatics (information retrieval from big data). Besides and due to those problems, we have been working in theoretical statistics, most in robust and nonparametric statistics, as well as in multivariate statistics and time series analysis. Now we briefly outline our vision of the topic of this book to indicate its place in the general context of statistical data analysis with its philosophy and ideological environment.

The reader should only remember that any classification is a convention, such are the forthcoming ones.

1.2.1   Forms of data representation

The customary forms of data representation are as follows (Shevlyakov and Vilchevski 2002, 2011):

- as a sample $\{x_1, \cdots, x_n\}$ of real numbers x_i being the most convenient form to deal with;

- as a sample $\{\mathbf{x}_1, \cdots, \mathbf{x}_n\}$ of real-valued vectors $\mathbf{x}_i = (x_{i1}, \cdots, x_{ip})^T$ of dimension p;

- as an observed realization $x(t)$, $t \in [0, T]$ of a real-valued continuous process (function);

- as a sample of "non-numerical nature" data representing qualitative variables;

- as the semantic type of data (statements, texts, pictures, etc.).

The first three possibilities mostly occur in the natural and technical sciences with the measurement techniques being well developed, clearly defined, and largely standardized. In the social sciences, the last forms are relatively common.

To summarize: in this book we deal mostly with the first three forms and, partially, with the fourth.

1.2.2   Types of data statistics

The experience of treating various statistical problems shows that practically all of them are solved with the use of only a few qualitatively different types of data statistics. Here we do not discuss how to use them in solving statistical problems: only note that their solutions result in computing some of those statistics, and final decision making essentially depends on their values (Mosteller and Tukey 1977; Tukey 1962).

These data statistics may be classified as follows:

- measures of location (central tendency, mean values),

- measures of scale (spread, dispersion, scatter),

- measures of correlation (interdependence, association),

- measures of extreme values,

- measures of a data distribution shape,

- measures of data spectrum.

To summarize: in this book we mainly focus on the measures of correlation, however dealing if needed with the other types of data statistics.

1.2.3   Principal aims of statistical data analysis

These aims can be formulated as follows:

(A1) compact representation of data,

(A2) estimation of model parameters explaining and/or revealing data structure,

(A3) prediction.

A human mind cannot efficiently work with large volumes of information, since there exist natural psychological bounds on the perception ability (Miller 1956). Thus it is necessary to provide a compact data output of information for expert analysis: only in this case we may expect a satisfactory final decision. Note that data processing often begins and ends with the first item *(A1)*.

The next step *(A2)* is to propose an explanatory underlying model for the observed data and phenomena. It may be a regression model, or a distribution model, or any other, desirably a low-complexity one: an essentially multiparametric model is usually a "bad" model; nevertheless, we should recall a cute note of George Box: "*All models are wrong, but some of them are useful*" (Box and Draper 1987). However, parametric models are the first to consider and examine.

Finally, the first two aims are only the steps to the last aim *(A3)*: here we have to state that this aim remains a main challenge to statistics and to science as a whole.

To summarize: in this book we pursue aims (A1) and (A2).

1.2.4   Prior information about data distributions and related approaches to statistical data analysis

The need for stability in statistical inference directly leads to the use of robust statistical methods. It may be roughly stated that, with respect to the level of prior information about underlying data distributions, robust statistical methods occupy the intermediate place between classical parametric and nonparametric methods.

In parametric statistics, the shape of an underlying data distribution is assumed known up to the values of unknown parameters. In nonparametric statistics, it is supposed that the underlying data distribution belongs to some sufficiently "wide" class of distributions (continuous, symmetric, etc.). In robust statistics, at least within Huber's minimax approach (Huber 1964), we also consider distribution classes but with more detailed information about the underlying distribution, say, in the form of a neighborhood of the normal distribution. The latter peculiarity allows the efficiency of robust procedures to be raised as compared with nonparametric methods, simultaneously retaining their high stability.

At present, there exist two main approaches in robustness:

- Huber's minimax approach — quantitative robustness (Huber 1981; Huber and Ronchetti 2009).

- Hampel's approach based on influence functions — qualitative robustness (Hampel 1968; Hampel et al. 1986).

In Chapter 3, we describe these approaches in detail. Now we classify the existing approaches in statistics with respect to the level of prior information about the underlying data distribution $F(x; \theta)$ in the case of point parameter estimation:

- A given data distribution $F(x; \theta)$ with a random parameter θ — the Bayesian statistics (Berger 1985; Bernardo and Smith 1994; Jaynes 2003).

- A given data distribution $F(x; \theta)$ with an unknown parameter θ — the classical parametric statistics (Fisher 1922; Kendall and Stuart 1963).

- A data distribution $F(x; \theta)$ with an unknown parameter θ belongs to a distribution class $\mathcal{F}$, usually a neighborhood of a given distribution, e.g., normal — the robust statistics (Hampel et al. 1986; Huber 1981; Kolmogorov 1931; Tukey 1960).

- A data distribution $F(x; \theta)$ with an unknown parameter θ belongs to some general distribution class $\mathcal{F}$ — the classical nonparametric statistics (Hettmansperger and McKean 1998; Kendall and Stuart 1963; Wasserman 2007).

- A data distribution $F(x; \theta)$ does not exist in the case of unique samples and frequency instability — the probability-free approaches to data analysis: fuzzy (Zadeh 1975), exploratory (Bock and Diday 2000; Tukey 1977), interval probability (Kuznetsov 1991; Walley 1990), logical-algebraic, geometrical (Billard and Diday 2003; Diday 1972).

Note that the upper and lower levels of this hierarchy, namely the Bayesian and the probability-free approaches, are being intensively developed at present.

To summarize: in this book we mainly use Huber's and Hampel's robust approaches to statistical data analysis.

References

Berger JO 1985 *Statistical Decision Theory and Bayesian Analysis,* Springer.

Bernardo JM and Smith AFM 1994 *Bayesian Theory,* Wiley.

Billard L and Diday E 2003 From the statistics of data to the statistics of knowledge: symbolic data analysis. *J. Amer. Statist. Assoc.* **98**, 991–999.

Bock HH and Diday E (eds) 2000 *Analysis of Symbolic Data: Exploratory Methods for Extracting Statistical Information from Complex Data,* Springer.

Box GEP and Draper NR 1987 *Empirical Model-Building and Response Surfaces,* Wiley.

Bravais A 1846 Analyse mathématique sur les probabilités des erreurs de situation d'un point. *Mémoires presents par divers savants l'Académie des Sciences de l'Institut de France. Sciences Mathématiques et Physiques* **9**, 255–332.

Diday E 1972 Nouvelles Méthodes et Nouveaux Concepts en Classification Automatique et Reconnaissance des Formes. These de doctorat d'état, Univ. Paris IX.

Fisher RA 1922 On the mathematical foundations of theoretical statistics. *Philosophical Transactions of the Royal Society,* A **222**, 309–368.

Galton F 1885 Regression towards mediocrity in hereditary stature. *Journal of Anthropological Institute* **15**, 246–263.

Galton F 1888 Co-relations and their measurement, chiefly from anthropometric data. *Proceedings of the Royal Society of London* **45**, 135–145.

Galton F 1894 *Natural Inheritance,* Macmillan, London.

Gauss CF 1809 *Theoria Motus Corporum Celestium, Perthes, Hamburg; English translation: Theory of the Motion of the Heavenly Bodies Moving about the Sun in Conic Sections.* New York: Dover, 1963.

Hampel FR 1968 *Contributions to the Theory of Robust Estimation.* PhD thesis, University of California, Berkeley.

Hampel FR, Ronchetti E, Rousseeuw PJ, and Stahel WA 1986 *Robust Statistics. The Approach Based on Influence Functions,* Wiley.

Hettmansperger TP and McKean JW 1998 *Robust Nonparametric Statistical Methods. Kendall's Library of Statistics,* Edward Arnold, London.

Huber PJ 1964 Robust estimation of a location parameter. *Ann. Math. Statist.* **35**, 73–101.

Huber PJ 1981 *Robust Statistics,* Wiley.

Huber PJ and Ronchetti E (eds) 2009 *Robust Statistics,* 2nd edn, Wiley.

Jaynes AT 2003 *Probability Theory. The Logic of Science,* Cambridge University Press.

Kendall MG 1938 A new measure of rank correlation. *Biometrika* **30**, 81–89.

Kendall MG and Stuart A 1962 *The Advanced Theory of Statistics. Distribution Theory,* vol. 1, Griffin, London.

Kendall MG and Stuart A 1963 *The Advanced Theory of Statistics. Inference and Relationship,* vol. 2, Griffin, London.

Kendall MG and Stuart A 1968 *The Advanced Theory of Statistics. Design and Analysis, and Time Series*, vol. 3, Griffin, London.

Kolmogorov AN 1931 On the method of median in the theory of errors. *Math. Sbornik* **38**, 47–50.

Kuznetsov VP 1991 *Interval Statistical Models*, Radio i Svyaz, Moscow (in Russian).

Legendre AM 1805 *Nouvelles methods pour la determination des orbits des cometes*, Didot, Paris.

Miller GA 1956 The magical number seven, plus or minus two: Some limits on our capacity for processing information. *Psychological Review* **63**, 81–97.

Mosteller F and Tukey JW 1977 *Data Analysis and Regression*, Addison–Wesley.

Pearson K 1895 Contributions to the mathematical theory of evolution. *Philosophical Transactions of the Royal Society of London* A **186**, 343–414.

Pearson K 1920 Notes on the history of correlations. *Biometrika* **13**, 25–45.

Rudwick MJS 1997 *Georges Cuvier, Fossil Bones, and Geological Catastrophes*, University of Chicago Press.

Shevlyakov GL and Vilchevski NO 2002 *Robustness in Data Analysis: criteria and methods*, VSP, Utrecht.

Shevlyakov GL and Vilchevski NO 2011 *Robustness in Data Analysis*, De Gruyter, Boston.

Stigler SM 1986 *The History of Statistics: The Measurement of Uncertainty before 1900*. Belknap Press/Harvard University Press.

Tukey JW 1960 A survey of sampling from contaminated distributions. In *Contributions to Probability and Statistics*. (ed. Olkin I). pp. 448–485. Stanford Univ. Press.

Tukey JW 1962 The future of data analysis. *Ann. Math. Statist.* **33**, 1–67.

Tukey JW 1977 *Exploratory Data Analysis*, Addison–Wesley.

Walley P 1990 *Statistical Reasoning with Imprecise Probabilities*, Chapman & Hall.

Wasserman L 2007 *All of Nonparametric Statistics*, Springer.

Zadeh LA 1975 Fuzzy logic and approximate reasoning. *Synthese* **30**, 407–428.

2

Classical Measures of Correlation

In this chapter we define several conventional measures of correlation, focusing most on Pearson's correlation coefficient and closely related to it constructions, enlist their principal properties and computational peculiarities.

2.1 Preliminaries

Here we comment on the requirements that should be imposed on the measures of correlation to distinguish them from the measures of location and scale (Renyi 1959; Schweizer and Wolff 1981).

Let $corr(X, Y)$ be a measure of correlation between any random variables X and Y. Here we consider both positive–negative correlation

$$-1 \leq corr(X, Y) \leq 1$$

and positive correlation

$$0 \leq corr(X, Y) \leq 1.$$

It is natural to impose the following requirements on $corr(X, Y)$:

(R1) Symmetry: $corr(X, Y) = corr(Y, X)$.

(R2) Invariancy to linear transformations of random variables:

$$corr(aX + b, cY + d) = \text{sgn}(ac)\, corr(X, Y), \quad ac \neq 0.$$

Robust Correlation: Theory and Applications, First Edition. Georgy L. Shevlyakov and Hannu Oja.
© 2016 John Wiley & Sons, Ltd. Published 2016 by John Wiley & Sons, Ltd.
Companion Website: www.wiley.com/go/Shevlyakov/Robust

(R3) Attainability of the limit values 0, +1, and −1:

(a) for independent X and Y, $corr(X, Y) = 0$;

(b) $corr(X, X) = 1$;

(c) $corr(X, aX + b) = \text{sgn}(a)$ for positive-negative correlation.

(R4) Invariancy to strictly monotonic transformations of random variables:

$$corr(g(X), h(Y)) = \pm corr(X, Y)$$

for strictly monotonic functions $g(X)$ and $h(Y)$.

(R5) $corr(g(X), h(X)) = \pm 1$.

Requirement (R1) holds almost for all known measures of correlation, being a natural assumption for correlation analysis as compared to regression analysis when it is not known which variables are dependent and which not.

Requirement (R2) makes a measure of correlation independent of the chosen measures of location and scale, since each of them reflects qualitatively different data characteristics.

Requirements (R3a), (R3b), and (R3c), on the one hand, are merely technical: it is practically and theoretically convenient to deal with a bounded scaleless measure of correlation; on the other hand, they refer to the correspondence of the limit values of $corr(X, Y)$ to the limit cases of association between random variables X and Y: the relation $corr(X, Y) = 0$ may mean independence of X and Y whereas the relation $|corr(X, Y)| = 1$ indicates the functional dependence between X and Y.

The first three requirements hold for almost all known measures of correlation. This is not so with the relation $corr(X, Y) = 0$, which does not guarantee independence of X and Y for several measures of correlation, for example, for Pearson's product-moment, the Spearman rank correlation, and for a few others.

However, this property holds for the maximal correlation coefficient $S(X, Y)$ defined as

$$S(X, Y) = \sup_{g, h} \rho(g(X), h(Y)),$$

where $\rho(X, Y)$ is Pearson's product-moment correlation coefficient (1.1), $g(x)$ and $h(x)$ are Borel-measurable functions such that $\rho(g(X), h(Y))$ has sense (Gebelein 1941). The independence of random variables X and Y also follows from the null value of Sarmanov's correlation coefficient (also called the maximal correlation coefficient) (Sarmanov 1958): in the case of a continuous symmetric bivariate distribution of X and Y, its value is reciprocal to the minimal eigenvalue of some integral operator. Apparently, Gebelein's and Sarmanov's correlation coefficients are rather complicated in their usage.

Recently, in Székely *et al.* (2007), a distance correlation has been proposed: its equality to null implies independence, but like Gebelein's and Sarmanov's correlation coefficients, it is much more complicated in computing than classical measures of correlation.

Requirements (R4) and (R5) refer to the rank measures of correlation, for example, to the Spearman and Kendall τ-rank correlation coefficients.

Now we enlist the well-known seven postulates of Renyi (1959), formulated for a measure of dependence $corr(X, Y)$ defined in the segment [0, 1]:

(P1) $corr(X, Y)$ is defined for any pair of random variables X and Y, neither of them being constant with probability 1.

(P2) $corr(X, Y) = corr(Y, X)$.

(P3) $0 \leq corr(X, Y) \leq 1$.

(P4) $corr(X, Y) = 0$ if and only if X and Y are independent.

(P5) $corr(X, Y) = 1$ if there is a strict dependence between X and Y, that is, either $X = g(Y)$ or $Y = h(X)$, where $g(x)$ and $h(X)$ are Borel-measurable functions.

(P6) If the Borel-measurable functions $g(x)$ and $h(x)$ map the real axis in a one-to-one way on to itself, $corr(g(X), h(Y)) = corr(X, Y)$.

(P7) If the joint distribution of X and Y is normal, then $corr(X, Y) = |\rho(X, Y)|$, where $\rho(X, Y)$ is Pearson's product-moment correlation coefficient.

This set of postulates is more restrictive than the proposed set (R1)–(R5) mostly because of the chosen range [0, 1] and the last postulate (P7) that yield the absolute value of Pearson's correlation $|\rho|$. Later we return to this set when considering informational measures of correlation. Moreover, in what follows, we generally focus on the conventional tools of correlation analysis based on Pearson's correlation coefficient and closely related to it measures, implicitly using Renyi's postulates.

2.2 Pearson's Correlation Coefficient: Definitions and Interpretations

Below we represent a series of different conceptual and computational definitions of the population and sample Pearson's correlation coefficients ρ and r, respectively. Each definition indicates a different way of thinking about this measure within different statistical contexts by using algebraic, geometric, and trigonometric settings (Rogers and Nicewander 1988).

2.2.1   Introductory remarks

The traditional for introductory statistics textbooks definitions (1.1) and (1.2) for Pearson's ρ and r can be evidently rewritten as

$$\rho = \frac{cov(X, Y)}{\sigma_X \, \sigma_Y}$$

and

$$r = \frac{1}{n} \sum_{i=1}^{n} \frac{x_i - \bar{x}}{s_x} \frac{y_i - \bar{y}}{s_y}, \tag{2.1}$$

where s_x and s_y are the mean squared errors.

Equation (2.1) for the sample correlation coefficient r can be regarded as the sample covariance of the standardized random variables, namely $(X - \mu_X)/\sigma_X$ and $(Y - \mu_Y)/\sigma_Y$.

Pearson's correlation coefficient possesses the properties (R1) and (R2) with the bounds $-1 \leq \rho, r \leq 1$: the cases $|\rho| = 1$ and $|r| = 1$ correspond to the linear dependence between variables.

Thus, Pearson's correlation coefficient is a measure of the linear interrelationship between random variables. Furthermore, the relations $\rho = 0$ and $r = 0$ do not induce independence of random variables. The typical shapes of correlated data clusters are exhibited in Figs 2.1 to 2.5.

2.2.2   Correlation via regression

The problem of estimation of the correlation coefficient ρ is directly related to the linear regression problem of fitting the straight line of the conditional expectation

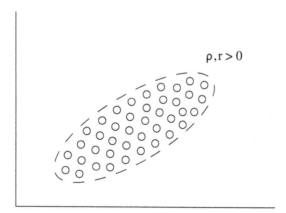

Figure 2.1   Data with positive correlation.

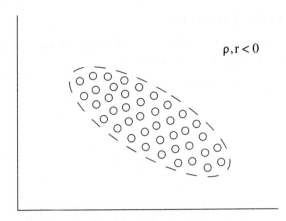

Figure 2.2   Data with negative correlation.

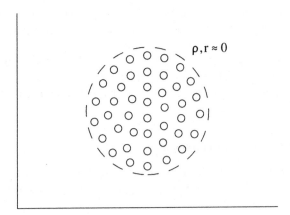

Figure 2.3   Data with approximately zero correlation.

(Kendall and Stuart 1963)

$$E(X|Y = y) = \mu_X + \beta_X(y - \mu_Y),$$
$$E(Y|X = x) = \mu_Y + \beta_Y(x - \mu_X). \tag{2.2}$$

For the bivariate normal distribution, the following relations hold (Kendall and Stuart 1963)

$$\rho = \beta_X \frac{\sigma_Y}{\sigma_X} = \beta_Y \frac{\sigma_X}{\sigma_Y},$$
$$\rho^2 = \beta_X \beta_Y. \tag{2.3}$$

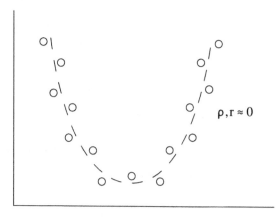

Figure 2.4   Approximately nonlinear dependent data correlation.

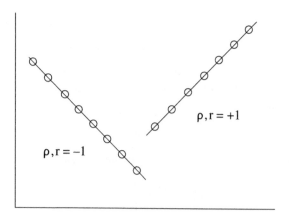

Figure 2.5   Linear dependent data correlation.

If the parameters of linear regression are estimated by the least squares (*LS*) method

$$(\widehat{\alpha}_X, \widehat{\beta}_X) = \arg \min_{\alpha_X, \beta_X} \sum_{i=1}^{n} (x_i - \alpha_X - \beta_X y_i)^2, \tag{2.4}$$

$$(\widehat{\alpha}_Y, \widehat{\beta}_Y) = \arg \min_{\alpha_Y, \beta_Y} \sum_{i=1}^{n} (y_i - \alpha_Y - \beta_Y x_i)^2, \tag{2.5}$$

then the sample correlation coefficient *r* can be expressed in terms of the *LS* estimates of slope $\widehat{\beta}_X$ and $\widehat{\beta}_Y$ as the geometric mean between the two regression slopes

(Kendall and Stuart 1963)

$$r = \widehat{\beta}_X \, \frac{s_y}{s_x} = \widehat{\beta}_Y \, \frac{s_x}{s_y}, \tag{2.6}$$

$$r^2 = \beta_X \beta_Y \quad \text{and} \quad r = \text{sgn}(\widehat{\beta}_Y) \sqrt{\widehat{\beta}_X \widehat{\beta}_Y}, \tag{2.7}$$

where s_x and s_y are the mean squared deviations.

2.2.3   Correlation via the coefficient of determination

Consider the simple linear regression model (2.2) with the response variable Y and the regressor X

$$E(Y|X) - \mu_Y = \beta_Y \, (X - \mu_X). \tag{2.8}$$

Squaring (2.8) and taking the expectation of it with respect to X, we get that

$$var(Y|X) = E_X((E(Y|X) - \mu_Y)^2)$$

$$= \beta_Y^2 \sigma_X^2 = \frac{cov(X, Y)^2}{\sigma_X^2}. \tag{2.9}$$

Define now *the coefficient of determination or the correlation ratio* as the ratio of the variance of the mean values of regression Y-sections and the variance of variable Y (Kendall and Stuart 1963)

$$\eta_Y^2 = \frac{var(Y|X)}{\sigma_Y^2}. \tag{2.10}$$

From (2.8) it follows that if regression is really linear then

$$\eta_Y^2 = \frac{cov(X, Y)^2}{\sigma_X^2 \, \sigma_Y^2} = \rho^2. \tag{2.11}$$

In the case of regression with the response variable X and the regressor Y, the coefficient of determination is defined similarly to (2.10):

$$\eta_X^2 = \frac{var(X|Y)}{\sigma_X^2}.$$

In the case of a simple linear regression, we do not distinguish the coefficients of determination η_X^2 and η_Y^2 as $\eta_X^2 = \eta_Y^2 = \eta^2 = \rho^2$.

The sample version of (2.11) is given by the following relation (Kendall and Stuart 1963):

$$r^2 = \frac{\sum_{i=1}^{n} (\widehat{y}_i - \bar{y})^2}{\sum_{i=1}^{n} (y_i - \bar{y})^2} = \frac{SS_{REG}}{SS_{TOT}}, \tag{2.12}$$

where SS_{REG} is the regression sum of squares (the explained sum of squares) with the expected values

$$\hat{y}_i = \hat{\alpha}_Y + \hat{\beta}_Y\, x_i, \quad i = 1, \ldots, n.$$

Here, $\hat{\alpha}_Y$ and $\hat{\beta}_Y$ are the LS estimates of intercept and slope, respectively, defined by (2.5); SS_{TOT} is the total sum of squares proportional to the sample variance.

Since the total sum of squares SS_{TOT} is partitioned into SS_{REG} and the residual sum of squares $SS_{ERR} = \sum (y_i - \hat{y}_i)^2$:

$$SS_{TOT} = SS_{REG} + SS_{ERR},$$

(2.12) can be rewritten as follows:

$$r^2 = 1 - \frac{SS_{ERR}}{SS_{TOT}}. \tag{2.13}$$

The coefficient of determination $\hat{\eta}^2 = r^2$ is a measure of the goodness of fit of linear regression to the real data: the relation $r^2 = 1$ indicates that the regression line perfectly fits the data points ($SS_{ERR} = 0$). Thus, the representation of correlation via the coefficient of determination directly exploits the interrelationship between linear dependence and correlation.

In the general case of nonlinear regression, the coefficients of determination η_X^2 and η_Y^2 can be used as the measures of functional dependence between random variables X and Y (for details, see Kendall and Stuart 1963).

Below we represent the main results of the interrelationship between the coefficients of correlation and determination (since the presented results hold for both η_X^2 and η_Y^2, we use the notation η^2 for them):

- $0 \leq \rho^2 \leq \eta^2 \leq 1$;

- $\rho^2 = 0$ if X and Y are independent, but the inverse is not true;

- $\rho^2 = \eta^2 = 1$ if and only if there is a linear dependence between X and Y;

- $\rho^2 < \eta^2 = 1$ if and only if there is a nonlinear dependence between X and Y;

- $\rho^2 = \eta^2 < 1$ if and only if there is a linear regression but there is no functional dependence between X and Y;

- $\rho^2 < \eta^2 < 1$ indicates that there is no functional dependence and that some nonlinear regression curve fits the data better than the best straight line.

Figures 2.6 to 2.9 illustrate these cases.

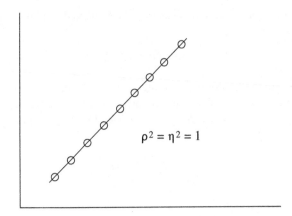

Figure 2.6   Linear dependent data correlation and determination coefficients.

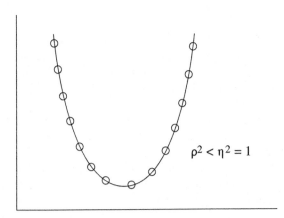

Figure 2.7   Nonlinear dependent data correlation and determination coefficients.

2.2.4   Correlation via the variances of the principal components

Consider the following identity for the correlation coefficient ρ between random variables X and Y (Gnanadesikan and Kettenring 1972)

$$\rho = \frac{var(\widetilde{X} + \widetilde{Y}) - var(\widetilde{X} - \widetilde{Y})}{var(\widetilde{X} + \widetilde{Y}) + var(\widetilde{X} - \widetilde{Y})},   \quad (2.14)$$

where $\widetilde{X}$ and $\widetilde{Y}$ are standardized:

$$\widetilde{X} = \frac{X - \mu_X}{\sigma_X}, \qquad \widetilde{Y} = \frac{Y - \mu_Y}{\sigma_Y}.   \quad (2.15)$$

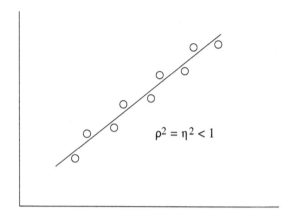

Figure 2.8   Approximately linear dependent data correlation and determination coefficients.

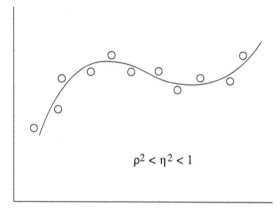

Figure 2.9   Approximately nonlinear dependent data correlation and determination coefficients.

Then, if we substitute the population means, variances and standard deviations in (2.14) and (2.15) by their sample analogs, namely, by the sample means, variances, and mean squared deviations, we get the representation (2.1) for Pearson's correlation

$$r = \frac{s^2_{\tilde{x}+\tilde{y}} - s^2_{\tilde{x}-\tilde{y}}}{s^2_{\tilde{x}+\tilde{y}} + s^2_{\tilde{x}-\tilde{y}}} \tag{2.16}$$

with

$$\tilde{x} = \frac{x - \bar{x}}{s_x}, \quad \tilde{y} = \frac{y - \bar{y}}{s_y}.$$

Further, identity (2.14) will be used for constructing robust minimax (in the Huber sense) estimates of the correlation coefficient.

Finally, we note that for any random pair (X, Y) the linear transformation of their standardized versions

$$U = \widetilde{X} + \widetilde{Y}, \quad V = \widetilde{X} - \widetilde{Y}$$

gives the uncorrelated random principal variables (U, V), actually independent for the bivariate normal distribution. Vice versa, for any two uncorrelated random variables U and V with variances $var(U) = \sigma_U^2$ and $var(V) = \sigma_V^2$, the correlation coefficient ρ of their linear combinations

$$X = U + V, \quad Y = U - V$$

is equal to (Shevlyakov and Vilchevski 2002, 2011)

$$\rho = \frac{var(U) - var(V)}{var(U) + var(V)}. \tag{2.17}$$

REMARK *Identities (2.14) and (2.17) are closely related to the geometric parameters of the bivariate normal ellipses of equal probability (Rogers and Nicewander 1988).*

Consider the standard bivariate normal distribution density

$$N(x, y; 0, 0, 1, 1, \rho) = \frac{1}{2\pi\sqrt{1 - \rho^2}} \exp\left\{-\frac{x^2 - 2\rho xy + y^2}{2(1 - \rho^2)}\right\}. \tag{2.18}$$

Let D and d be the lengths of the major and minor axes of the ellipse of equal probability $x^2 - 2\rho xy + y^2 = C$, where $C > 0$ is an arbitrary positive constant (see Fig. 2.10).

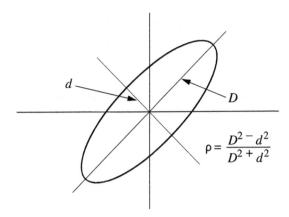

$$\rho = \frac{D^2 - d^2}{D^2 + d^2}$$

Figure 2.10   Ellipse of equal probability for the standard bivariate normal distribution with the major and minor diameters dependent on the correlation coefficient.

Then it can be shown that (Schilling 1984)

$$\rho = \frac{D^2 - d^2}{D^2 + d^2} \tag{2.19}$$

with

$$D \propto \sqrt{1+\rho}, \quad d \propto \sqrt{1-\rho} \quad \text{for} \quad \rho > 0,$$

and

$$D \propto \sqrt{1-\rho}, \quad d \propto \sqrt{1+\rho} \quad \text{for} \quad \rho < 0.$$

2.2.5   Correlation via the cosine of the angle between the variable vectors

This definition directly follows from the elementary analytic geometry formula for the angle α between two vectors $\mathbf{a} = (a_1, \dots, a_n)^T$ and $\mathbf{b} = (b_1, \dots, b_n)^T$ in $\mathbb{R}^n$:

$$r = \cos(\alpha) = \frac{\mathbf{a} \cdot \mathbf{b}}{\|\mathbf{a}\| \, \|\mathbf{b}\|}, \tag{2.20}$$

where the scalar product of the vectors $\mathbf{a}$ and $\mathbf{b}$ stands in the nominator being scaled by their Euclidean norms (see Fig. 2.11). Using the centered observations $\{x_i - \bar{x}\}_1^n$ and $\{y_i - \bar{y}\}_1^n$ as vector components, we immediately arrive at definition (1.2) for Pearson's r.

When the angle is 0 or π, $\cos(\alpha) = \pm 1$, and the vectors lie on the same line. When the angle is $\pi/2$, the vectors are perpendicular, and $\cos(\alpha) = 0$. Thus, collinearity of vectors means maximal correlation, positive or negative; orthogonality means zero

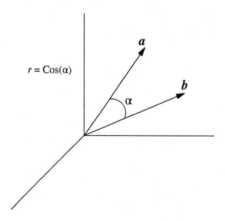

Figure 2.11 *The sample correlation coefficient as the cosine of the angle between the variable vectors.*

correlation – here we observe the concise geometrical interpretation of the values of r. Moreover, visually it is much easier to "see the size" of correlation by observing an angle than by looking at how data points cluster on the plane.

This definition of the correlation coefficient is widely used in information retrieval studies (Manning *et al.* 2008).

2.2.6 Correlation via the ratio of two means

This interpretation of the population correlation coefficient ρ was first proposed by Galton (1885, 1888). Moreover, for Galton it was natural to focus on correlation as a ratio of means, because he was very much interested to understand the relations between different kinds of means, for example, between the average stature of sons of unusually tall fathers as compared to the average stature of their fathers. Numerous problems of this kind arise in genetics.

Consider the following setting (Rogers and Nicewander 1988): let X be a random variable denoting a parent's feature, and let Y denote this feature of the oldest offspring. Next assume that the population means μ_X and μ_Y are 0 and that the standard deviations σ_X and σ_Y are unity.

Now select an arbitrarily large value of X, say x_c, and compute the mean feature for parents with feature values greater than x_c: $E(X|X > x_c)$. Next, average the feature values Y of the oldest offspring of these exceptional parents. Denote this mean by $E(Y|X > x_c)$.

Then under the bivariate normality of standardized X and Y, it can be shown that

$$\rho = \frac{E(Y|X > x_c) - \mu_Y}{E(X|X > x_c) - \mu_X} = \frac{E(Y|X > x_c)}{E(X|X > x_c)}. \tag{2.21}$$

The proof of (2.21) is straightforward: the nominator and denominator of the ratio (2.21) are equal to $\rho \exp(-x_c^2/2)/\sqrt{2\pi}$ and $\exp(-x_c^2/2)/\sqrt{2\pi}$, respectively.

Now we show that the ratio (2.21) can be derived without any assumption of normality, only correlation between X and Y is needed. Define a random variable Y as follows:

$$Y = \rho X + \sqrt{1 - \rho^2} Z,$$

where X and Z are independent random variables with zero means. Then we get

$$var(Y) = \rho^2 var(X) + (1 - \rho^2) var(Z),$$

and if $var(X) = var(Z) = 1$ then $var(Y) = 1$. Further, $E(XY) = \rho$, and

$$E(Y|X > x_c) = \rho E(X|X > x_c) + \sqrt{1 - \rho^2} E(Z|X > x_c)$$

with a zero last term by independence of X and Z. Thus, we arrive at (2.21).

Note that $E(X|X > x_c)$ and $E(Y|X > x_c)$ are the increasing functions of x_c. The interpretation of the ratio (2.21) suggests that explicit selection on one variable also implicitly selects on a second variable in case of correlation between them.

2.2.7   Pearson's correlation coefficient between random events

The correlation coefficient between two random events has a long history (Bernstein 1946; Pearson 1920). We are going to consider this measure in detail, since it is closely related to several measures of correlation.

Consider random events A and B such that $P(A)P(B) > 0$: the dependence between them can be measured by the following conditional probabilities:

$$P(A|B) = \frac{P(AB)}{P(B)} \quad \text{and} \quad P(B|A) = \frac{P(AB)}{P(A)}.$$

However, each of these characteristics is asymmetric with respect to events A and B.

It is desirable to have a symmetric measure of correlation between A and B constructed of the probabilities $P(A)$, $P(B)$, and $P(AB)$ that characterizes dependence or independence and is convenient to use. Pearson's correlation coefficient between two random events is such a measure

$$\rho_{AB} = \frac{P(AB) - P(A)P(B)}{\sqrt{P(A)P(\overline{A})P(B)P(\overline{B})}}. \tag{2.22}$$

It can be easily shown that this measure is the classical Pearson's correlation coefficient ρ defined just between the indicators of events I_A and I_B:

$$E(I_A) = P(A), \quad var(I_A) = P(A)P(\overline{A}),$$

$$E(I_B) = P(B), \quad var(I_B) = P(B)P(\overline{B}).$$

Analogously,

$$cov(I_A, I_B) = E(I_A I_B) - E(I_A)E(I_B) = P(AB) - P(A)P(B).$$

Thus, we get that

$$\rho_{AB} = \frac{cov(I_A, I_B)}{\sqrt{var(I_A)var(I_B)}} = \frac{P(AB) - P(A)P(B)}{\sqrt{P(A)P(\overline{A})P(B)P(\overline{B})}}.$$

By definition, Pearson's correlation between random events inherits some properties of Pearson's correlation between random variables; the essential difference is that the equality $\rho_{AB} = 0$ induces independence of random events. Below we enlist the properties of ρ_{AB}:

- $-1 \leq \rho_{AB} \leq 1$.

- $\rho_{AB} = 0$ if and only if the events A and B are independent. This assertion follows immediately from the necessary and sufficient condition of independence of random events A and B: $P(AB) = P(A)P(B)$.

- $\rho_{AB} = 1$ if and only if $P(A) = P(B)$, i.e., $A = B$ up to the set of measure null (Bernstein 1946). Analogously, $\rho_{AB} = -1$ if and only if $P(A) = P(\overline{B})$, i.e., $A = \overline{B}$.

- From (2.22) it follows that if $\rho_{AB} < 0$ then $P(A|B) < P(A)$ and if $\rho_{AB} > 0$ then $P(A|B) > P(A)$.

Example 2.2.1. *Take the following events as A and B: $A = \{X > \text{Med } X\}$ and $B = \{Y > \text{Med } Y\}$, where Med X and Med Y are the population medians. Then for symmetric distributions of X and Y, $P(A) = P(\overline{A}) = P(B) = P(\overline{B}) = 1/2$, and hence*

$$\rho_{AB} = 4P(X > \text{Med } X, Y > \text{Med } Y) - 1.$$

In the particular case of the standard bivariate normal distribution density (2.18)

$$N(x, y; 0, 0, 1, 1, \rho) = \frac{1}{2\pi\sqrt{1 - \rho^2}} \exp\left\{ -\frac{x^2 - 2\rho xy + y^2}{2(1 - \rho^2)} \right\},$$

the correlation coefficient between the events A and B takes the form

$$\rho_{AB} = 4 \int_0^\infty \int_0^\infty N(x, y; 0, 0, 1, 1, \rho)\, dxdy - 1 = \frac{2}{\pi} \arcsin \rho. \tag{2.23}$$

The correlation coefficient (2.23) arises in Chapter 5 as the expectation of the quadrant correlation coefficient r_Q at the standard bivariate normal distribution.

2.3   Nonparametric Measures of Correlation

2.3.1   Introductory remarks

The nonparametric approach to statistical data analysis is based on the *nonparametric facts of statistics*, or in other words, on the existence of distribution-free and weakly dependent on the underlying distribution statistical methods (Kendall and Stuart 1963). As a rule, most of those methods use the signs and the ranks of observations. In correlation analysis, they are represented by the quadrant (sign) correlation coefficient (Blomqvist 1950), and the Spearman and the Kendall τ-rank correlation coefficients (Kendall 1938; Spearman 1904).

Recall that the rank $R(x_i)$ of an observation x_i is defined as the location of the corresponding order statistic in the variational series $x_{(1)} \leq \cdots \leq x_{(n)}$, that is, if an observation x_i from the initial sample $x_1, \ldots, x_n$ is transformed to the order statistic $x_{(j)}$, then $R(x_i) = j$, $1 \leq j \leq n$.

The usage of observation signs and ranks instead of observations is grounded by the high correlation between observations and their signs, as well as between observations and their ranks.

Table 2.1   Pearson's correlation between a random observation X and its sign S.

Distribution F	U-shape	Uniform	Normal	Laplace
$\rho_{XS}(F)$	0.97	0.87	0.80	0.71

Let X be a random variable with a distribution function $F(x)$, symmetric about zero. Now compute Pearson's correlation between X and its sign $S = \text{sgn}(X)$:

$$\rho_{XS}(F) = \frac{\int x\,\text{sgn}(x)\,dF(x)}{\left[\int x^2 dF(x)\,\int dF(x)\right]^{1/2}} = \frac{\int |x|\,dF(x)}{\sigma_X}. \qquad (2.24)$$

The numerical values of $\rho_{XS}(F)$ are exhibited in Table 2.1 for the following distributions: U-shape with density

$$f(x) = \begin{cases} 3/2\,x^2 & \text{for} \quad |x| \le 1, \\ 0 & \text{for} \quad |x| > 1, \end{cases}$$

uniform, normal, and Laplace with $f(x) = 0.5\exp(-|x|)$.

Pearson's correlation between a random variable X and its rank $R = R(X)$ is given by the following expression (Shulenin 2012):

$$\rho_{XR}(F) = \frac{\sqrt{3}\,\Delta_X}{2\,\sigma_X}, \qquad (2.25)$$

where Δ_X is the Gini mean deviation

$$\Delta_X = 2\int_{-\infty}^{\infty} F(x)(1 - F(x))\,dx.$$

The numerical values of $\rho_{XR}(F)$ are represented in Table 2.2 for the following distributions: uniform, normal, logistic with density

$$f(x) = \frac{\exp(-x)}{[1 + \exp(-x)]^2}$$

and Laplace.

Table 2.2   Pearson's correlation between a random observation X and its rank R.

Distribution F	Uniform	Normal	Logistic	Laplace
$\rho_{XR}(F)$	1.00	0.98	0.95	0.92

Note that the heavier distribution tails, the smaller are $\rho_{XR}(F)$ and $\rho_{XR}(F)$. From Tables 2.1 and 2.2 it also follows that correlation between observations and their ranks is noticeably stronger that between observations and their signs.

2.3.2   The quadrant correlation coefficient

This measure of correlation is defined as follows (Blomqvist 1950):

$$q = \frac{1}{n} \sum_{i=1}^{n} \operatorname{sgn}(x_i - \operatorname{med} x) \operatorname{sgn}(y_i - \operatorname{med} y), \qquad (2.26)$$

where med x and med y are the sample medians and

$$\operatorname{sgn}(x) = \begin{cases} +1 & \text{for} \quad x \geq 0, \\ -1 & \text{for} \quad x < 0. \end{cases} \qquad (2.27)$$

The chosen form of the signum function (2.27), not the conventional choice

$$\operatorname{sign}(x) = \begin{cases} +1 & \text{for} \quad x > 0, \\ 0 & \text{for} \quad x = 0, \\ -1 & \text{for} \quad x < 0, \end{cases} \qquad (2.28)$$

provides the proper values $q = +1$ and $q = -1$ in the cases of autocorrelation when $\{y_i = x_i\}_1^n$ and $\{y_i = -x_i\}_1^n$, respectively. Further, it can be directly checked that the sign correlation coefficient is just Pearson's correlation coefficient between the signs of deviations from medians when $n = 2k$.

Consider another way to define the quadrant correlation coefficient (see Fig. 2.12): divide the plane into the four quadrants with the center at the coordinate-wise median

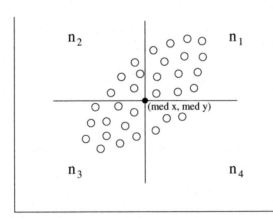

Figure 2.12   The quadrant correlation coefficient.

(med x, med y) and set n_1, n_2, n_3, and n_4 as the numbers of observations in quadrants 1, 2, 3, and 4, respectively (the observations falling on the axes should be attributed to the upper and right quadrants). Then we arrive at the following definition:

$$q = \frac{(n_1 + n_3) - (n_2 + n_4)}{(n_1 + n_3) + (n_2 + n_4)},$$

$$n = n_1 + n_2 + n_3 + n_4,$$

$$-1 \leq q \leq 1.$$

2.3.3 The Spearman rank correlation coefficient

This measure of correlation (Spearman 1904) is Pearson's correlation coefficient between the observation ranks $\{R(x_i)\}_1^n$ and $\{R(y_i)\}_1^n$

$$\hat{\rho}_S = \frac{\sum_{i=1}^n [R(x_i) - \bar{R}_x][R(y_i) - \bar{R}_y]}{\left(\sum_{i=1}^n [R(x_i) - \bar{R}_x]^2 \sum_{i=1}^n [R(y_i) - \bar{R}_y]^2\right)^{1/2}}. \tag{2.29}$$

Since the ranks are integers, for computing it is more convenient to use the transformed version of (2.29) (Kendall and Stuart 1963)

$$\hat{\rho}_S = 1 - \frac{6S(d^2)}{n(n^2 - 1)}, \quad S(d^2) = \sum_{i=1}^n [R(x_i) - R(y_i)]^2, \tag{2.30}$$

from which it follows that for computing the Spearman correlation coefficient it suffices to get the squared rank deviations.

Now we show this by applying the following formulas of elementary mathematics:

$$\sum R(x_i) = \sum R(y_i) = \sum_{i=1}^n i = \frac{n(n+1)}{2} = n\bar{R}_x = n\bar{R}_y,$$

where

$$n\bar{R}_x = n\bar{R}_y = \frac{1}{n}\sum_{i=1}^n i = \frac{n+1}{2},$$

$$\sum R^2(x_i) = \sum R^2(y_i) = \sum_{i=1}^n i^2 = \frac{n(n+1)(2n+1)}{6}.$$

Further,

$$\sum [R(x_i) - \bar{R}_x]^2 = \sum [R(x_i)]^2 - 2\bar{R}_x \sum R(x_i) + n[\bar{R}_x]^2$$

$$= \sum [R(x_i)]^2 - n[\bar{R}_x]^2 = \frac{n(n+1)(2n+1)}{6} - n\frac{(n+1)^2}{4} = \frac{n^3 - n}{12},$$

$$\sum [R(y_i) - \bar{R}_y]^2 = \frac{n^3 - n}{12}.$$

Consider now the sum of squared rank deviations:

$$\sum d_i^2 = \sum [R(x_i) - R(y_i)]^2 = \sum [(R(x_i) - \overline{R}_x) - (R(y_i) - \overline{R}_y)]^2$$
$$= \sum (R(x_i) - \overline{R}_x)^2 + \sum (R(y_i) - \overline{R}_y)^2 - 2\sum (R(x_i) - \overline{R}_x)(R(y_i) - \overline{R}_y)$$
$$= 2\frac{n^3 - n}{12} - \sum (R(x_i) - \overline{R}_x)(R(y_i) - \overline{R}_y).$$

Hence,

$$\sum (R(x_i) - \overline{R}_x)(R(y_i) - \overline{R}_y) = \frac{n^3 - n}{12} - \frac{1}{2}\sum d_i^2.$$

Combining all the above results, we arrive at (2.30):

$$r_S = \frac{\sum [R(x_i) - \overline{R}_x][R(y_i) - \overline{R}_y]}{\left(\sum [R(x_i) - \overline{R}_x]^2 \sum [R(y_i) - \overline{R}_y]^2\right)^{1/2}}$$
$$= \frac{\frac{n^3-n}{12} - \frac{1}{2}\sum d_i^2}{\frac{n^3-n}{12}} = 1 - \frac{6\sum d_i^2}{n(n^2 - 1)}.$$

Since the Spearman correlation coefficient is the particular case of Pearson's one, it inherits all its properties, e.g., $-1 \le \hat{\rho}_S \le 1$. If the direct rank correspondence is observed, then $d_i = 0$, $i = 1, \ldots, n$, and thus $\hat{\rho}_S = 1$. On the contrary, if the reverse rank correspondence occurs, then $R(x_i) = n + 1 - R(y_i)$, $i = 1, \ldots, n$, and $\hat{\rho}_S = -1$.

In practice it happens that two or more sample elements cannot be distinguished from each other. In this case, to each of those elements is assigned the rank equal to the average of their initial order numbers (Kendall and Stuart 1963).

2.3.4   The Kendall τ-rank correlation coefficient

Let $(x_1, y_1), \ldots, (x_n, y_n)$ be a sample of observations of the joint random variables X and Y, such that all the values of x_i and y_i are unique. Two pairs of observations (x_i, y_i) and (x_j, y_j) are said to be *concordant* if their ranks agree: that is, if both $x_i > x_j$ and $y_i > y_j$ or if both $x_i < x_j$ and $y_i < y_j$. They are said to be *discordant* if $x_i > x_j$ and $y_i < y_j$ or if $x_i < x_j$ and $y_i > y_j$. If $x_i = x_j$ or $y_i = y_j$, the pair is neither concordant nor discordant.

Denote the numbers of concordant and discordant pairs as n_+ and n_-, respectively. Then the Kendall τ-rank correlation coefficient is defined as follows (Kendall 1938):

$$\tau = \frac{n_+ - n_-}{n(n-1)/2}. \tag{2.31}$$

Since the denominator is the total number of pair combinations, $-1 \le \tau \le 1$.

Another way to define τ is given by the following construction:

$$\tau = \frac{2}{n(n-1)} \sum_{i<j} \text{sgn}(x_i - x_j)\,\text{sgn}(y_i - y_j). \tag{2.32}$$

The perfect agreement between two rankings, i.e., when they are the same, implies $\tau = 1$. From the perfect disagreement between two rankings, i.e., when one ranking is the reverse of the other, it follows that $\tau = -1$. If X and Y are independent, then we would expect τ to be approximately zero.

2.3.5 Concluding remark

Unlike Pearson's correlation coefficient, which is a measure of a linear dependence between random variables X and Y, the introduced nonparametric measures of correlation, namely, the quadrant, Spearman and Kendall correlations, are the measures of monotonic dependencies between random variables. For instance, the following property holds for them: $corr\,(g(X), h(Y)) = corr\,(X, Y)$ with strictly monotonic functions $g(x)$ and $h(x)$ (see Section 2.1).

2.4 Informational Measures of Correlation

Pearson's correlation coefficient is a well-defined measure of the linear dependence between continuous random variables X and Y. This partially refers to closely related rank measures as the quadrant, Spearman and Kendall correlation coefficients. However, if one is interested either in processing discrete data or in revealing the possible nonlinear relationship between random variables, then difficulties may arise both in implementation of these classical measures as well as in their interpretation.

In the literature, several proposals were made to solve these problems, for instance, the aforementioned Gebelein's (1941), and Sarmanov's (1958) correlation coefficients, and the distance correlation coefficient (Székely *et al.* 2007). The other proposals are based on the informational measures of association between random variables (Shannon 1948).

Joe's dependence measure (Joe 1989) exploits the concept of the relative entropy that measures the similarity of two random variables with the distributions $p(x)$ and $q(x)$ in the discrete case

$$D(p\|q) = \sum_x p(x) \log \frac{p(x)}{q(x)}.$$

Silvey (1964) used the measure of dependence between two random variables defined by the ratio of their joint density and the product of their marginal densities

$$\phi(x, y) = \frac{p(x, y)}{p(x)p(y)}.$$

The introduced measure is defined as follows:

$$\Delta = E(d(x))$$

where

$$d(x) = \int_{y:\ \phi(x,y)>1} (p(y|x) - p(y))\, dy.$$

Thus, it can be rewritten as

$$\Delta = \int \int_{(x,y):\ \phi(x,y)>1} (p(x,y) - p(x)p(y))\, dx\, dy.$$

Granger *et al.* (2004) introduced another formula for the measure of dependence

$$S_p = \frac{1}{2} \int \int \left([p(x,y)]^{1/2} - [p(x)p(y)]^{1/2} \right)^2 dx\, dy.$$

Unfortunately none of these formulas satisfy the Renyi postulate (P7). Furthermore, Joe's dependence measure is not symmetrical, and, in addition, Silvey's and Granger's measures do not satisfy postulate (P6), and they are very hard to compute.

Linfoot (1957) proposed the informational correlation coefficient based on the mutual information (Shannon 1948). Mutual information for any pair of discrete random variables X and Y is defined by

$$I(X,Y) = \sum_{x,y} p(x,y) \log \frac{p(x,y)}{p(x)p(y)}$$

and by

$$I(X,Y) = \int \int f(x,y) \log \frac{p(x,y)}{p(x)p(y)}\, dx\, dy \tag{2.33}$$

in the continuous case. Mutual information satisfies some desired postulates for the measures of dependence, for example, (P4) and (P5), but fails to satisfy (P3), (P6), and (P7).

The informational correlation coefficient, firstly introduced by (Linfoot 1957) solves these problems in the continuous case. The informational correlation coefficient (ICC) is defined as follows:

$$r_{ICC}(X,Y) = \sqrt{1 - e^{-2I(X,Y)}}. \tag{2.34}$$

Formally, Equation (2.34) is obtained by the substitution of the bivariate normal joint distribution density $p(x,y)$ into Equation (2.33). This measure of dependence is well-defined only in the continuous case—in the discrete case postulate (P5) does not hold.

2.5   Summary

In this chapter we most focus on Pearson's correlation coefficient and its numerous forms of representation and various interpretations some of which will be used in what follows. For instance, the representation of Pearson's correlation through the principal component variances is intensively exploited for designing its robust counterparts in Chapter 5.

The other groups of the classical measures of correlation comprise both the non-parametric measures based on data ranks (the quadrant, Spearman's, and Kendall's correlation coefficients) which naturally possess the property of robustness and thus they are in the scope of our work, and the classical measures of association aimed at the processing of categorical data such as Yule's coefficient of colligation, Pearson's contingency coefficient, and many others, which we only mention here for they deserve a serious separate study with respect to their robustness. For instance, a thorough general survey of the statistical methods of categorical data analysis is given in Kendall and Stuart (1963).

The methods based on the concepts of information theory constitute the last represented group of the classical measures of correlation and association.

References

Bernstein SN 1946 *Probability Theory*, OGIZ, Moscow-Leningrad (in Russian).

Blomqvist N 1950 On a measure of dependence between two random variables. *Ann. Math. Statist.* **21**, 593–600.

Galton F 1885 Regression towards mediocrity in hereditary stature. *Journal of Anthropological Institute* **15**, 246–263.

Galton F 1888 Co-relations and their measurement, chiefly from anthropometric data. *Proceedings of the Royal Society of London* **45**, 135–145.

Gebelein H 1941 Das statistische problem der korrelation als variations und eigenwerproblem und sein zuzammenhang mit der ausgleichsrechnung. *Z. Angewandte Math. Mech.* **21**, 364–379.

Gnanadesikan R and Kettenring JR 1972 Robust estimates, residuals, and outlier detection with multiresponse data. *Biometrics* **28**, 81–124.

Granger CW, Maasoumi and Racine J 2004 A dependence metric for possibly nonlinear processes. *J. of Time Series Analysis* **25**, 649–669.

Joe H 1989 Relative entropy measures of multivariate dependence. *Journal of the American Statistical Association* **84**, 157–164.

Kendall MG 1938 A new measure of rank correlation. *Biometrika* **30**, 81–89.

Kendall MG and Stuart A 1963 *The Advanced Theory of Statistics. Inference and Relationship*, vol. 2, Griffin, London.

Linfoot EH 1957 An informational measure of correlation. *Information and Control* **1**, 85–89.

Manning CD, Raghavan P, and Schutze H 2008 *Introduction to Information Retrieval*, Cambridge Univ. Press.

Pearson K 1920 Notes on the history of correlations. *Biometrika* **13**, 25–45.

Renyi A 1959 On measures of dependence. *Acta Mathematica Academiae Scieniarum Hungaricae* **10**, 441–451.

Rogers JL and Nicewander WA 1988 Thirteen ways to look at the correlation coefficient. *The American Statistician* **42**, 59–66.

Sarmanov OV 1958 Maximum correlation coefficient (symmetric case). *Doklady Akad. Nauk SSSR* **120**, 715–718.

Schilling MF 1984 Some remarks on quick estimation of the correlation coefficient. *The American Statistician* **38**, 330.

Schweizer B and Wolff EF 1981 On nonparametric measures of dependence for random variables. *Ann. Statist.* **9**, 879–885.

Shannon CE 1948 A mathematical theory of communication. *Bell Syst. Tech. J.* **7**, 329–423, 623–656.

Shevlyakov GL and Vilchevski NO 2002 *Robustness in Data Analysis: criteria and methods*, VSP, Utrecht.

Shevlyakov GL and Vilchevski NO 2011 *Robustness in Data Analysis*, De Gruyter, Boston.

Shulenin VP 2012 *Nonparametric Statistics*, Tomsk University (in Russian).

Silvey SD 1964 On a measure of association. *Ann. Math. Statist.* **35**, 1157–1166.

Spearman C 1904 The proof and measurement of association between two things. *Amer. J. Psychol.* **15**, 88–93.

Székely GJ, Rizzo ML and Bakirov NK 2007 Measuring and testing dependence by correlation of distances. *Ann. Statist.* **35**, 2769–2794.

3

Robust Estimation of Location

In this chapter we consider the principal for all further constructions the problem of robust estimation of location.

3.1 Preliminaries

Historically first robust procedures based on rejection of outliers appear in the eighteenth century, namely, they come from Boscovich (1757) and Daniel Bernoulli (1777). Several outstanding scientists of the late nineteenth and early twentieth century (the astronomer Newcomb (1886), the chemist Mendeleyev (1895), the astrophysicist Eddington (1914), and the geophysicist Jeffreys (1932) amongst them) saw the drawbacks of the standard estimates under heavy-tailed error distributions and proposed some robust alternatives to them (for details, see Stigler 1973). Here, it is also noteworthy to name the work of Kolmogorov (1931), in which he compared the performances of the sample mean and the sample median with the recommendation to use the latter under heavy-tailed distributions.

The strong arguments for robust statistics are given in Hampel *et al.* (1986), Huber (1981), and Tukey (1960). In what follows, we also refer to an overview of robust statistics represented in Shevlyakov and Vilchevski (2002).

In general, robust statistics deals with the consequences of possible perturbations of the assumed probabilistic model and suggests the methods protecting statistical procedures against such perturbations. Thus the probabilistic models used in robust statistics are chosen so as to account for possible departures from the assumptions about the underlying distribution. For description of these departures, several forms of neighborhoods of the underlying model are constructed with the use of an appropriately chosen metric, for example, the Kolmogorov, Prokhorov, or Lévy

Robust Correlation: Theory and Applications, First Edition. Georgy L. Shevlyakov and Hannu Oja.
© 2016 John Wiley & Sons, Ltd. Published 2016 by John Wiley & Sons, Ltd.
Companion Website: www.wiley.com/go/Shevlyakov/Robust

(Hampel *et al.* 1986; Huber 1981). Hence the initial model (basic or ideal) is extended up to the so-called *supermodel* that describes both the ideal distribution model and the departures from it.

Designing a robust procedure, it is useful to answer the following three questions:

- Robustness of what?

- Robustness against what?

- Robustness in what sense?

The first answer defines the type of statistical procedure (point or interval estimation, hypotheses testing, regression parameters estimation, etc.); the second specifies the supermodel, and the third introduces the criterion of quality of a statistical procedure and some related requirements towards its performance. The wide spectrum of the problems observed in robust statistics can be explained by the fact that there exist a great variety of answers to each of the above questions.

Now we briefly enlist main supermodels used in robust statistics (Bickel 1976; Hampel *et al.* 1986; Huber 1981). Here, we are most interested in the supermodels describing possible changes of the distribution shape.

For supermodels, we may distinguish the following two types: local and global (Bickel 1976). A local type suggests setting an ideal model and then the related supermodel is defined as a neighborhood of this ideal model. A global supermodel represents some class $\mathcal{F}$ of distributions with given properties that also comprises an ideal model.

Hodges and Lehmann (1963) consider a supermodel in the form of all absolutely continuous symmetric distributions. Birnbaum and Laska (1967) propose a supermodel as a finite collection of distribution functions: $\mathcal{F} = \{F_1, F_2, \ldots, F_k\}$. Andrews *et al.* (1972) examine statistical procedures in the supermodels containing distributions with heavier tails than the normal. In particular, they use Tukey's supermodel based on the quantile function, the inverse to the distribution function. This supermodel comprises rather accurate approximations to the normal, Laplace, logistic, Cauchy, and Student distributions.

Various supermodels are used to study deviations from normality: the family of power-exponential or generalized Gaussian distributions with the normal, Laplace, and uniform distributions as particular cases; the family of the Student t-distributions with the normal and Cauchy distributions; also the influence of non-normality can be studied with the use of the measures of asymmetry and kurtosis, where the positive values of the latter indicate gross errors and heavy tails.

For describing gross errors and outliers, Tukey's supermodel in the form of the mixture of normal distributions with different location and scale parameters is used (Tukey 1960)

$$\mathcal{F} = \left\{ F : F(x) = (1 - \varepsilon)\Phi(x) + \varepsilon\Phi\left(\frac{x - \theta}{k}\right), 0 \leq \varepsilon \leq 1, k \geq 1 \right\}. \qquad (3.1)$$

The generalization of this supermodel is given by

$$\mathcal{F} = \{F : F(x) = (1 - \varepsilon)F_0(x) + \varepsilon H(x), 0 \leq \varepsilon \leq 1\}, \qquad (3.2)$$

where F_0 is a given distribution (the ideal model) and H is an arbitrary continuous distribution is considered in Huber (1964). The supermodel (3.1) has the following natural interpretation: the parameter ε is the probability of gross errors ($k \gg 1$) in the data. On the contrary, in the supermodel (3.2) the parameter ε describes the level of uncertainty of the ideal distribution shape.

Huber (1964) also considers the Kolmogorov distance ε-neighborhood of the normal distribution

$$\mathcal{F} = \left\{ F : \sup_x |F(x) - \Phi(x)| < \varepsilon \right\}.$$

In general, a supermodel can be defined by the use of some suitable metric $d(F_0, F)$ in the space of all distributions: $\mathcal{F} = \{F : d(F_0, F) \leq \varepsilon)\}$. The Prokhorov metric (Prokhorov 1956) and its particular case, the Lévy metric, are rather convenient choices, since the supermodels based on them describe simultaneously the effects of gross errors, grouping, and rounding-off in the data (for details, see Huber and Ronchetti 2009).

The use of other metrics for constructing supermodels is discussed in Bickel (1976). The analysis of relations between various metrics can be found in Huber and Ronchetti (2009) and Zolotarev (1997).

Summarizing the above, we answer the second question: "Robustness against what?" as follows: "Robustness against the extension of ideal models to supermodels."

Now we are in a position partially to answer the third question: "Robustness in what sense?" by tackling the problem of robust estimation of a location parameter that is basic for all further considerations.

3.2   Huber's Minimax Approach

3.2.1   Introductory remarks

The first general approach to robust estimation is based on the minimax principle (Huber 1964; Huber and Ronchetti 2009). The minimax approach aims at the worst situation for which it suggests the best solution. Thus, in some sense, this approach provides a guaranteed result, perhaps too pessimistic. However, being applied to the problem of parameter estimation, it yields a robust modification of the principle of maximum likelihood (Huber 1964). We describe this approach in detail, as in the sequel, we exploit it to design robust estimates of the correlation coefficient.

3.2.2   Minimax variance M-estimates of location

Generalities

Let $x_1, \ldots, x_n$ be i.i.d. observations from a distribution with density $f(x, \theta)$ depending on the scalar parameter θ to be estimated. Estimating equations of the form

$$\sum_{i=1}^{n} \psi(x_i, \hat{\theta}_n) = 0 \tag{3.3}$$

were studied by Godambe (1960), who showed that under regularity conditions the maximum likelihood (*ML*) score function

$$\psi_{ML}(x, \theta) \propto -\frac{\partial \log(f(x, \theta))}{\partial \theta}$$

is the best choice.

Huber (1964) showed that if the model $f(x, \theta)$ is only approximately true, the optimality is not even approximately true. He also derived optimal minimax ψ-functions and, because estimates different from the maximum likelihood choice appeared to be useful, called the estimates of form (3.3) as M-estimates.

Here we restrict ourselves to the problem of estimating the location parameter θ of a symmetric density $f(x - \theta)$ in which case the ψ-functions take the form $\psi(x - \theta)$. Henceforth, without any loss of generality, we set $\theta = 0$.

Under rather general conditions of regularity imposed on ψ and f, M-estimates are consistent and asymptotically normally distributed with asymptotic variance

$$V(\psi, f) = \frac{A(\psi, f)}{B^2(\psi, f)}, \tag{3.4}$$

where

$$A(\psi, f) = \int_{-\infty}^{\infty} \psi^2(x) f(x) \, dx,$$

$$B(\psi, f) = \int_{-\infty}^{\infty} \psi'(x) f(x) \, dx. \tag{3.5}$$

As pointed out by Godambe (1960), for a given $f \in C^1(\mathbb{R})$, the variance is minimized when the ψ-function is equal to $\psi_{ML}(x) = -f'(x)/f(x)$. In this case, the minimal variance is equal to

$$\frac{1}{A(\psi_{ML}, f)} = \frac{1}{B(\psi_{ML}, f)} = \frac{1}{I(f)},$$

where

$$I(f) = \int_{-\infty}^{\infty} \left(\frac{f'(x)}{f(x)}\right)^2 f(x) \, dx \tag{3.6}$$

is Fisher information for location.

The asymptotic efficiency of any M-estimate of location is commonly defined as follows:

$$\text{eff}(\psi,f) = \frac{V(\psi_{ML},f)}{V(\psi,f)} = \frac{1}{I(f)V(\psi,f)} \ . \tag{3.7}$$

REMARK *Since efficiency (3.7) can be rewritten as*

$$\text{eff}(\psi,f) = \frac{B^2(\psi,f)}{I(f)A(\psi,f)} = \frac{\left[\int \psi'(x)f(x)\,\mathrm{d}x\right]^2}{\int \psi_{ML}^2(x)f(x)\,\mathrm{d}x \int \psi^2(x)f(x)\,\mathrm{d}x} \ ,$$

and then integrating the nominator of the last ratio by parts

$$B(\psi,f) = \int \psi'(x)f(x)\,\mathrm{d}x = -\int \psi(x)f'(x)\,\mathrm{d}x \ ,$$

we arrive at the squared Pearson's correlation coefficient between the scores of the maximum likelihood and M-estimates as a measure of efficiency

$$corr^2(\psi_{ML},\psi) = \frac{\left[\int \psi_{ML}(x)\psi(x)f(x)\,\mathrm{d}x\right]^2}{\int \psi_{ML}^2(x)f(x)\,\mathrm{d}x \ \int \psi^2(x)f(x)\,\mathrm{d}x} \ .$$

Compare this result with the result of Van Eeden (1963) on the relation between Pitman's asymptotic relative efficiency of two tests and the correlation coefficient between their test statistics.

In cases where a ψ-function is not smooth or a distribution density contains point masses, the integrals in (3.5) must be interpreted with care. Given a class Ψ of ψ-functions and a class $\mathcal{F}$ of densities f (various suggestions can be found in Deniau *et al.* (1977a,b,c), Hampel *et al.* (1986), Huber (1964), and Huber and Ronchetti (2009), it is often possible to identify a minimax estimate score function ψ^*.

These minimax M-estimates satisfy the property of a guaranteed accuracy for any density $f \in \mathcal{F}$ in the sense that there exists a worst-case density f^* such that

$$V(\psi^*,f) \leq V(\psi^*,f^*) = \inf_{\psi \in \Psi} \sup_{f \in \mathcal{F}} V(\psi,f). \tag{3.8}$$

This least favorable (informative) density f^* minimizes Fisher information for location over the class $\mathcal{F}$

$$f^* = \arg \min_{f \in \mathcal{F}} I(f) \ . \tag{3.9}$$

The minimax score function ψ^* is equal to the maximum likelihood choice for the least informative density f^*, i.e., $\psi^*(x) = -f^{*\prime}(x)/f^*(x)$. The upper bound $V(\psi^*,f^*)$ on the asymptotic variance of the optimal estimate in (3.8) strongly depends on the characteristics of the chosen class of distribution densities $\mathcal{F}$.

For most of our aims, the following regularity conditions defining the classes $\mathcal{F}$ and Ψ are sufficient (for details, see Hampel *et al.* 1986, pp. 125–127):

($\mathcal{F}1$) f is twice continuously differentiable and satisfies $f(x) > 0$ for all x in $\mathbb{R}$.

($\mathcal{F}2$) Fisher information for location satisfies $0 < I(f) < \infty$.

($\Psi1$) ψ is well-defined and continuous on $\mathbb{R} \backslash C(\psi)$, where $C(\psi)$ is finite. At each point of $C(\psi)$ there exist finite left and right limits of ψ that are different. Moreover, $\psi(-x) = -\psi(x)$ if $(-x, x) \subset \mathbb{R} \backslash C(\psi)$ and $\psi(x) \geq 0$ for $x \geq 0$ not belonging to $C(\psi)$.

($\Psi2$) The set $D(\psi)$ of points at which ψ is continuous but in which ψ' is not defined or not continuous is finite.

($\Psi3$) $\int \psi^2 \, dF < \infty$.

($\Psi4$) $0 < \int \psi'(x) \, dF(x) = - \int \psi(x) f'(x) \, dx < \infty$.

The key-point of Huber's approach is the solution of the variational problem (3.9). The surveys of the least informative distributions for different distribution classes can be found in Huber (1981), Huber and Ronchetti (2009), and Shevlyakov and Vilchevski (2002).

In what follows, we specify the methods of searching for an optimal solution f^* in the class of distribution densities $\mathcal{F}$.

Searching for a least informative distribution density

Now we consider the restrictions defining the classes of distribution densities $\mathcal{F}$. Generally, these restrictions are of the following forms:

$$\int_{-\infty}^{\infty} s_k(x) f(x) \, dx \leq \alpha_k, \quad k = 1, \dots, m, \tag{3.10}$$

$$f(x) \geq \varphi(x). \tag{3.11}$$

In particular, the normalization condition

$$\int f(x) \, dx = 1$$

when $s(x) = 1$ and the restriction on the variance

$$\int x^2 f(x) \, dx \leq \overline{\sigma}^2$$

when $s(x) = x^2$ are referred to (3.10); the conditions of non-negativeness $f(x) \geq 0$ and of the approximate normality (3.2)

$$f(x) \geq (1 - \varepsilon)N(x; 0, \sigma_N)$$

are described by (3.11), etc.

The variational problem of minimization of Fisher information under conditions (3.10) and (3.11)

$$\text{Minimize} \quad I(f) \quad \text{with} \quad f \in \mathcal{F},$$

$$\mathcal{F} = \left\{ f : \int_{-\infty}^{\infty} s_k(x)f(x)\, dx \leq \alpha_k, \; k = 1, \ldots, m, \; f(x) \geq \varphi(x) \right\} \tag{3.12}$$

is non-standard, and at present there are no general methods of its solution.

Nevertheless, using heuristic considerations, it is possible to find a candidate for the optimal solution of the variational problem (3.12), and then to check its validity. Certainly, such a reasoning must be grounded on the classical results of the calculus of variations. In general, it may be described as follows:

- first, use the restrictions of form (3.10), solve the Euler equation and determine the family of extremals;

- second, try to satisfy the restrictions of form (3.11) by gluing the pieces of free extremals with the constraints $\varphi(x)$;

- and, finally, verify the obtained solution.

Now we describe a procedure of searching for an eventual candidate for the solution of problem (3.12) proposed in Tsypkin (1984).

Consider the classes only with the restrictions of form (3.10). In this case, the Lagrange functional is composed as

$$L(f, \lambda_1, \ldots, \lambda_m) = I(f) + \sum_{k=1}^{m} \lambda_k \left(\int_{-\infty}^{\infty} s_k(x)f(x)\, dx - \alpha_k \right),$$

where $\lambda_1, \ldots, \lambda_m$ are the Lagrange multipliers. Taking the variation of this functional and equating it to zero, we obtain the Euler equation in the form

$$-2\frac{f''(x)}{f(x)} + \left(\frac{f'(x)}{f(x)}\right)^2 + \sum_{k=1}^{m} \lambda_k s_k(x) = 0. \tag{3.13}$$

Euler equation (3.13), as a rule, cannot be solved in a closed form. Hence one should use numerical methods. However, there is a serious obstacle in satisfying the restrictions of the form $f(x) \geq \varphi(x)$.

In what follows, we consider some classes $\mathcal{F}$ with analytical solutions for the least informative density.

Another approach is based on the direct application of numerical methods to variational problem (3.12). These methods are associated with some approximation to the distribution density $f(x)$ followed by the subsequent solution of the problem of mathematical programming.

Thus, if to approximate $f(x)$ by a piecewise linear finite function with step h, integrals by sums and derivatives by differences, we arrive at the problem of nonlinear programming with linear restrictions

$$\text{Minimize} \quad \sum_{i=1}^{N} \frac{(f_{i+1} - f_i)^2}{f_i}$$

$$f_i = f(ih), \quad i = 1, \dots, N;$$

$$\sum_{i=1}^{N} s_k(ih) f_i h \le \alpha_k, \quad k = 1, \dots, m; \quad f_i \ge \varphi(ih), \quad i = 1, \dots, N.$$

For this problem of nonlinear programming, there exist well-elaborated methods of its solution (Ben-Tal *et al.*, 1999).

Verifying optimality

Now we assume that there exists an analytical solution for the least informative density $f^*(x)$. Huber (1981) proposed a direct method for final checking of the eventual candidate for the least informative density.

Assume that $\mathcal{F}$ is convex, $0 < I(f) < \infty$, and the set where density f^* is strictly positive is convex. Set also the variation of f^* in the form of the mixture of densities

$$f_t = (1 - t)f^* + tf_1, \qquad 0 \le t \le 1,$$

where $f_1 \in \mathcal{F}$. Under these assumptions, Huber (1981) shows that density f^* minimizes Fisher information if and only if the inequality

$$\left. \frac{d}{dt} I(f_t) \right|_{t=0} \ge 0 \tag{3.14}$$

holds for any distribution density $f_1 \in \mathcal{F}$.

The inequality (3.14) can be rewritten in the convenient form for verifying the optimal solution f^*

$$\int_{-\infty}^{\infty} (2\psi^{*\prime} - \psi^{*2})(f_1 - f^*) \, dx \ge 0, \tag{3.15}$$

where $\psi^*(x) = f^{*\prime}(x)/f^*(x)$ is the optimal score function for any $f_1 \in \mathcal{F}$.

REMARK *This is a direct and explicit rule for verifying the earlier obtained optimal solution. For distribution classes with the restrictions of the inequality form given by (3.10) and (3.11), the experience of application of the inequality rule (3.15) in the verifying procedure yields the corresponding inequality of a distribution density class.*

Extremals of the basic variational problem

In order to get a candidate for the least informative density, one needs to have an idea about the possible structure of an optimal solution. Now we consider the family of extremals whose constituents would have the minimized Fisher information for location with the only side normalization condition

$$\text{Minimize} \quad I(f) = \int_{-\infty}^{\infty} \left(\frac{f'(x)}{f(x)} \right)^2 f(x) \, dx$$

$$\text{under the condition} \quad \int_{-\infty}^{\infty} f(x) \, dx = 1. \tag{3.16}$$

Set $\sqrt{f(x)} = g(x) \geq 0$ and rewrite the minimization problem (3.16) as follows:

$$\text{Minimize} \quad I(f) = 4 \int_{-\infty}^{\infty} (g'(x))^2 \, dx$$

$$\text{under the condition} \quad \int_{-\infty}^{\infty} g^2(x) \, dx = 1. \tag{3.17}$$

Using the Lagrange multiplier λ together with the normalization condition, we obtain the differential equation

$$4g''(x) + \lambda g(x) = 0. \tag{3.18}$$

The general solutions of (3.18) are of the following possible forms depending on the sign of λ:

- the exponential form
$$g(x) = C_1 e^{kx} + C_2 e^{-kx}; \tag{3.19}$$

- the cosine form
$$g(x) = C_1 \sin kx + C_2 \cos kx; \tag{3.20}$$

- the linear form
$$g(x) = C_1 + C_2 x,$$

where $k = \sqrt{\pm\lambda}/2$.

In what follows, the exponential and cosine forms and their combinations are used in the structures of optimal solutions for various classes of distribution densities.

Example 3.2.1. *Recall the well-known solution for the class of ε-contaminated normal distributions (Huber 1964)*

$$\mathcal{F}_H = \{f : f(x) = (1 - \varepsilon)\varphi(x) + \varepsilon h(x), \quad 0 \le \varepsilon < 1\}, \qquad (3.21)$$

where $\varphi(x) = (2\pi)^{-1/2} \exp(-x^2/2)$ is the standard normal density, $h(x)$ is an arbitrary density, and ε is a contamination parameter describing the uncertainty of our knowledge about the true underlying distribution and giving the aforementioned probability of gross outliers in the data.

As the restriction of the class $\mathcal{F}_H$ can be rewritten in the inequality form

$$f(x) \ge (1 - \varepsilon)\varphi(x),$$

we can foresee the qualitative structure of the least informative density: there should be free extremals of the exponential form (3.19) smoothly sewed with the constraint $(1 - \varepsilon)\varphi(x)$.

Thus the least informative density is of the following structure:

$$f_H^*(x) = \begin{cases} (1 - \varepsilon)\varphi(x) & \text{for } |x| \le k, \\ (1 - \varepsilon)(2\pi)^{-1/2} \exp(-k|x| + k^2/2) & \text{for } |x| > k, \end{cases}$$

where $k = k(\varepsilon) > 0$ satisfies

$$2\varphi(k)/k - 2\Phi(-k) = \varepsilon/(1 - \varepsilon), \quad \Phi(x) = \int_{-\infty}^{x} \varphi(t)\, dt, \qquad (3.22)$$

and its values are tabulated in Huber (1981, p. 87). The optimal score function is a truncated linear function $\psi_H^*(x) = \max[-k, \min(x, k)]$ (see Fig. 3.1).

The minimax variance M-estimate $\widehat{\theta}_n$ of location is defined by the solution of the following equation:

$$\sum_{i=1}^{n} \psi_H^*(x_i - \widehat{\theta}_n) = 0.$$

The particular cases of this solution for $\varepsilon = 0$ and $\varepsilon \to 1$ are given by the sample mean and the sample median, respectively.

The Huber score function yields a robust version of the *MLE*: in the central zone $|x_i| \le k$, the data are processed by the *ML* method and they are trimmed within

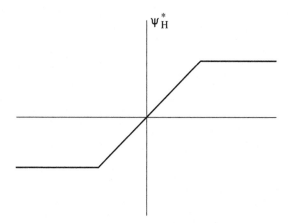

Figure 3.1   Huber's optimal score function

distribution tails. In the limit case of a completely unknown density as $\varepsilon \to 1$, the minimax variance M-estimate of location is the sample median.

It is likely that one could construct classes of distribution densities in which the minimax estimate would be a bad choice because it would attempt to ensure an adequate variance for some abstruse density. In the above example and in other cases discussed in the literature on robustness this does, however, not happen. It is, for example, quite surprising to note that the least informative distribution has exponential rather than the more severe Pareto-like tails. Under Pareto-like tails, quite extreme outliers would be expected to be found among the data. Such gross errors are, however, not very difficult to spot and, as Huber's result shows, the estimate that protects against exponential tails already does a sufficiently good job of it. If f^* had Pareto tails, it would follow that $\lim_{|x| \to \infty} \psi^*(x) = 0$, that is, ψ^* would be redescending. Large outliers have no or negligible effect on the estimate computed with a redescender. By Huber's result, such estimates pay the price of an increased asymptotic variance at intermediate tail indices.

Within the minimax approach, robustness of estimation is measured in terms of efficiency, namely by the supremum of asymptotic variance in the supermodel $\mathcal{F}$: $\sup_{f \in \mathcal{F}} V(\psi, f)$. Apparently, the smaller this characteristic, the more robust an M-estimate is.

3.2.3   Minimax bias M-estimates of location

Another measure of robustness is given by the supremum of an asymptotic bias

$$\sup_{f \in \mathcal{F}} |b(\psi, f)|$$

under asymmetric distributions (Huber 1981; Huber and Ronchetti 2009; Rychlik 1987; Zieliński 1987), where

$$b(\psi, f) = \lim_{n \to \infty} E(\hat{\theta}_n - \theta) = \frac{E_F(\psi)}{E_F(\psi')} = \frac{\int \psi(x)\, dF(x)}{\int \psi'(x)\, dF(x)}.$$

In particular, for the normal density $f_0(x) = N(x; \theta, \sigma)$ and asymmetric contaminating density $h(x)$ in the supermodel of gross errors (3.2), the minimax bias M-estimate is determined by the following condition

$$\psi^* = \arg \inf_{\psi} \sup_{h} |b(\psi, f)|\ ,$$

and hence it is given by the sample median (Huber 1981).

3.2.4 *L*-estimates of location

L-estimates are the linear combinations of order statistics defined as

$$\hat{\theta}_n = \sum_{i=1}^{n} C_i x_{(i)}, \quad \sum_{i=1}^{n} C_i = 1\ .$$

The normalization condition $\sum_{i=1}^{n} C_i = 1$ provides the equivariancy of L-estimates under translation. The trimmed mean estimate

$$\bar{x}_{tr}(k) = \frac{1}{n - 2k} \sum_{i=k+1}^{n-k} x_{(i)} \tag{3.23}$$

belongs to this class. In asymptotics, the fraction α of censored observations is used: $k = [\alpha n]$.

    L-estimates were proposed by Daniel (1920) and since then they have been forgotten for thirty years, being revived in robustness studies. The description of L-estimates can be formalized with a *weight function* $h(t)$. Let a function $h(t)$ of bounded variation on $[0, 1]$ satisfy the following conditions:

$$h(t) = h(1 - t) \quad \text{for all } t \in [0, 1],$$

$$\int_0^1 h(t)\, dt = 1.$$

Then

$$\hat{\theta}_n = \frac{1}{n} \sum_{i=1}^{n} h\left(\frac{i}{n+1}\right) x_{(i)}$$

is called an *L*-estimate with the weight function $h(t)$. Under regularity conditions (Huber 1981), *L*-estimates are consistent and asymptotically normal with the asymptotic variance

$$n^{1/2} var(\widehat{\theta}_n) = A_L(h, F) = \int_0^1 K^2(t)\, dt,$$

where

$$K(t) = \int_{1/2}^t \frac{h(u)}{f(F^{-1}(u))}\, du,$$

$$\int_0^1 K(t)\, dt = 0.$$

3.2.5   *R*-estimates of location

R-estimates proposed in Hodges and Lehmann (1963) are based on rank tests. There are several methods of their construction. Now we briefly describe one of those (Azencott, 1977a,b; Huber 1981).

Let $y_1, \ldots, y_n$ and $z_1, \ldots, z_n$ be independent samples from the distributions $F(x)$ and $F(x - \theta)$, respectively. For testing the hypothesis $\theta = 0$ against the alternative $\theta > 0$, the following statistic is used:

$$W_n(y_1, \ldots, y_n, z_1, \ldots, z_n) = \sum_{i=1}^n J\left(\frac{R_i}{2n+1}\right),$$

where R_i is the rank of y_i, $i = 1, \ldots, n$, in the united sample of size $2n$.

Let a function $J(t)$, $0 \le t \le 1$ satisfy the following conditions:

$J(t)$ is increasing; $J(t) + J(1 - t) = 0$ for all $t \in [0, 1]$;

$J'(t)$ is defined on $(0, 1)$;

the functions J' and $f(F^{-1})$ are of bounded variation on $[0, 1]$,

and $\int_0^1 J'(t) f(F^{-1}(t))\, dt \ne 0$.

Under these conditions, the test with the critical region $W_n > c$ is optimal in power (Hájek and Šidák 1967).

The *R*-estimate $\widehat{\theta}_n$ based on this test is defined as a solution to the following equation

$$W_n(x_1 - \widehat{\theta}_n, \ldots, x_n - \widehat{\theta}_n, -(x_1 - \widehat{\theta}_n), \ldots, -(x_n - \widehat{\theta}_n)) = 0.$$

Under the above conditions, $\widehat{\theta}_n$ is consistent and asymptotically normal with the asymptotic variance

$$n^{1/2} \text{var}(\widehat{\theta}_n) = A_R(J, F)$$

$$= \frac{\int_0^1 J^2(t)\, dt}{\left[\int J'(F(x))f^2(x)\, dx\right]^2}.$$

For any given function $F(x)$, it is possible to find the function $J(t)$ minimizing the asymptotic variance $A_R(J, F)$. The test based on such function $J(t)$ also has optimal properties for given F.

In particular, the logistic distribution $F(x) = (1 + e^{-x})^{-1}$ leads to the Wilcoxon test. The corresponding estimate of location is the Hodges–Lehmann median

$$\widehat{\theta}_n = \text{med}\left\{\frac{x_{(i)} + x_{(k)}}{2}\right\}, \quad 1 \le i \le k \le n. \tag{3.24}$$

3.2.6   The relations between M-, L- and R-estimates of location

In Jaeckel (1971) the asymptotic equivalence of these estimates is established. Let F and ψ be given. Set

$$h(t) = \frac{\psi'(F^{-1}(t))}{\int \psi'(x)f(x)\, dx}, \quad t \in [0, 1]$$

and

$$J(t) = \psi(F^{-1}(t)), \quad t \in [0, 1].$$

Then it holds that

$$V(\psi, f) = A_L(h, F) = A_R(J, F).$$

However, M-estimates are most convenient for analysis and L-estimates are the simplest in computation.

Example 3.2.2. *The optimal robust minimax variance L-estimate of location is the two-sided trimmed mean (3.23) (Huber 1981)*

$$\bar{x}_{tr}(n^*) = \frac{1}{n - 2n^*} \sum_{i=n^*+1}^{n-n^*} x_{(i)},$$

where $n* = [\alpha n]$, $\alpha = F_H^*(-k)$, and k is given by (3.22); the values $F_H^*(-k)$ are tabulated in Huber and Ronchetti (2009, p. 85).

For instance, the fraction values α of censored observations are as follows:

$$\alpha = 0.031 \quad with \quad \varepsilon = 0.01, \quad \alpha = 0.102 \quad with \quad \varepsilon = 0.05,$$

$$\alpha = 0.164 \quad with \quad \varepsilon = 0.1, \quad \alpha = 0.256 \quad with \quad \varepsilon = 0.2.$$

3.2.7   Concluding remarks

In this section we formulate the basic steps of the historically first Huber's mini-max approach to robustness, outline the concepts and techniques of the derivation of least informative distributions minimizing Fisher information for location, and present the classical Huber minimax M-estimate of a location parameter for the class of ε-contaminated normal distributions. Finally, the asymptotic equivalency of M-, L-, and R-estimates is established.

3.3   Hampel's Approach Based on Influence Functions

3.3.1   Introductory remarks

The main advantage of robust methods is their lower sensitivity to possible variations of data distributions as compared to conventional statistical methods. Thus it is necessary to have specific mathematical tools that allow analysis of the sensitivity of estimates to outliers, rounding-off errors, etc. On the other hand, such tools make it possible to solve the inverse problem: to design estimates with the required sensitivity. Now we introduce the above-mentioned apparatus, namely, the sensitivity curves and the influence functions.

3.3.2   Sensitivity curve

Let $\{T_n\}$ be some sequence of statistics, where $T_n(X)$ is the statistic from $\{T_n\}$ on the sample $X = (x_1, \ldots, x_n)$, and let $T_{n+1}(x, X)$ denote the same statistic on the sample $(x_1, \ldots, x_n, x)$. Then the function

$$SC_n(x; T_n, X) = (n+1)[T_{n+1}(x, X) - T_n(X)]$$

characterizes the sensitivity of T_n to the addition of one observation at x, and it is called the *sensitivity curve* for this statistic (Tukey 1977).

In particular, it holds that

$$SC_n(x; \bar{x}, X) = x - \frac{1}{n} \sum_{i=1}^{n} x_i = x - \bar{x}$$

for the sample mean $\bar{x}$ and

$$SC_n(x; \text{med } x, X) = \begin{cases} 0.5(n+1)[x_{(k)} - x_{(k+1)}] & \text{for} \quad x \le x_{(k)}, \\ 0.5(n+1)[x - x_{(k+1)}] & \text{for} \quad x_{(k)} \le x \le x_{(k+2)}, \\ 0.5(n+1)[x_{(k)} - x_{(k+1)}] & \text{for} \quad x \ge x_{(k+2)} \end{cases}$$

for the sample median med x with $n = 2k + 1$ (see Figs 3.2 and 3.3).

We can see that the sensitivity curve of the sample mean is unbounded, and hence only one extreme observation can completely destroy this estimate. On

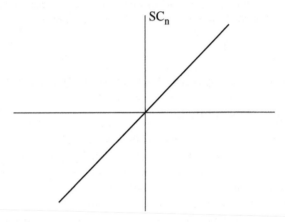

Figure 3.2   Tukey's sensitivity curve for the sample mean

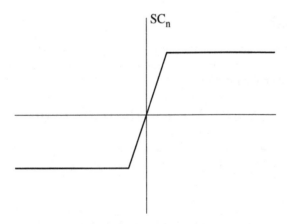

Figure 3.3   Tukey's sensitivity curve for the sample median

the contrary, the sensitivity curve of the sample median is bounded; however, the sample median is sensitive to the occurrence of new sample elements in the interval $(x_{(k)}, x_{(k+2)})$.

There exist some other characteristics describing the influence of the data perturbations on estimates. It is desirable to have such a characteristic that does not depend on the specific sample X. One of most convenient for asymptotic analysis is the influence function (curve) introduced by Hampel (1974).

3.3.3   Influence function and its properties

Hampel's method consists in constructing an estimate with a predetermined influence function, which in its turn determines the qualitative robustness properties of an estimation procedure such as its sensitivity to outliers in the data, to rounding off, etc.

Let F be a distribution corresponding to $f \in \mathcal{F}$, the class of distribution densities, and let $T(F)$ be a functional defined in a subset of all distribution functions. The natural estimate defined by T is $T_n = T(F_n)$, i.e., the functional computed at the sample distribution function F_n.

The influence function $IF(x; T, f)$ of this functional at one of the model distribution densities is defined as

$$IF(x; T, f) = \lim_{t \to 0} \frac{T((1 - t)F + t\Delta_x) - T(F)}{t}, \qquad (3.25)$$

where Δ_x is the degenerate distribution taking mass 1 at x (Hampel 1968, 1974; Hampel *et al.* 1986).

The influence function measures the impact of an infinitesimal contamination at x on the value of an estimate, formally being the Gâteaux derivative of the functional $T(F)$. For an M-estimate with a score function ψ, the influence function takes the following form (Hampel 1974; Hampel *et al.* 1986) (see Figs 3.4 and 3.5):

$$IF(x; \psi, f) = \frac{\psi(x)}{B(\psi, f)} . \qquad (3.26)$$

From the above result and (3.4) it follows that the asymptotic variance of M-estimates is equal to

$$V(\psi, f) = \int_{-\infty}^{\infty} IF^2(x; \psi, f) f(x) \, \mathrm{d}x. \qquad (3.27)$$

Furthermore, under regularity conditions, Equation (3.27) holds for more general estimates $T_n = T(F_n)$ than M-estimates (Hampel *et al.* 1986).

For M-estimates, the relation between the influence function and the score function is the simplest. This allows M-estimates to be applied to solve some specific extremal problems of maximization of the estimate efficiency subject to their bounded

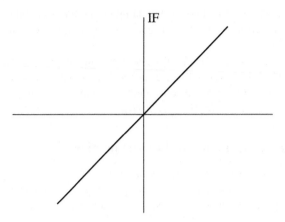

Figure 3.4   Influence function for the population mean

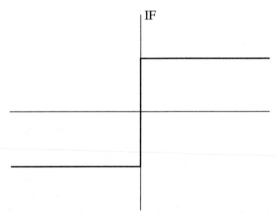

Figure 3.5   Influence function for the population median

sensitivity to outliers, in other words, the so-called problems of optimal Huberization (Deniau *et al.* 1977d; Hampel *et al.* 1986; Huber 1972).

The influence function, which measures the impact of an infinitesimal contamination at x on the value of an estimate, however, is a useful tool having deep intrinsic relations with other important statistical concepts (Hampel 1968 1974; Hampel *et al.* 1986; Huber 1981; Huber and Ronchetti 2009). For instance, with the use of $IF(x; T, f)$, the functional $T(F)$ can be linearized in the neighborhood of the ideal model F_0 (f_0) as

$$T(F) = T(F_0) + \int IF(x; T, f_0) \, [f(x) - f_0(x)] \, dx + \text{remainder}.$$

Further, $\sqrt{n}\,[T_n(F_n) - T(F)]$ tends to $\int IF(x; T,f)\,dF_n(x)$ in probability so that

$$T_n(F_n) = T(F) + \int IF(x; T,f)\,dF_n(x) + \text{remainder.}$$

Next,

$$\sqrt{n}(T_n - T(F)) = \frac{1}{\sqrt{n}} \sum_{i=1}^{n} IF(x_i; T,f) + \text{remainder.}$$

Since in most cases the remainder is negligible as $n \to \infty$, $\sqrt{n}\,T_n$ is asymptotically normal with asymptotic variance

$$V(T,f) = \int IF^2(x; T,f)\,f(x)\,dx. \tag{3.28}$$

This line of reasoning is accurately verified in Fernholz (1983).

3.3.4   Local measures of robustness

From the influence function, the following robustness measures are defined as follows (Hampel 1968, 1974):

- *Supremum of the influence function absolute value*

$$\gamma^*(T,f) = \sup_x |IF(x; T,f)| \tag{3.29}$$

is called *the gross-error sensitivity* of T at f. This sensitivity gives an upper bound upon the asymptotic estimate bias and measures the worst influence of an infinitesimal contamination on the value of an estimate. Minimizing the asymptotic variance under the condition of a finite gross-error sensitivity leads formally to the same estimate as Huber's minimax approach. The use of the gross-error sensitivity as a leading indicator of robustness excludes redescenders, because redescendency has no effect on this indicator.

- *Local-shift sensitivity*

$$\lambda^*(T,f) = \sup_{x \neq y} \frac{|IF(y; T,f) - IF(x; T,f)|}{|y - x|}$$

accounts for the effects of rounding-off and grouping of the observations.

- *Rejection point*

$$\rho^*(T,f) = \inf\{r > 0 : IF(x; T,f) = 0, \quad \text{where} \quad |x| > r\},$$

defines the observations to be rejected completely.

By analogy with the influence function, the change-of-variance function and its sensitivity are defined as

$$CVF(x;T,f) = \lim_{t \to 0} \left[V\left(T,(1-t)F + \frac{t}{2}(\Delta_x + \Delta_{-x})\right) - V(T,f) \right] \Big/ t$$

and

$$\kappa^*(T,f) = \sup_x \frac{CVF(x;T,f)}{V(T,f)} \;.$$

At present, the influence function is the main heuristic tool for designing esti-mates with given robustness properties (Hampel *et al.* 1986; Huber 1981; Rey 1978. For example, slightly changing the maximum likelihood estimate, it is possible to improve considerably its sensitivity to gross errors by lessening γ^* and its sensi-tivity to local effects of rounding-off and grouping types by bounding the slope of $IF(x;T,f)$ (i.e., λ^*) above. Setting $IF(x;T,f)$ tending to zero as $n \to \infty$ leads to the stabilization of the asymptotic variance while bounding the slope above stabilizes the asymptotic bias.

3.3.5   *B*- and *V*-robustness

The following concepts turned out to be useful in robustness studies.

Definition 3.3.1. *An estimate T_n of $T(F)$ is called B-robust (Rousseeuw 1981) if its gross-error sensitivity is bounded: $\gamma^*(T,f) < \infty$, and those for which there exists a positive minimum of γ^* are the most B-robust estimates (Hampel et al. 1986).*

Definition 3.3.2. *An estimate $T_n = T(F_n)$ of $T(F)$ is called V-robust if its change-of-variance sensitivity is bounded: $\kappa^*(T,f) < \infty$ (Hampel et al. 1986).*

3.3.6   Global measure of robustness: the breakdown point

All the above-introduced measures of robustness based on the influence function and its derivatives are of a local character, being evaluated at the model distribution F. Hence it is desirable to have a measure of *the global robustness* of an estimate over the chosen class of distributions, in other words, in the chosen supermodel $\mathcal{F}$.

Since the general definition of a supermodel is based on the concept of a dis-tance (Kolmogorov, Lévy, Prokhorov) in the space of all distributions (for details, see (Hampel *et al.* 1986; Huber and Ronchetti 2009) the same concept is involved in the construction of a measure of the global robustness.

Let d be such a distance. Then the *breakdown point* ε^* of the estimate $T_n = T(F_n)$ for the functional $T(F)$ at F is defined by

$$\varepsilon^*(T, F) = \sup \left\{ \varepsilon \leq 1: \sup_{F:d(F,F_0)<\varepsilon} |T(F) - T(F_0)| < \infty \right\}.$$

The breakdown point characterizes the maximal deviation (in the sense of a chosen metric) from the ideal model F_0 that provides the boundedness of the estimate bias.

For our further aims, the concept of *the gross-error breakdown point* suffices:

$$\varepsilon^*(T, F) = \sup \left\{ \varepsilon: \sup_{F:F=(1-\varepsilon)F_0+\varepsilon H} |T(F) - T(F_0)| < \infty \right\}.$$

This notion defines the largest fraction of gross errors that still keeps the bias bounded.

Example 3.3.1. *The sample median possesses the following properties: the Huber minimax variance, the Huber minimax bias, B-robustness, V-robustness, and global robustness with the maximal value of the breakdown point* $\varepsilon^* = 1/2$.

The basic relations between Huber's and Hampel's approaches are thoroughly analyzed in Hampel *et al.* (1986, pp. 172–178), and between the concepts of continuity and qualitative robustness in Hampel (1971). In particular, for sufficiently small ε, the Huber minimax solution ψ^* minimizing the asymptotic variance $V(\psi,f)$ of M-estimates in the gross-error supermodel turns out to be optimal V-robust minimizing $V(\psi,f)$ under the condition that κ^* is an upper bound of the change-of-variance sensitivity (Rousseeuw 1981). A similar assertion holds for the Huber minimax bias solution and the optimal B-robust estimate, namely the sample median.

3.3.7   Redescending M-estimates

By Monte Carlo modeling of a variety of M-estimates in a variety of distribution models (Andrews *et al.* 1972, p. 126), it was found that the redescending M-estimate dubbed 25A overall performed better than the others. Redescending ψ-functions are conventionally designed to vanish outside some central region, in other words, the following class (Hampel *et al.* 1986) is considered:

$$\Psi_r = \{\psi(x): \quad \psi(x) = 0 \quad \text{for} \quad |x| \geq r\}, \tag{3.30}$$

where $0 < r < \infty$ is fixed.

Some examples of redescending estimates are given by

- the aforementioned Hampel's 25A estimate from Andrews *et al.* (1972) with the redescending three-part score function

$$\psi_{25A}(x) = \begin{cases} x & \text{for} \quad 0 \le |x| \le a, \\ a \ \text{sgn}(x) & \text{for} \quad a \le |x| \le b, \\ a \ \dfrac{r - |x|}{r - b} \text{sgn}(x) & \text{for} \quad a \le |x| \le b, \\ 0 & \text{for} \quad r \le |x|, \end{cases}$$

with $0 < a \le b < r < \infty$ (see Fig. 3.6),

- the Huber skipped mean with ψ-function (see Fig. 3.7)

$$\psi_{sk(r)}(x) = x \ 1_{[-r,r]}(x),$$

- the biweight estimate (Mosteller and Tukey 1977, p. 205) (see Fig. 3.8)

$$\psi_{bi(r)}(x) = x(r^2 - x^2)^2 \ 1_{[-r,r]}(x), \tag{3.31}$$

and a compromise between the last two,

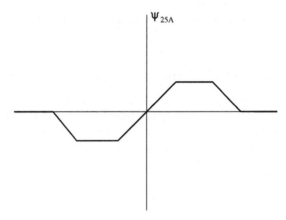

Figure 3.6   Hampel's ψ_{25A} score function

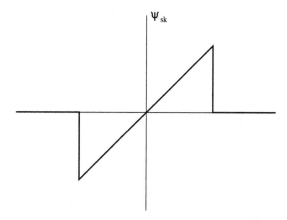

Figure 3.7   Huber's skipped mean score function

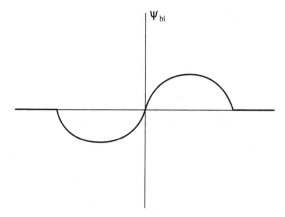

Figure 3.8   Mosteller-Tukey's biweight score function

- Smith's estimate (Stigler 1981)

$$\psi_{Sm(r)}(x) = x(r^2 - x^2)\, 1_{[-r,r]}(x), \tag{3.32}$$

where $1_{[-r,r]}(x)$ is the indicator function taking unit value in $[-r, r]$ and zero otherwise.

Optimality of redescending estimates can be derived by implementing either Huber's or Hampel's program with the "hard rejection" condition (3.30). However, given their success in finite sample simulation studies, the following two questions are of interest: first, is it possible to derive redescendency without simply imposing it via the rejection point r and, second, is it possible to weaken the condition imposed by the rejection point by considering score functions tending to zero for $|x| \to \infty$ as in Holland and Welsch (1977)? The answers to these questions are given in the next section.

3.3.8   Concluding remark

In this section we give a brief overview of Hampel's historical second general approach to robustness with the definitions of the related concepts, notions, local and global measures of robustness – most of them will be used in sequel.

3.4   Robust Estimation of Location: A Sequel

3.4.1   Introductory remarks

The theory of robustness is definitely confined to distribution models (supermodels) in the form of neighborhoods of a given distribution, usually normal, as in Huber's approach (Huber 1981; Huber and Ronchetti 2009), and/or to the impact on statistical inferences caused by the specific forms of perturbations of a given distribution, also a neighborhood, as in Hampel's approach (Hampel et al. 1986). This robustness "main-stream" justly follows the general mathematical concept of stability in the Lyapunov sense, that is, roughly continuity, within which robustness of a statistical procedure means that "small perturbations" lead to "small impacts".

However, this is not the only possibility and distribution neighborhoods are not the only models of interest. In applications rather often there exists prior information about the dispersion of a distribution, about its central part and/or its tails, about the moments and/or subranges of a distribution. The empirical distribution function and corresponding estimates of a distribution shape (quantile functions and their approximations, histograms, kernel estimates) along with their confidence boundaries give other examples (Shevlyakov and Vilchevski 2002). It seems reasonable to use such information within Huber's minimax setting by introducing the corresponding classes $\mathcal{F}$ of distribution densities $f(x)$ in order to increase the efficiency of robust minimax variance estimates. As before, here we confine ourselves to a robust estimation of a location parameter.

Another point of interest is the stability of robust solutions – evaluation of their sensitivity to possible violations of the conditions under which they are derived. In what follows, this question is partially considered with respect to the stability of least informative distributions within the minimax approach to robust estimation of location.

3.4.2 Huber's minimax variance approach in distribution density models of a non-neighborhood nature

Recall that the key point of Huber's minimax approach is the solution of the variational problem of minimization of Fisher's information for location over a given distribution density class $\mathcal{F}$

$$f^* = \arg \min_{f \in \mathcal{F}} I(f) \,.$$

The classes of distribution densities

Now we enlist some typical examples of distribution density classes that seem most natural and convenient for the description of prior knowledge about data distributions.

The conditions of symmetry $f(-x) = f(x)$, non-negativeness $f(x) \geq 0$, and normalization $\int_{-\infty}^{\infty} f(x)\,dx = 1$ are common for all the classes.

- The class of nondegenerate distribution densities (Polyak and Tsypkin 1976):

$$\mathcal{F}_1 = \left\{ f : f(0) \geq \frac{1}{2a} > 0 \right\}. \tag{3.33}$$

It is one of the most wide classes: any distribution density with a nonzero value at the center of symmetry belongs to it. The parameter a of this class characterizes the dispersion of the central part of a data distribution, in other words, a is an upper bound for that dispersion. The condition of belonging to this class is very close to the complete lack of information about a data distribution.

- The class of distribution densities with a bounded variance (Kagan et al. 1973):

$$\mathcal{F}_2 = \left\{ f : \sigma^2(f) = \int_{-\infty}^{\infty} x^2 f(x)\,dx \leq \bar{\sigma}^2 \right\}. \tag{3.34}$$

All distributions with variances bounded from above are members of this class. Apparently, the heavy-tailed Cauchy-type distributions do not belong to it.

- The class of finite distribution densities (Shevlyakov and Vilchevski 2002):

$$\mathcal{F}_3 = \left\{ f : \int_{-l}^{l} f(x)\,dx = 1 \right\}. \tag{3.35}$$

The restriction on this class defines the boundaries of the data (i.e., $|X| \leq l$ holds with probability one) and there is no more information about the distribution except the conditions of regularity at the boundaries $\pm l$

$$f(\pm l) = 0, \qquad f'(\pm l) = 0.$$

- The class of approximately finite distribution densities or the class with a given subrange:

$$\mathcal{F}_4 = \left\{ f : \int_{-l}^{l} f(x)\, dx = 1 - \beta \right\}. \tag{3.36}$$

The parameters l and β, $0 \le \beta < 1$, are given; the latter characterizes the degree of closeness of $f(x)$ to a finite distribution density. The restriction on this class means that the inequality $|X| \le l$ holds with probability $1 - \beta$. Obviously, the class of finite distributions $\mathcal{F}_3$ is a particular case of the class $\mathcal{F}_4$; these classes are considered in Huber (1981), Sacks and Ylvisaker (1972), and Shevlyakov and Vilchevski (2002).

Example 3.4.1. *Assuming symmetry and setting $\beta = 1/2$, we can rename the class*

$$\mathcal{F}_4 = \left\{ f : \int_{-l}^{l} f(x)\, dx = 1/2 \right\}$$

as the class with a given semi-interquartile distribution range SIQR(f)

$$\mathcal{F}_4 = \{ f : SIQR(f) = l \}.$$

The least informative distribution densities and optimal score functions

Now we write out the least informative distribution densities and the corresponding score functions for the aforementioned distribution classes.

- In the class of nondegenerate distribution densities $\mathcal{F}_1$, the least informative distribution density is the Laplace or the double-exponential one (Polyak and Tsypkin 1976)

$$f_1^*(x) = L(x; 0, a) = \frac{1}{2a} \exp\left(-\frac{|x|}{a} \right) \tag{3.37}$$

with the optimal sign score function $\psi_1^*(x) = \text{sign}(x)$ (see Fig. 3.9). This result is shown to be quite natural if we recall the exponential form (3.19) of free extremals for the basic variational problem.

- In the class of distribution densities with a bounded variance, the least informative distribution density is normal (Kagan *et al.* 1973)

$$f_2^*(x) = N(x; 0, \overline{\sigma}) = \frac{1}{\overline{\sigma}\sqrt{2\pi}} \exp\left(-\frac{x^2}{2\overline{\sigma}^2} \right) \tag{3.38}$$

with the optimal linear score function $\psi_2^*(x) = x$ (see Fig. 3.10).

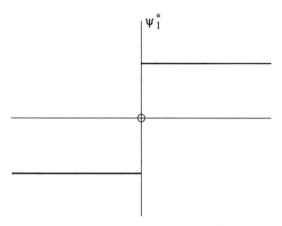

Figure 3.9   Optimal score function for the class of nondegenerate distribution densities $\mathcal{F}_1$

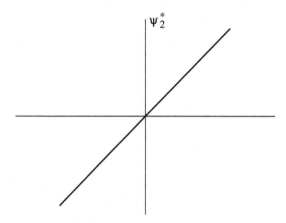

Figure 3.10   Optimal score function for the class of distribution densities with a bounded variance $\mathcal{F}_2$

- In the class of finite distribution densities, the least informative distribution density has the cosine form (Shevlyakov and Vilchevski 2002)

$$
f_3^*(x) = \begin{cases} \dfrac{1}{l}\cos^2\left(\dfrac{\pi x}{2l}\right) & \text{for } |x| \leq l, \\ 0 & \text{for } |x| > l. \end{cases} \tag{3.39}
$$

The optimal score function is unbounded: $\psi_3^*(x) = \tan(\pi x / 2l)$ for $|x| \leq l$ (see Fig. 3.11). In this case, the cosine form (3.61) of free extremal works.

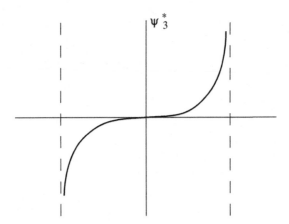

Figure 3.11   Optimal score function for the class of finite distribution densities $\mathcal{F}_3$

- In the class of approximately finite distribution densities, the least informative distribution density has two branches: the cosine one in the central part and the exponential at the tails (here both the exponential (3.19) and the cosine (3.61) forms of free extremals work) (Huber 1981; Shevlyakov and Vilchevski 2002)

$$
f_4^*(x) = \begin{cases} A_1\cos^2(B_1x) & \text{for } |x| \le l, \\[2mm] A_2\exp(-B_2|x|) & \text{for } |x| > l, \end{cases} \tag{3.40}
$$

where the constants A_1, A_2, B_1, and B_2 are determined from the simultaneous equations characterizing the restrictions of the class $\mathcal{F}_4$, namely, the conditions of normalization and approximate finiteness, and the conditions of continuity and continuous differentiability of the optimal solution at $x = l$ (for details, see Shevlyakov and Vilchevski (2002, p. 64).

The solution of those equations is given by the following formulas:

$$
A_1 = \frac{(1-\beta)\omega}{l(\omega+\sin(\omega))}, \qquad B_1 = \frac{\omega}{2l},
$$

$$
A_2 = \frac{\beta\lambda}{2l}e^\lambda, \qquad B_2 = \frac{\lambda}{l},
$$

where the parameters ω and β satisfy the equations

$$
\frac{2\cos^2(\omega/2)}{\omega\tan(\omega/2)+2\sin^2(\omega/2)} = \frac{\beta}{1-\beta}, \qquad 0 < \omega < \pi,
$$

and $\lambda = \omega\tan(\omega/2)$.

The optimal score function $\psi_4^*(x)$ is bounded (see Fig. 3.12)

$$\psi_4^*(x) = \begin{cases} \tan(B_1 x) & \text{for} \quad |x| \leq l, \\ \\ \tan(B_1 l)\text{sign}(x) & \text{for} \quad |x| > l. \end{cases}$$

REMARK 1 *The least informative density f_4^* also minimizes Fisher information in the class with the restriction of an inequality form*

$$\int_{-l}^{l} f(x)\, dx \geq 1 - \beta, \qquad 0 \leq \beta < 1,$$

or as a class with an upper bound on the distribution subrange $SR\beta(f)$ of level β

$$SR\beta(f) \leq b = 2l, \qquad 0 \leq \beta < 1.$$

REMARK 2 *The least informative density in the class F_1 of nondegenerate distribution densities is the special case of the optimal solution in the class of approximate finite distributions as*

$$l \to 0, \quad 1 - \beta \to 0, \quad \frac{1 - \beta}{2l} \to \frac{1}{2a}.$$

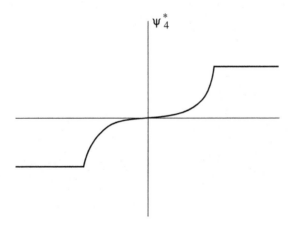

Figure 3.12   Optimal score function for the class of approximate finite distribution densities $\mathcal{F}_4$

3.4.3   Robust estimation of location in distribution models with a bounded variance

Here, due both to their importance for further reasoning and for practical applications, we consider the classes of distribution densities with a bounded variance (Shevlyakov and Vilchevski 2002).

The Weber–Hermite distribution densities

In Shevlyakov and Vilchevski (2002, p. 66) it is shown that a family of the Weber–Hermite distribution densities arises as the extremals of the variational problem of minimization of Fisher information $I(f)$ for location, namely, the following result holds.

Theorem 3.4.1. *Under the restrictions of non-negativeness, symmetry, normalization, and under a given variance d*

$$f(x) \geq 0, \qquad f(-x) = f(x),$$

$$\int_{-\infty}^{\infty} f(x)\, dx = 1,$$

$$\int_{-\infty}^{\infty} x^2 f(x) dx = d^2,$$

the extremals of the variational problem of minimization of Fisher information for location

$$f^*(x) = \arg \min_f \int_{-\infty}^{\infty} \left(\frac{f'(x)}{f(x)} \right)^2 f(x)\, dx \tag{3.41}$$

are of the following form:

$$f(x; v, d) = \frac{\Gamma(-v)\sqrt{2v+1+1/S(v)}}{d\sqrt{2\pi}S(v)} D_v^2 \left(\frac{|x|}{d} \sqrt{2v+1+1/S(v)} \right), \tag{3.42}$$

where

- $D_v(\cdot)$ *are the Weber–Hermite functions or the functions of the parabolic cylinder (Abramowitz and Stegun 1972);*
- *the parameter v takes its values in the interval* $(-\infty, 0]$;
- $S(v) = [\psi(1/2 - v/2) - \psi(-v/2)]/2,$
- $\psi(x) = \frac{d \ln \Gamma(x)}{dx}$ *is the digamma function.*

In this case, Fisher information for location is given by

$$I(v, d) = \frac{1}{d^2} \left[(2v + 1)^2 + \frac{4(2v + 1)}{S(v)} + \frac{3}{S^2(v)} \right].$$ (3.43)

The proof can be found in Shevlyakov and Vilchevski (2002, pp. 81–84).

We call the family of distribution densities (3.42) the family of the Weber–Hermite distribution densities. This family includes:

- the normal distribution density with $v = 0$

$$f(x; 0, d) = \frac{1}{d\sqrt{2\pi}} \exp\left(-\frac{x^2}{2 d^2}\right);$$

- the family of the $(k + 1)$-modal Hermite distribution densities with $v = k$, $k = 0, 1, \ldots,$

$$f(x; k, d) = \frac{\sqrt{2k + 1}}{d\sqrt{2\pi} \, k! \, 2^k} H_k^2 \left(\frac{|x|}{d} \sqrt{k + 1/2}\right) \exp\left(-\frac{(2k + 1)x^2}{2d^2}\right),$$

  where $H_k(x) = (-1)^k e^{x^2} \frac{d^k(e^{-x^2})}{dx^k}$ are the Hermite polynomials;

- the Laplace distribution density as $v \to -\infty$

$$f(x; -\infty, d) = L(x; 0, d/\sqrt{2}) = \frac{1}{d\sqrt{2}} \exp\left(-\frac{|x|\sqrt{2}}{d}\right);$$

- the unimodal Weber–Hermite densities with $-\infty < v < 0$ that are intermediate between the normal and Laplace densities.

REMARK *The Weber–Hermite densities (3.42) have two free parameters d and v; thus they arise in the solutions of the variational problems with two restrictions (one of them should be imposed upon a distribution variance).*

The least informative distribution density and the optimal score function in the class $\mathcal{F}_{12}$

Now we exhibit the minimax variance robust estimate of location in the class with the restrictions of an inequality form upon the values of a distribution density at the center of symmetry and a distribution variance

$$\mathcal{F}_{12} = \left\{ f : f(0) \geq \frac{1}{2a} > 0, \quad \sigma^2(f) = \int_{-\infty}^{\infty} x^2 f(x) \, dx \leq \overline{\sigma}^2 \right\}.$$ (3.44)

Here, the following result holds (Shevlyakov and Vilchevski 2002).

Theorem 3.4.2. *In the class $\mathcal{F}_{12}$, the least informative density is of the form*

$$
f_{12}^*(x) = \begin{cases} f_2^*(x) & \text{for} \quad \overline{\sigma}^2/a^2 < 2/\pi, \\ f(x; v, \overline{\sigma}) & \text{for} \quad 2/\pi \le \overline{\sigma}^2/a^2 \le 2, \\ f_1^*(x) & \text{for} \quad \overline{\sigma}^2/a^2 > 2, \end{cases} \tag{3.45}
$$

where $f(x; v, \overline{\sigma})$ are the Weber–Hermite distribution densities with $v \in (-\infty; 0]$ determined from the equation

$$
\frac{\overline{\sigma}}{a} = \frac{\sqrt{2v + 1 + 1/S(v)}\,\Gamma^2(-v/2)}{\sqrt{2\pi}\,2^{v+1}\,S(v)\,\Gamma(-v)}. \tag{3.46}
$$

The proof is based on checking of the inequalities (3.14) and (3.15); it can be found in Shevlyakov and Vilchevski (2002, pp. 84–85).

The optimal solution has the three branches: they arise with respect to the degree in which the constraints of the class $\mathcal{F}_{12}$ are taken into account:

- if $\overline{\sigma}^2 \le 2a^2/\pi$, that is, with relatively small variances and/or light distribution tails, only the restriction upon a variance matters (it takes the equality form $\sigma^2(f_{12}^*) = \overline{\sigma}^2$ whereas the restriction upon the value of a density at the center of symmetry is of the form of the strict inequality $f_{12}^*(0) > 1/2a$ and thus it can be neglected;

- vice versa, if $\overline{\sigma}^2 > 2a^2$, that is, with relatively large variances and/or heavy distribution tails, only the restriction on the density value $f_{12}^*(0) = 1/2a$ is essential whereas the restriction upon the variance $\sigma^2(f_{12}^*) < \overline{\sigma}^2$ does not matter;

- finally, in the middle zone $2a^2/\pi < \overline{\sigma}^2 \le 2a^2$ both restrictions work, taking the form of equalities $f_{12}^*(0) = 1/2a$ and $\sigma^2(f_{12}^*) = \overline{\sigma}^2$.

The optimal minimax variance algorithm of robust estimation of a location parameter is defined by the maximim likelihood score function for the least informative distribution density

$$
\psi_{12}^*(z) = \begin{cases} z/\overline{\sigma}^2 & \text{for} \quad \overline{\sigma}^2/a^2 \le 2/\pi, \\ -f'(z; v, \overline{\sigma})/f(z; v, \overline{\sigma}) & \text{for} \quad 2/\pi < \overline{\sigma}^2/a^2 \le 2, \\ a^{-1}\text{sign}(z) & \text{for} \quad \overline{\sigma}^2/a^2 > 2. \end{cases} \tag{3.47}
$$

From Equation (3.47) it follows that

- with relatively small variances and/or light distribution tails, the normal distribution density and the corresponding least squares (*LS*) method with the linear score function $\psi(z) \propto z$ is optimal;

- with relatively large variances and/or heavy distribution tails, the Laplace distribution density and the corresponding least absolute values (*LAV*) method with the sign score function $\psi(z) \propto \mathrm{sgn}(z)$ is optimal;

- with relatively moderate variances, a compromise between the *LS* and *LAV* methods with the score function $\psi_{12}^*(z)$ is the best: its behavior is displayed in Fig. 3.13.

This minimax variance robust *M*-estimate of location qualitatively differs from the Huber minimax variance robust *M*-estimate that is optimal in the class $\mathcal{F}_H$ (see subsection 3.2.1), even though they both have the *LS* and *LAV* estimates as the limiting cases.

Robust estimation of location in distribution models with bounded subranges

Similar to the above, results can be obtained in distribution models with bounded subranges.

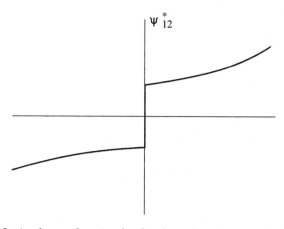

Figure 3.13   Optimal score function for the class of nondegenerate distribution densities with a bounded variance $\mathcal{F}_{12}$ in the intermediate between the classes $\mathcal{F}_1$ and $\mathcal{F}_2$ zone

Consider the class of approximately finite distribution densities

$$\mathcal{F}_5 = \left\{ f: \int_{-l}^{l} f(x)\, dx \geq 1 - \beta, \quad 0 < \beta < 1 \right\} \tag{3.48}$$

or the class with the restriction on the distribution mass in the central zone, where the constraint on the distribution mass can be rewritten as the constraint on the distribution subrange $SR\beta$

$$\mathcal{F}_5 = \{ f: SR\beta \leq b, \quad 0 < \beta < 1 \}, \tag{3.49}$$

where $b = 2l$.

Recall that in this case the least informative density has the *cosine*-central part and the exponential tails

$$f^*(x;\ A, B, C, D, b) = \begin{cases} A \cos^2(Bx), & |x| \leq b/2, \\[2mm] C \exp(-D|x|), & |x| > b/2, \end{cases} \tag{3.50}$$

where the values $A = A(\beta, b)$, $B = B(\beta, b)$, $C = C(\beta, b)$, and $D = D(\beta, b)$ are chosen to satisfy the conditions:

- the normalization condition

$$\int_{-\infty}^{\infty} f^*(x;\ A, B, C, D, b)\, dx = 1;$$

- the characterization condition of the class $\mathcal{F}_5$

$$\int_{-b/2}^{b/2} f^*(x;\ A, B, C, D, b)\, dx = 1 - \beta;$$

- the conditions of smoothness at $x = b/2$

$$f^*(b/2 - 0;\ A, B, C, D, b) = f^*(b/2 + 0;\ A, B, C, D, b),$$
$$f^{*\prime}(b/2 - 0;\ A, B, C, D, b) = f^{*\prime}(b/2 + 0;\ A, B, C, D, b).$$

The exponential tails of the least informative density and the corresponding form of the robust minimax contrast function $\rho^* = |x|$ for $|x| > b/2$ imply that the observed data with $|x| > b/2$ are simply ignored (rejected) when we apply this method. The smaller b, the more data are rejected.

Consider now the class $\mathcal{F}_{55}$ with the inequality constraints on distribution subranges

$$\mathcal{F}_{55} = \{ f: SR\beta_1 \leq b_1, \quad SR\beta_2 \leq b_2 \} \tag{3.51}$$

with $0 \leq \beta_2 \leq \beta_1 < 1, b_1 \leq b_2$. The following result holds in this case (Shevlyakov and Vilchevski 2002).

Theorem 3.4.3. *In the class* $\mathcal{F}_{55}$*, the least informative density is of the form*

$$f_{55}^*(x) = \begin{cases} f^*(x; \ A_2, B_2, C_2, D_2, b_2), \ b_2/b_1 \leq k_1, \\ f^*(x; \ A^*, B^*, C^*, D^*, b^*), \ k_1 < b_2/b_1 \leq k_2, \\ f^*(x; \ A_1, B_1, C_1, D_1, b_1), \ b_2/b_1 > k_2, \end{cases} \quad (3.52)$$

where

- *the function* $f^*(x; \ A, B, C, D, b)$ *is defined by Equation (3.50);*

- *the values of the parameters* $A_1, \dots, D_1$ *are set to* $A_1 = A(\beta_1, b_1)$, $B_1 = B(\beta_1, b_1)$, $C_1 = C(\beta_1, b_1)$, $D_1 = D(\beta_1, b_1)$;

- *the values of the parameters* $A_2, \dots, D_2$ *are set to* $A_2 = A(\beta_2, b_2)$, $B_2 = B(\beta_2, b_2)$, $C_2 = C(\beta_2, b_2)$, $D_2 = D(\beta_2, b_2)$;

- *the values of the parameters* A^*, B^*, C^*, D^**, and* b^**,* $b_1 \leq b^* \leq b_2$*, are determined from the equations including:*

the normalization condition

$$\int_{-\infty}^{\infty} f^*(x; \ A^*, B^*, C^*, D^*, b^*) \, dx = 1;$$

the characterization conditions of the class

$$\int_{-b_1/2}^{b_1/2} f^*(x; \ A^*, B^*, C^*, D^*, b^*) \, dx = 1 - \beta_1,$$

$$\int_{-b_2/2}^{b_2/2} f^*(x; \ A^*, B^*, C^*, D^*, b^*) \, dx = 1 - \beta_2;$$

the conditions of smoothness at $x = b^*$

$$f^*(b^* - 0; A^*, B^*, C^*, D^*, b^*) = f^*(b^* + 0; A^*, B^*, C^*, D^*, b^*),$$

$$f^{*\prime}(b^* - 0; A^*, B^*, C^*, D^*, b^*) = f^{*\prime}(b^* + 0; A^*, B^*, C^*, D^*, b^*);$$

- *the switching parameters k_1 and k_2 of solution (3.52) are derived from the equations*

$$\int_0^{b_2/2k_2} f^*(x; A_1, B_1, C_1, D_1, b_1)\, dx = (1 - \beta_1)/2,$$

$$\int_0^{k_1 b_1/2} f^*(x; A_2, B_2, C_2, D_2, b_2)\, dx = (1 - \beta_2)/2.$$

Three branches of the optimal solution (3.52) are connected with the degree in which the constraints are taken into account:

- in the first zone when $b_2/b_1 \leq k_1$, only the second restriction matters;

- in the third zone when $b_2/b_1 > k_2$, only the first restriction is substantial;

- in the intermediate zone, both restrictions are used.

From Equation (3.52) we can conclude that

- for a relatively small distribution dispersion (in the first zone), the "mild" robust algorithm based on $f^*(x; A_2, B_2, C_2, D_2, b_2)$ is optimal;

- for a relatively large distribution dispersion (in the third zone), the hard robust algorithm (with the hard rejection of sample elements) based on $f_1^*(x; A_1, B_1, C_1, D_1, b_1)$ is optimal,

- in the middle zone, a compromise between these algorithms is the best solution.

REMARK 1 *Given β_1 and β_2, the parameters b_1 and b_2 can be roughly estimated from the empirical distribution function; one of the possible choices is to set $\beta_1 = 0.9$ and $\beta_2 = 0.1$.*

REMARK 2 *The important feature of the obtained results on the least informative distributions is that the characterization conditions of the considered distribution classes are given in the inequality form, e.g., as the bounded variance and sub-ranges constraints. This allows both the application areas of the optimal solutions as compared to the equality form restrictions to be widened and to obtain fruitful three-branch analytical solutions.*

3.4.4 On the robustness of robust solutions: stability of least informative distributions

Generalities

This section is concerned with the stability of the least informative distributions minimizing Fisher information for location in a given class of distributions. Generally, the solutions of variational problems essentially depend on the regularity conditions of the assumed functional class. Here, the stability of these optimal solutions with respect to the violations of regularity conditions is studied under lattice distributions.

The discrete analogs of Fisher information for location are obtained: they have the form of the Hellinger metrics while estimating a real continuous location parameter and the form of the χ^2 metrics while estimating an integer discrete location parameter. The analytical expressions for the corresponding least informative discrete distributions are derived in some classes of lattice distributions by means of generating functions and the Bellman recursive functional equations of dynamic programming. These classes include the class of nondegenerate distributions with a restriction on the value of the density at the center of symmetry, the class of finite distributions, and the class of contaminated distributions.

The obtained least informative lattice distributions preserve the form of their prototypes in the continuous case. These results show the stability of robust minimax structures under different types of transitions from the continuous to be widened distribution to the discrete one (Shevlyakov and Vilchevski 2002).

As shown before, the form of the solution obtained by the minimax approach substantially depends on the characteristics of the distribution class. As a rule, the classes of continuous and symmetric distributions are considered. In many real-life problems of data processing, the results of measurements include groups of equal values. Furthermore, the results of measurements usually come rounded off in accordance with the scale of the measurement device. Thus, in these cases, the use of continuous distribution models does not seem adequate to the original problem of data processing, and for applications, it is quite important to design robust methods for discrete distribution models corresponding to the real nature of data.

Here we describe the analogs of Fisher information for the discrete distribution classes while considering:

- the direct discretization procedure of Fisher information functional in the problem of estimation of a continuous location parameter;

- the discrete analog of the Rao–Cramér inequality in the problem of estimation of a discrete location parameter.

In the latter case, the obtained form of the Rao–Cramér inequality is similar to the Chapman–Robbins inequality (Chapman and Robbins 1951).

The derived terms corresponding to the Fisher information functional are quite different in the above cases, but the solutions of the variational problems of minimization of these functionals (the least informative distributions) are the same.

Moreover, they demonstrate a remarkable correspondence with their continuous analogs. Thus we can conclude that the structure of robust minimax procedures is rather stable to deviations from the assumptions of regularity of the distribution classes.

Discrete analogs of Fisher information

Consider the class of lattice distributions

$$f_l(x) = \sum_i p_i \delta(x - i\Delta), \quad \sum_i p_i = 1, \tag{3.53}$$

where $\delta(\cdot)$ is the Dirac delta-function and Δ is the step of the lattice distribution. Henceforth, we set $\Delta = 1$.

We consider the following two cases:

- the location parameter is continuous with $\theta \in \mathbb{R}$;

- the location parameter is discrete with $\theta \in \mathbb{Z}$.

In the first case, the following result is true.

Theorem 3.4.4. *In the class of lattice distributions with a continuous parameter θ, the variational problem of minimization of Fisher information for location is equivalent to the following optimization problem:*

$$\sum (\sqrt{p_{i+1}} - \sqrt{p_i})^2 \to \min; \tag{3.54}$$

The proof of this and all further results can be found in Shevlyakov and Vilchevski (2002, p. 105–111).

In the second case with a discrete location parameter θ, the following analog of the Rao–Cramér inequality holds.

Theorem 3.4.5. *Let $x_1, \ldots, x_n$ be i.i.d. random variables with distribution density $f_l(x - \theta)$ (3.53), and $p_i > 0$, $i \in \mathbb{Z}$, $x_1, \ldots, x_n, \theta \in \mathbb{Z}$. Let $\widehat{\theta}_n = \widehat{\theta}_n(x_1, \ldots, x_n)$ be a discrete unbiased estimate of the discrete location parameter θ*

$$E(\widehat{\theta}_n) = \theta, \quad \widehat{\theta}_n \in \mathbb{Z}.$$

Then the variance of this estimate satisfies the Rao–Cramér inequality

$$var(\widehat{\theta}_n) \geq \left[\left(\sum_{i\in\mathbb{Z}} \frac{(p_{i-1} - p_i)^2}{p_i} + 1 \right)^n - 1 \right]^{-1}. \tag{3.55}$$

REMARK *The key feature of the obtained result is that in the discrete case the lower boundary of the estimate's variance decreases exponentially as $n \to \infty$, providing the corresponding efficiency of estimation is much greater than in the continuous case.*

Example 3.4.6. *Consider the double-geometric lattice distribution*

$$p_{-i} = p_i = \alpha^i p_0, \quad \alpha = \frac{1 - p_0}{1 + p_0}, \quad i = 0, 1, \ldots, \tag{3.56}$$

where $0 < p_0 < 1$.
In this case, the Rao–Cramér inequality has the form

$$var(\widehat{\theta}_n) \geq \frac{1}{\left(1 + \frac{4p_0^2}{1-p_0} \right)^n - 1}. \tag{3.57}$$

For sufficiently small p_0, the principal part of the denominator of the right-hand side of Equation (3.57) (the analog of Fisher information for the double-geometric lattice distribution) is equal to $4p_0^2$. Furthermore, the Rao–Cramér inequality takes the form

$$var(\widehat{\theta}_n) \geq \frac{1}{4p_0^2\, n},$$

and yields the classical order $1/n$ of decreasing asymptotic variance. It is noteworthy that the obtained lower bound upon the asymptotic variance is similar to that for the sample median if $p_0 = f(Me)$: $var(\text{med}) \geq [4f^2(Me)\, n]^{-1}$ is set.

In the class of lattice distributions (3.53) with a discrete parameter θ, the problem of minimization of Fisher information is equivalent to the optimization problem

$$\sum_{i\in\mathbb{Z}} \frac{p_{i-1}^2}{p_i} \to \min. \tag{3.58}$$

Least informative lattice distributions

Now we consider the discrete analogs of the least informative distributions for the classes of continuous distributions $\mathcal{F}_1$, $\mathcal{F}_3$, and $\mathcal{F}_H$.

Let $\mathcal{P}_1$ be the class of lattice symmetric nondegenerate distributions

$$\mathcal{P}_1 = \left\{ p_i, i \in \mathbb{Z} : p_i > 0, p_0 \geq \gamma_0 > 0, p_{-i} = p_i, \sum p_i = 1 \right\}.$$

Theorem 3.4.7. *In the class $\mathcal{P}_1$ of lattice distributions, the solutions of optimization problem (3.54) and (3.58) are given by*

$$p_{-i}^* = p_i^* = \alpha^i \gamma_0, \quad \alpha = \frac{1 - \gamma_0}{1 + \gamma_0}, \quad i = 0, 1, \ldots . \tag{3.59}$$

REMARK *The least informative lattice distribution $f_{\mathfrak{l}1}^*$ (3.53) with the geometric progression of p_i^* is the direct discrete analog of the least informative Laplace density f_1^* for the distribution class $\mathcal{F}_1$ of nondegenerate distributions (compare with the corresponding Rao–Cramér inequality (3.57)).*

Now consider the discrete analog of the class $\mathcal{F}_H$ of ε-contaminated distributions with the restrictions on the values of p_i in the central zone:

$$\mathcal{P}_H = \{ p_i, i \in \mathbb{Z} : p_i > 0, p_{-i} = p_i, p_i \geq \gamma_i > 0, i = 0, 1, \ldots, k; \} \tag{3.60}$$
$$\sum p_i = 1.$$

Theorem 3.4.8. *In the class $\mathcal{P}_H$ of lattice distributions with the additional restrictions on γ_i,*

$$\gamma_i^{1/2} - \gamma_{i-1}^{1/2} \leq \frac{(1 - \alpha^{1/2})^2}{2\alpha^{1/2}} \sum_{j=0}^{i-1} \gamma_j^{1/2},$$

the solution of variational problem (3.54) is of the form

$$p_{-i}^* = p_i^* = \begin{cases} \gamma_i, & i = 0, 1, \ldots, s^*, \ s^* \leq k, \\ \alpha^{i-s^*} \gamma_{s^*}, & i > s^*, \end{cases}$$

where

$$\alpha = \frac{1 - \gamma_0 - 2 \sum_{i=0}^{s^*} \gamma_i}{1 - \gamma_0 - 2 \sum_{i=0}^{s^*} \gamma_i + 2\gamma_{s^*}};$$

the sewing number s^* is determined by the maximum value of s satisfying

$$2(\gamma_{s-1}\gamma_s)^{1/2} + \left(1 - \gamma_0 - 2\sum_{i=0}^{s-1}\gamma_i\right)^{1/2}$$

$$\times\left[\left(1 - \gamma_0 - 2\sum_{i=0}^{s}\gamma_i\right)^{1/2} - \left(1 - \gamma_0 - 2\sum_{i=0}^{s-2}\gamma_i\right)^{1/2}\right] > 0.$$

The connection of this result with the Huber least informative density f_H^* (see Section 3.2) is obvious.

Finally, consider the discrete analog of the class of finite distributions $\mathcal{F}_3$:

$$\mathcal{P}_3 = \left\{p_i, i \in \mathbb{Z} : p_{-i} = p_i > 0, i = 0, 1, \dots, n; \ p_i = 0, i > n; \ \sum p_i = 1\right\}.$$

Theorem 3.4.9. *In the class $\mathcal{P}_3$ of lattice distributions, the solution of optimization problem (3.54) has the cosine form*

$$p_{-i}^* = p_i^* = \frac{1}{n+1}\cos^2\left(\frac{i\pi}{2(n+1)}\right), \qquad i = 0, \dots, n. \tag{3.61}$$

3.4.5   Concluding remark

The obtained results manifest the stability of the forms of robust minimax solutions to violations of regularity conditions caused by different types of transitions from the continuous to the discrete case.

3.5   Stable Estimation

3.5.1   Introductory remarks

The basis of what follows below is a reversal of the conventional setting described above: the maximum of some measure of sensitivity is minimized under the guaranteed value of the estimate variance or efficiency in a given class of distributions. The conventional point-wise local measures of sensitivity such as the influence and change-of-variance functions are not appropriate for this purpose – a global indicator of sensitivity is desirable. We show that a novel global indicator of robustness proposed by Shurygin (1994a,b, 2000a,b), namely, the estimate stability based on the variance sensitivity notion, is closely related, firstly, to the change-of-variance function, and, secondly, to the classical variation of the functional of the estimate asymptotic variance. Although a general theory has been developed for stable estimation of an arbitrary parameter of the underlying distribution, here we focus on estimating a parameter of location.

3.5.2   Variance sensitivity

A new measure of an estimate sensitivity derived from the asymptotic variance $V(\psi, f)$ was introduced by Shurygin (1994a,b) (see also Shevlyakov *et al.* 2008). Formally, it is derived as follows:

$$
\begin{aligned}
\frac{\partial V(\psi, f)}{\partial f} &= \frac{\partial}{\partial f} \frac{\int \psi^2 f \, dx}{\left(\int \psi' f \, dx\right)^2} \\
&= \frac{\int \psi^2 \, dx}{\left(\int \psi' f \, dx\right)^2} - 2 \frac{\int \psi' \, dx \int \psi^2 f \, dx}{\left(\int \psi' f \, dx\right)^3} \\
&= \frac{\int \psi^2 \, dx}{B(\psi, f)^2} - 2 \frac{\int \psi' \, dx \, A(\psi, f)}{B(\psi, f)^3}.
\end{aligned}
\tag{3.62}
$$

In order for this to make mathematical sense, the density f and the function ψ must be smooth. Equation (3.62) defines a global measure of the stability of an estimate under an improper model where the outliers occur with equal chance anywhere on the real line. Since the existence of the asymptotic variance (3.4) is guaranteed by the existence and positiveness of the integrals $A(\psi, f)$ and $B(\psi, f)$ in (3.5), finiteness of the Lagrange derivative (3.62) holds under the condition of square integrability, i.e., $\psi \in L^2(\mathbb{R})$. Such ψ-functions are automatically redescending, because the integral only converges if $\psi(x) \to 0$ for $|x| \to \infty$. From this it follows that $\int \psi' \, dx = 0$, so that for estimates with a finite Lagrange functional derivative, the second term summand in (3.62) vanishes and the Lagrange derivative of the asymptotic variance can in fact be simplified.

Definition 3.5.1. *The scalar*

$$
VS(\psi, f) = \frac{\partial V(\psi, f)}{\partial f} = \frac{\int \psi^2 \, dx}{\left(\int \psi' f \, dx\right)^2}
\tag{3.63}
$$

is called the variance sensitivity of the M-estimate defined by the score function ψ.

The variance sensitivity is an extremely strict indicator of stability of an estimate: its finiteness is equivalent to redescendency. Since the square of the influence function averaged with respect to the model density is equal to the asymptotic variance (3.27), it is not surprising that

$$
VS(\psi, f) = \int IF^2(x; \psi, f) \, dx.
$$

The following connection of the variance sensitivity with the change-of-variance function is of importance. For M-estimates that satisfy $\psi \in C^1(\mathbb{R})$, the change-of-variance function is given by (Hampel *et al.* 1986)

$$CVF(x;\psi,f) = \frac{A(\psi,f)}{B^2(\psi,f)} \left(1 + \frac{\psi^2(x)}{A(\psi,f)} - 2\frac{\psi'(x)}{B(\psi,f)} \right), \tag{3.64}$$

where $A(\psi,f)$ and $B(\psi,f)$ are given by (3.5). Thus, up to an additive constant, the variance sensitivity is merely equal to the integral of the *CVF*, if it exists.

Now we dwell on the relation of the Lagrange derivative (3.62) to the variation $\delta V(\psi,f)$ of the asymptotic variance $V(\psi,f)$, finding its principal part with respect to $\|\delta f\|$:

$$\begin{aligned}
V(\psi,f+\delta f) &= \frac{\int \psi^2(f+\delta f)\,\mathrm{d}x}{\left(\int \psi'(f+\delta f)\,\mathrm{d}x \right)^2} \\[2mm]
&= \frac{\int \psi^2 f\,\mathrm{d}x + \int \psi^2 \delta f\,\mathrm{d}x}{\left(\int \psi'f\,\mathrm{d}x \right)^2 + 2\int \psi'\delta f\,\mathrm{d}x \int \psi'\delta f\,\mathrm{d}x + \left(\int \psi'\delta f\,\mathrm{d}x \right)^2} \\[2mm]
&= \frac{\int \psi^2 f\,\mathrm{d}x + \int \psi^2 \delta f\,\mathrm{d}x}{\left(\int \psi'f\,\mathrm{d}x \right)^2} \left(1 - 2\frac{\int \psi'\delta f\,\mathrm{d}x}{\int \psi'f\,\mathrm{d}x} \right) + o(\|\delta f\|) \\[2mm]
&= V(\psi,f) + \frac{\int \psi'f\,\mathrm{d}x \int \psi^2 \delta f\,\mathrm{d}x - 2\int \psi^2 f\,\mathrm{d}x \int \psi'\delta f\,\mathrm{d}x}{\left(\int \psi'f\,\mathrm{d}x \right)^3} + o(\|\delta f\|).
\end{aligned}$$

Finally we get

$$\begin{aligned}
\delta V(\psi,f) &= V(\psi,f+\delta f) - V(\psi,f) \\[2mm]
&= \frac{B(\psi,f)\int \psi^2 \delta f\,\mathrm{d}x - 2A(\psi,f)\int \psi'\delta f\,\mathrm{d}x}{B(\psi,f)^3} + o(\|\delta f\|) \\[2mm]
&= \int \left(\frac{\psi(x)^2}{B(\psi,f)^2} - 2\frac{A(\psi,f)\psi'(x)}{B(\psi,f)^3} \right) \delta f(x)\,\mathrm{d}x + o(\|\delta f\|).
\end{aligned}$$

Taking into account the analog of the Lagrange mean theorem (Bohner and Guseinov 2003), we arrive at the following relation for the sought variation:

$$\delta V(\psi,f) = \delta f(x^*) \int \left(\frac{\psi(x)^2}{B(\psi,f)^2} - 2\frac{A(\psi,f)\psi'(x)}{B(\psi,f)^3} \right) \mathrm{d}x$$

where $x^* \in \mathbb{R}$, and thus its principal part is proportional to the Lagrange derivative (3.62)

$$\delta V(\psi,f) \propto \frac{\partial V(\psi,f)}{\partial f}.$$

In other words, the Lagrange derivative (3.62) reflects the variation of the asymptotic variance $V(\psi,f)$ corresponding to the uncontrolled variation of a distribution density f.

REMARK   *The variance sensitivity of any M-estimate with nondecreasing ψ-function is infinite. The mean, the trimmed means, and the Huber estimates all have infinite variance sensitivity. Writing the median as a limit of such estimates shows that even the median has infinite sensitivity, which comes as a surprise. In fact, from the point of view of the bias, this estimate is the most robust possible.*

3.5.3   Estimation stability

In what follows, the following problem is considered: what is the least variance sensitive ψ-function for a given distribution F with density $f \in C^1(\mathbb{R})$? This is a sensible problem, because any estimate with low variance sensitivity has an asymptotic variance that is little affected if the assumed model is only approximately true. Robustness with regard to uncertainty in the model density has thus been taken care of. Under this point of view, exploring the range of the variance sensitivity and of the efficiency at a given density is all that remains to be done. As shown in Shurygin (1994a,b), the solution of the problem is given by

$$\psi_{MVS}(x) = \arg \min_{\psi \in C^1(\mathbb{R})} VS(\psi,f) = -f'(x). \qquad (3.65)$$

This and further similar results are obtained by the calculus of variations techniques through writing out the Euler–Lagrange equations for the appropriate functionals. In the case of minimization of the variance sensitivity, the problem is reduced to minimization with respect to ψ of the numerator of the fraction (3.63) subject to its bounded denominator, i.e., the functional $J(\psi) = \int (\psi^2 + \lambda \psi'f) \, dx$, where λ is the Lagrange multiplier corresponding to the aforementioned condition. The Euler–Lagrange equation has the form $2\psi - \lambda f' = 0$, giving the required result $\psi_{MVS}(x) = -f'(x)$.

The estimate with the score function (3.65) is called the estimate of *minimum variance sensitivity* with $VS_{\min} = VS(\psi_{MVS},f)$. It is easy to check that the minimum sensitivity functional takes the form

$$VS_{\min} = VS(\psi_{MVS},f)$$

$$= \left(\int \psi_{MVS}^2 dx \right)^{-1} = \left(\int (f'(x))^2 dx \right)^{-1}. \qquad (3.66)$$

Definition 3.5.2. *The stability of any M-estimate is defined as the ratio of the minimum variance and estimate variance sensitivities*

$$\text{stb}(\psi,f) = \frac{VS_{\min}(f)}{VS(\psi,f)}. \qquad (3.67)$$

Similarly to estimate efficiency, the estimate stability values lie in the segment [0, 1].

Setting different weights for efficiency and stability, various optimality criteria can be constructed (Shurygin 1994a,b 2000a). In particular, the structure of redescenders is specified by the analog of Hampel's lemma (Hampel *et al.* 1986): the maximum efficiency under the guaranteed stability

$$\max_{\psi \in C^1(\mathbb{R})} \mathrm{eff}(\psi, f), \quad \mathrm{stb}(\psi, f) \geq \mathrm{stb}, \quad 0 \leq \mathrm{stb} < 1$$

is provided by the redescending M-estimate called *conditionally optimal* with the following score function:

$$\psi_{c.opt}(x) = \frac{\psi_{ML}(x)}{1 + \lambda/f(x)}, \tag{3.68}$$

where $\psi_{ML} = -f'/f$ is the maximum likelihood score function and λ is the Lagrange multiplier corresponding to the restriction upon stability. From (3.68) it follows that if the restriction upon stability does not matter, that is, when stb = 0 and therefore $\lambda = 0$, then $\psi_{c.opt}(x) = \psi_{ML}(x)$; otherwise $\psi_{c.opt}(x)$ is redescending, i.e., $\psi_{c.opt}(x) \to 0$ as $|x| \to \infty$. Obviously, the conditionally optimal estimate (3.68) also maximizes stability under guaranteed efficiency.

In practice, the freedom in choosing the level of guaranteed stability or efficiency may be inconvenient. In this case, a reasonable choice can be made by equating efficiency and stability, $\mathrm{eff}(\psi) = \mathrm{stb}(\psi)$: this estimate is called *radical*, and the score function of the radical M-estimate is given by the maximum likelihood score function ψ_{ML} multiplied by the weight function $\sqrt{f(x)}$

$$\psi_{rad}(x) = \psi_{ML}(x) \sqrt{f(x)} = -\frac{f'(x)}{\sqrt{f(x)}}. \tag{3.69}$$

For distributions from the exponential family, the estimates of minimum sensitivity and the radical estimates belong to the class of M-estimates with the exponentially weighted maximum likelihood score functions previously proposed on intuitive grounds by Meshalkin (1971).

In Table 3.1, efficiency and stability of the maximum likelihood estimate with the score function ψ_{ML}, the sample mean and median, the estimate of minimum variance sensitivity ψ_{MVS}, the radical estimate ψ_{rad}, as well as the minimax variance of Huber's and redescending Hampel's M-estimates with the score functions ψ_H and ψ_{25A}, respectively, are computed at the following distribution densities:

- the standard normal $\varphi(x) = (2\pi)^{-1/2} \exp(-x^2/2)$;
- the Laplace $L(x) = 2^{-1} \exp(-|x|)$;
- the Cauchy $C(x) = \pi^{-1}(1 + x^2)^{-1}$;
- the Cauchy contaminated normal $f(x) = 0.9\, \varphi(x) + 0.1\, C(x)$.

Table 3.1   Efficiency and stability of M-estimates of location

Score function		Normal	Contaminated normal	Laplace	Cauchy
ψ_{ML}	eff	1	1	1	1
	stb	0	0.38	0	0.50*
ψ_{mean}	eff	1	0	0.50	0
	stb	0	0	0	0
ψ_{med}	eff	0.64	0.69	1	0.89*
	stb	0	0	0	0
ψ_H	eff	**0.92**	0.88*	0.67	0.75
	stb	0	0	0	0
ψ_{25A}	eff	0.91*	**0.89**	0.64	0.81
	stb	0.69*	0.62*	0.15*	0.49
ψ_{MVS}	eff	0.65	0.71	0.75*	0.80
	stb	1	1	1	1
ψ_{rad}	eff	0.84	**0.89**	**0.89**	**0.92**
	stb	**0.84**	**0.89**	**0.89**	**0.92**

Table 3.1 exhibits the results (the best are boldfaced, next to them are starred) for four distributions, two of them, Cauchy and Cauchy contaminated normal, with extremely heavy tails. Since the corresponding score functions are not square integrable on the real line, the maximum likelihood estimates of location for the normal and Laplace distributions, namely, the mean and median, as well as Huber's M-estimate, have zero stability. The redescending maximum likelihood score functions for the heavy-tailed Cauchy and Cauchy contaminated normal distributions, on the other hand, provide a reasonably moderate level of stability equal to 0.5 and 0.38, respectively. The redescending minimum variance sensitivity and radical estimates perform well both in efficiency and stability, especially the radical estimate.

3.5.4   Robustness of stable estimates

The above optimization problem is local, dealing with a given density f. What if the density f is unknown or it belongs to some class $\mathcal{F}$? As pointed out above, finite variance stability implies a high degree of robustness with regard to changes in the model density f. While it is of interest to study classes of densities, it is not necessary to look at full neighborhoods as in Huber's theory. Nor is it necessary to include heavy-tailed densities in the class.

The simplest and most direct way to examine the variability of the newly introduced characteristics, the variance sensitivity and estimate's stability, is to compare their values on some set of distributions. Partially, it was done above for the normal, Laplace, Cauchy, and Cauchy contaminated distributions. Now we enlarge that set of examples.

Example 3.5.1. *Consider the class of exponential-power densities*

$$f(x; q) \propto e^{-|x|^q}, \qquad 1 \le q < \infty,$$

and the corresponding radical estimate with the score function ψ_{rad} *(3.69).*
The efficiency and stability of these estimates are equal

$$\text{eff}(\psi_{rad}) = \text{stb}(\psi_{rad}) = \frac{64}{81} \left(\frac{9}{8}\right)^{1/q}.$$

It is decreasing in q attaining a maximal value at the Laplace distribution (q = 1):

$$\text{eff}(\psi_1) = \text{stb}(\psi_1) = 8/9 \approx 0.89$$

and a minimal value at q = ∞*:*

$$\text{eff}(\psi_\infty) = \text{stb}(\psi_\infty) = 64/81 \approx 0.79.$$

For the normal distribution (q = 2), the values of efficiency and stability are

$$\text{eff}(\psi_2) = \text{stb}(\psi_2) = 32/(27\sqrt{2}) \approx 0.84.$$

So, under exponential power distributions, we observe a rather low variability of the examined characteristics.

Example 3.5.2. *Let now f(x) be the Huber least favorable density (3.21) in the class of ε-contaminated normal distributions and consider the radical estimate of the center of symmetry θ = 0 with the score function (3.69).*
The efficiency and stability of the radical estimate are

$$\text{eff } \hat{\theta}(\varepsilon) = \text{stb } \hat{\theta}(\varepsilon) = \frac{32}{27\sqrt{2}} \frac{\Phi^2(\sqrt{3/2}\, k(\varepsilon))}{\Phi(k(\varepsilon))\, \Phi(\sqrt{2}\, k(\varepsilon))},$$

where $\Phi(x) = \int_0^x \varphi(t)\, dt$ *and k(ε) satisfies (3.22).*

The minimum and maximum values are attained at the normal distribution with
$\varepsilon = 0$ *and at the Laplace distribution with* $\varepsilon \to 1$, *respectively:*

$$\text{eff } \hat{\theta}(0) = \text{stb } \hat{\theta}(0) = 32/(27\sqrt{2}) \approx 0.84,$$

$$\lim_{\varepsilon \to 1} \text{eff } \hat{\theta}(\varepsilon) = \lim_{\varepsilon \to 1} \text{stb } \hat{\theta}(\varepsilon)) = 8/9 \approx 0.89.$$

*In this case, we have also a rather narrow range of the variability of estimate effi-
ciency and stability, both being relatively high.*

Example 3.5.3. *Consider the standard normal distribution density $f(x) = \varphi(x)$ and
the optimal V-robust redescending hyperbolic tangent estimate minimizing asymptotic variance $V(\psi, f)$ subject to the bounded sensitivity of the change-of-variance
function*

$$\kappa^*(\psi, f) = \sup_x \frac{CVF(x; \psi, f)}{V(\psi, f)} \le k,$$

with the score function (Hampel et al. 1986, pp. 160–167)

$$\psi_{\text{tanh}}(x) = \begin{cases} x, & 0 \le |x| \le p, \\ \sqrt{A(k-1)} \tanh\left[\sqrt{\frac{k-1}{A}} \frac{B(r-|x|)}{2}\right] \text{sgn}(x), & p \le |x| \le r, \\ 0, & r \le |x|, \end{cases}$$

*where the recommended choice $r = 4.0$ and $k = 4.5$ (Hampel et al. 1986, p. 163)
implies*

$$A = 0.804598, \quad B = 0.877210 \quad and \quad p = 1.634416.$$

The efficiency and stability of this estimate are

$$\text{eff}(\psi_{\text{tanh}}) = 0.96, \quad \text{stb}(\psi_{\text{tanh}}) = 0.53.$$

Here we observe a highly efficient estimate with an acceptable stability.

In Figs 3.14 to 3.16, efficiency and stability of the main M-estimates of location
defined by the following score functions are scattered (Shurygin 1994a):

- the maximum likelihood score $\psi_{ML}(x) = -f'(x)/f(x)$,

- the linear score $\psi_{mean}(x) = x$ with the sample mean as an estimate,

- the sign score $\psi_{med}(x) = \text{sign}(x)$ with the sample median as an estimate,

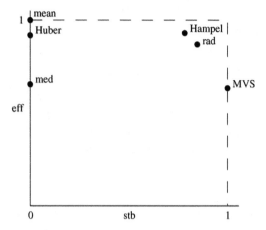

Figure 3.14   Efficiency and stability of the estimates of location at the normal distribution with density $N(x; \mu, 1)$

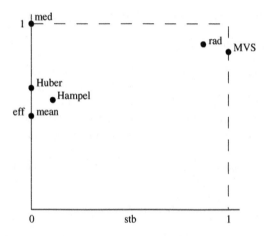

Figure 3.15   Efficiency and stability of the estimates of location at the Laplace distribution with density $L(x; \mu, 1)$

- Huber's linear bounded score (Huber 1964) with $k = 1.5$

$$\psi_H(x) = \max[-1.5, \min(x, 1.5)];$$

- Hampel's redescending three-part score $\psi_{25A}(x)$ of subsection 3.3.7 with the following values of the parameters: $a = 1.31, b = 2.039,$ and $r = 4$ (see Hampel *et al.* 1986, p. 167);

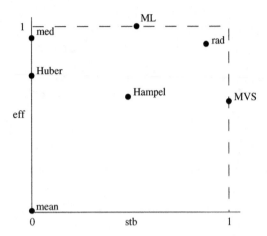

Figure 3.16   Efficiency and stability of the estimates of location at the Cauchy distribution with density $C(x; \mu, 1)$

- the minimum variance sensitivity score given by $\psi_{MVS}(x) = -f'(x)$,

- the radical score function given by $\psi_{rad}(x) = -f'(x)/\sqrt{f(x)}$.

Hampel's three-part M-estimate revealed its advantageous properties in the Princeton experiment (Andrews *et al.* 1972). Later, it was shown that the optimal V-robust redescending hyperbolic tangent M-estimate is very close to it in performance, being only about 1% more efficient (see Hampel *et al.* 1986, p. 167). Here we prefer Hampel's three-part score due to its much simpler implementation as compared to the optimal *tanh*-score.

In the comparative analysis of the estimates of a location parameter μ, the following noise distribution densities $f(x)$ are used:

- the normal distribution with density $N(x; \mu, 1)$,

- the Laplace distribution with density

$$L(x; \mu, 1) = \frac{1}{2} \exp\left[-|x - \mu|\right],$$

- the Cauchy distribution with density

$$C(x; \mu, 1) = \frac{1}{\pi[1 + (x - \mu)^2]}.$$

From Figs 3.14 to 3.16 it follows that, on the whole, the stable radical estimate outperforms the others including Huber's and Hampel's V-robust optimal estimates.

3.5.5   Maximin stable redescending M-estimates

From these examples, one may expect that the optimal M-estimates of location that maximize the minimum variance sensitivity will provide a high guaranteed level of stability of estimation over wide nonparametric classes of distributions.

Now we briefly describe this version of Huber's minimax approach (for details, see Shevlyakov *et al.* 2008). Consider the following maximin problem:

$$(\psi^*, f^*) = \arg \max_{f \in \mathcal{F}} \min_{\psi \in \Psi} VS(\psi, f), \tag{3.70}$$

where $\mathcal{F}$ and Ψ are some suitable classes of distribution densities and score functions, respectively. This setting is almost equivalent to Huber's setting of the problem of minimax estimation of location (Huber 1964) up to the substitution of the asymptotic variance $V(\psi, f)$ by the variance sensitivity $VS(\psi, f)$.

The solution of the inner minimization problem in (3.70) is given by the minimum variance sensitivity score function $\psi_{MVS}(x) = -f'(x)$ (3.65), and since the minimum sensitivity takes the form (3.66), the solution of problem (3.70) is reduced to the solution of the variational problem of minimization of the following functional:

$$f^* = \arg \min_{f \in \mathcal{F}} J(f), \qquad J(f) = \int (f'(x))^2 \, dx, \tag{3.71}$$

with the subsequent application of Equation (3.65) to the least informative density f^*

$$\psi^*(x) = -f^{*'}(x). \tag{3.72}$$

The optimal pair (ψ^*, f^*) provides the guaranteed stability of estimation over a chosen class of distribution densities: the variance sensitivity $VS(\psi^*, f)$ of the maximin M-estimate with the score function ψ^* does not exceed the value of $VS(\psi^*, f^*)$ for all $f \in \mathcal{F}$

$$VS(\psi^*, f) \le VS(\psi^*, f^*).$$

Similarly to the solution of Huber's minimax problem (Huber 1964), in this case the optimal maximin solution strongly depends on the characteristics of the chosen class of densities $\mathcal{F}$.

Under regularity conditions imposed on distribution densities and score functions (their specifics as compared with the standard ones described, for example, in Hampel *et al.* (1986) is that they allow densities with finite support, provided the existence of nonzero Fisher information), the following result holds (Shevlyakov *et al.* 2008): in the class of densities with a bounded variance

$$\mathcal{F} = \left\{ f : \ f(x) \ge 0, \int f(x) dx = 1, \sigma^2(f) = \int x^2 f(x) \, dx \le \overline{\sigma}^2 \right\}, \tag{3.73}$$

the maximin M-estimate of location is Smith's estimate with the score function $\psi_{Sm}^*(x)$ given by (3.32).

Example 3.5.4. *Let $\bar{\sigma} = 1$. In this case, the score function takes the form*

$$\psi_{Sm}^*(x) = x(7 - x^2)\, 1_{[-\sqrt{7},\sqrt{7}]}(x).$$

The efficiency and stability of the corresponding redescending Smith's M-estimate at the standard normal density $\varphi(x)$ are

$$\text{eff}(\psi_{Sm}^*, \varphi) = 0.835, \quad \text{stb}(\psi_{Sm}^*, \varphi) = 0.912.$$

Note the high levels of the estimate's efficiency and stability, naturally the latter higher than the former since the objective functional relates to variance stability. Furthermore, in Shevlyakov et al. (2008) it is also shown that in the class with a bounded fourth distribution moment the optimal M-estimate providing the maximum of the minimum variance sensitivity is just Tukey's biweight estimate with the score function $\psi_{bi}(x)$ given by (3.31).

3.5.6   Concluding remarks

The influence function is a basic tool for studying statistical functionals. Besides the smoothness and boundedness of an influence function, we introduce square integrability as another characteristic of interest. Square integrability implies stability of the asymptotic variance and it excludes estimates whose ψ-functions are not redescending. For such highly resistant estimates, it can be argued that optimality at an ideal model is all that is required.

Given an ideal model density f, redescending M-estimates naturally arise, either when minimizing the variance sensitivity or when maximizing the efficiency under a lower bound on the stability. Generally, the optimal score functions $\psi(x)$ are of the form of a product of a weight function $w(x)$ and the maximum likelihood score function $\psi_{ML}(x) = -f'(x)/f(x)$, that is,

$$\psi(x) = w(x)\, \psi_{ML}(x).$$

The weights are:

- $w(x) = f(x)$ for the estimate of minimum variance sensitivity (3.65),
- $w(x) = f(x)/[\lambda + f(x)]$ for the conditionally optimal estimate (3.68),
- $w(x) = \sqrt{f(x)}$ for the radical estimate (3.69).

Among these, we recommend the radical M-estimate, which is at the same time highly efficient and stable (see Table 3.1; Shevlyakov *et al.* 2008; Shurygin 1994a,b 2000a).

If no ideal model density is assumed and instead of it a class $\mathcal{F}$ of model densities is considered, we may maximize the minimum variance sensitivity. We show that such well-known redescending M-estimates as Smith's estimate and Tukey's biweight can be justified in this manner.

Here we mostly deal with the so-called V-robustness not touching at all the aspects of B-robustness. However, redescending M-estimates of location optimal in the V-robustness sense perform quite well and also with respect to bias (Hampel *et al.* 1986).

3.6   Robustness Versus Gaussianity

3.6.1   Introductory remarks

As emphasized in Section 3.1, the main argument pro robustness is that we never exactly but only approximately know the shape of an underlying data distribution; thus various neighborhood models of an ideal distribution (supermodels) are proposed to describe the observed uncertainty, for instance, the Tukey gross error model (Huber 1981) with a Gaussian as an ideal distribution. Hence, it can be roughly said that practically all the developed robust statistical methods finally can be reduced to the maximum likelihood approach applied to some non-Gaussian distribution in the neighborhood of a Gaussian or another ideal distribution, and thus, on the whole, robust approaches can be called contra-Gaussian.

At the same time, the Gaussian distribution models with the related statistical tools are successfully and widely used in almost all data and signal processing tasks. The reasons for that in no way can be reduced only to the central limit theorem (CLT) and its consequences.

In what follows, we briefly highlight those reasons and finally arrive at the following paradox: due to its essential properties, a Gaussian distribution with the related statistical methods is entirely connected with the essence of the robustness paradigm.

Henri Poincaré: *Physicists believe that the Gaussian law has been proved in mathematics while mathematicians think that it was experimentally established in physics.*

This witty remark of a great mathematician (Poincaré 1904) reflects the fact of the ubiquitous use and success of the Gaussian distribution law and at the same moment gives both a humorous and serious hint to explain this phenomenon. The majority of scientific community shares the common belief that it is due to the CLT. We will show that the CLT is not only a unique reason but perhaps it is even not the main one.

Here, we try to answer the question: "Why the ubiquitous use and success of the Gaussian distribution law?" The history of the Gaussian or normal distribution is rather long has lasted for nearly three hundred years since it was discovered by de Moivre in 1733 and the related literature is immense. An extended and thorough treatment of the topic and a survey of the works in the related area are given in the posthumously edited book of E.T. Jaynes (2003), and we partially follow this source, in particular while considering the history of the posed question. The important aspects of the general history of noise, especially of Brownian motion, are given in Cohen (2005). Our main contribution to the topic is concerned with highlighting the role of Gaussian models in signal processing based on the optimal property of the Gaussian distribution minimizing Fisher information over the class of distributions with a bounded variance.

In what follows, we deal only with the univariate Gaussian distribution, leaving the properties of the multivariate Gaussian distribution out of the scope of our consideration. First of all, we present the ideas of classical derivations of the Gaussian law. Then we consider its properties and characterizations including the CLT and minimization of the distribution entropy and Fisher information. Finally, we dwell on the connections between Gaussianity and robustness in signal processing.

The Gaussian or normal distribution density is defined as

$$N(x;\ \mu, \sigma) = \frac{1}{\sigma\sqrt{2\pi}} \exp\left[-\frac{(x-\mu)^2}{2\sigma^2}\right], \quad -\infty < x < \infty, \qquad (3.74)$$

where μ and σ are the parameters of location (mean) and scale (standard deviation), respectively. Its standard form is commonly denoted by $\varphi(x) = N(x;\ 0, 1)$ (see Fig. 3.17).

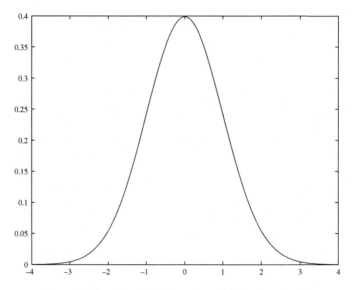

Figure 3.17   Standard Gaussian distribution density

Using Stirling's approximations for factorials, it can be shown that the Gaussian distribution is a limiting form of the binomial distribution (Feller 1971)

$$P_{n,k}(p) = C_n^k p^k q^{n-k}, \qquad k = 0, 1, \ldots, n,$$

$$0 < p < 1, \qquad q = 1 - p, \qquad C_n^k = \frac{n!}{k!(n-k)!},$$

$$P_{n,k}(p) \; \to \; \frac{1}{\sqrt{npq}} \; \varphi \left(\frac{k - np}{\sqrt{npq}} \right) \quad \text{as} \quad n, k \; \to \; \infty,$$

with $(k - np)/\sqrt{npq}$ finite.

In the particular case with $p = q = 1/2$, the Gaussian distribution had been found by de Moivre (1733), who did not recognize its significance. In the general case when $0 < p < 1$, Laplace (1781) had derived its main properties and suggested that it should be tabulated due to its importance. Gauss (1809) considered another derivation of this distribution (not as a limiting form of the binomial distribution); it became popularized by his work and thus his name was attached to it. The fundamental Boltzmann distribution of statistical mechanics (Boltzmann 1871), exponential in energies, is the Gaussian in velocities (Maxwell 1860).

It seems likely that the term "normal" is associated with a linear regression model $y = \Phi \, \beta + e$, where the vector y and the matrix Φ are known, the vector of parameters β and the noise vector e unknown; to solve this linear regression problem, Gauss (1823) suggested the least squares method and called the system of equations $\Phi' \Phi \, \beta = \Phi' \, y$, which gives the least square parameter estimates $\hat{\beta}$, the "normal equations".

A well-known historian of statistics, Stigler (1980) formulated a universal law of eponymy that "no discovery is named for its original discoverer". Jaynes (2003) truly notices that "the history of this terminology excellently confirms this law, since the fundamental nature of this distribution and its main properties were derived by Laplace when Gauss was six years old; and the distribution itself had been found by de Moivre before Laplace was born".

3.6.2   Derivations of the Gaussian distribution

Derivation of Gauss (1809)

Consider a sample of $n + 1$ independent observations $x_0, x_1, \ldots, x_n$ taken from the distribution with density $f(x; \theta)$, where θ is a parameter of location. Its maximum likelihood estimate $\hat{\theta}$ must satisfy

$$\frac{\partial}{\partial \theta} \log \left\{ \prod_{i=0}^{n} f(x_i; \theta) \right\} = \sum_{i=0}^{n} \frac{\partial}{\partial \theta} \log f(x_i; \theta) = 0.$$

Assuming differentiability of $f(x; \theta)$ and denoting

$$\log f(x; \theta) = g(x - \theta),$$

we have that the maximum likelihood estimate will be the solution of

$$\sum_{i=0}^{n} g'(x_i - \hat{\theta}) = 0. \tag{3.75}$$

Carl Friedrich Gauss (1809) asked the following question: "What would be a distribution density $f(x; \theta)$ for which the maximum likelihood estimate $\hat{\theta}$ is the sample mean

$$\hat{\theta} = \bar{x} = \frac{1}{n+1} \sum_{i=0}^{n} x_i \quad ? \quad "$$

Note that here we use the modern terminology adopted by the scientific community more than a century later (the method of maximum likelihood was proposed by Fisher (1922).

To answer the posed question, we apply the following method of functional equations (Aczel 2002). First, set $x_0 = x_1 = \cdots = x_n = 0$, then check that $\bar{x} = 0$, write out Equation (3.75), and get that

$$g'(0) = 0. \tag{3.76}$$

Second, set $x_0 = u, x_1 = -u, x_2 = 0, \ldots, x_n = 0$, check that $\bar{x} = 0$, and then from Equations (3.75) and (3.76) it follows that

$$g'(-u) = -g'(u). \tag{3.77}$$

Third, set $x_0 = -u\,(n+1), x_1 = 0, \ldots, x_n = 0$, get that $\bar{x} = -u$, and then Equations (3.75) and (3.77) imply

$$g'(-n\,u) + \sum_{i=1}^{n} g'(u) = 0,$$

and thus $g(u)$ must satisfy the functional equation

$$g'(n\,u) = n\,g'(u), \qquad n = 1, 2, 3, \ldots . \tag{3.78}$$

From Equation (3.78) it follows that

$$g'(1) = n\,g'\left(\frac{1}{n}\right) \qquad n = 1, 2, 3, \ldots ,$$

$$g'\left(\frac{m}{n}\right) = m\,g'\left(\frac{1}{n}\right) \qquad m, n = 1, 2, 3, \ldots ,$$

and we get the linear equation that holds for all rational numbers

$$g'\left(\frac{m}{n}\right) = a\,\frac{m}{n} \qquad m,n = 1,2,3,\dots\,, \tag{3.79}$$

where $a = g'(1)$. Since any real number can be arbitrarily accurately approximated by rational numbers, the linear equation (3.79) holds for real u

$$g'(u) = au,$$

with the corresponding quadratic form of $g(u)$:

$$g(u) = \frac{1}{2}\,au^2 + b,$$

and the Gaussian density

$$f(x;\theta) = N(x;\ \theta, 1/\sqrt{\alpha}), \alpha = -a > 0.$$

Let us look at this derivation from another point of view: Gauss assumed the sample mean (the estimate of the least squares method, the honor of inventing which he shares with Legendre 1805) due to its computational low-complexity and derived the Gaussian law. This line of reasoning is quite the opposite to the modern exposition in text books on statistics and signal processing, where the least squares method is derived from the assumed Gaussianity.

Derivation of Herschel (1850) and Maxwell (1860)

The astronomer John Herschel (1850) considered the two-dimensional probability distribution for errors in measuring the position of a star and, ten years later, James Clerk Maxwell (1860) gave a three-dimensional version of the same derivation for the probability distribution density for the velocities of molecules in a gas, which has become well-known to physicists as the *"Maxwellian velocity distribution law"*, fundamental in kinetic theory and statistical mechanics.

Here we consider the two-dimensional case. Let x be the error in the east–west direction and y the error in the north–south direction, and $f(x,y)$ be the joint probability distribution density. First, assume the independence and identity of coordinate error distributions

$$f(x,y)\,\mathrm{d}x\,\mathrm{d}y = f(x)\,\mathrm{d}x \times f(y)\,\mathrm{d}y. \qquad (A1)$$

Second, require that this distribution should be invariant to the rotation of the coordinate axes

$$f(x,y) = g(x^2 + y^2). \qquad (A2)$$

From assumptions $(A1)$ and $(A2)$ it immediately follows that

$$g(x^2) = f(x)f(0), \qquad g(y^2) = f(y)f(0),$$

yielding the functional equation

$$g(x^2 + y^2) \propto g(x^2)\, g(y^2),$$

with the exponential solution (Aczel 2002)

$$g(u^2) = \exp(\lambda\, u^2),\, \lambda < 0,$$

and the Gaussian law for the coordinate error distribution

$$f(x) = N(x;\; 0, 1/\sqrt{-2\lambda}).$$

Derivation of Landon (1941)

Vernon D. Landon (1941), an electrical engineer, considered the distribution density $p(x;\ \sigma^2)$ of the electrical noise voltage $x(t)$ observed in a circuit at time t, where σ is the standard deviation of the noise voltage. He suggested that this distribution is so universal that it must be determined theoretically: namely, that there exists an hierarchy of distribution densities $p(x;\ \sigma^2)$ of the same functional form characterized only by σ. Moreover, all the different levels of σ at which it occurs correspond to different noise environments, such as temperatures, amplifications, impedance levels, and even to different kinds of sources –natural or man-made industrial – resulting only in a new value of σ and preserving the functional shape. Landon's original derivation exploits the particular case of a sinusoidal noise amplitude (Landon 1941); in what follows, we use the generalization of Landon's approach proposed by Jaynes (2003).

Assume that the noise amplitude x has the distribution density $p(x;\ \sigma^2)$. Let it be incremented by a small extra contribution Δx so that $x' = x + \Delta x$, where Δx is small compared with σ, and let Δx have a distribution density $q(\Delta x)$ independent of $p(x;\ \sigma^2)$. Then, given a specific Δx, the probability for the new noise amplitude to have the value x' would be the previous probability that x should have the value $(x' - \Delta x)$. Next, by the product and sum rules of probability theory, the new distribution density is the convolution

$$f(x') = \int p(x' - \Delta x;\ \sigma^2)\, q(\Delta x)\, d(\Delta x). \qquad (3.80)$$

Expanding Equation (3.80) in powers of the small quantity Δx and dropping the prime, we get

$$f(x) = p(x; \sigma^2) - \frac{\partial p(x; \sigma^2)}{\partial x} \int \Delta x\, q(\Delta x)\, d(\Delta x)$$

$$+ \frac{1}{2} \frac{\partial^2 p(x; \sigma^2)}{\partial x^2} \int (\Delta x)^2\, q(\Delta x)\, d(\Delta x) + \cdots$$

or

$$f(x) = p(x;\ \sigma^2) - \overline{\Delta x} \frac{\partial p(x;\ \sigma^2)}{\partial x} + \frac{1}{2} \overline{\Delta x^2} \frac{\partial^2 p(x;\ \sigma^2)}{\partial x^2} + \cdots, \tag{3.81}$$

where $\overline{\Delta x}$ and $\overline{\Delta x^2}$ stand for the expectation and second moment of the increment Δx, respectively.

Since the increment is as likely to be positive as negative, assume that $\overline{\Delta x} = 0$. Moreover, assume also that the moments of order higher than 2 can be neglected, that is, $\overline{\Delta x^k} = o(\overline{\Delta x^2})$ for all $k > 2$. Then Equation (3.81) can be rewritten as follows:

$$f(x) = p(x;\ \sigma^2) + \frac{1}{2} \overline{\Delta x^2} \frac{\partial^2 p(x;\ \sigma^2)}{\partial x^2} + o\left(\overline{\Delta x^2}\right). \tag{3.82}$$

Further, the variance of x is increased to $\sigma^2 + var\ (\Delta x)$ and Landon's invariancy property requires that $f(x)$ should be equal to

$$f(x) = p(x;\ \sigma^2 + var\ (\Delta x)). \tag{3.83}$$

Expanding Equation (3.83) with respect to small $var\ (\Delta x)$, we get

$$f(x) = p(x;\ \sigma^2) + var\ (\Delta x) \frac{\partial p(x;\ \sigma^2)}{\partial(\sigma^2)} + o\left(\overline{\Delta x^2}\right). \tag{3.84}$$

Equating the principal parts of Equations (3.82) and (3.84), we obtain the following condition for this invariance:

$$\frac{\partial p(x;\ \sigma^2)}{\partial(\sigma^2)} = \frac{1}{2} \frac{\partial^2 p(x;\ \sigma^2)}{\partial x^2}$$

that is the "diffusion equation" (Cohen 2005), whose solution with the initial condition $p(x;\ \sigma^2 = 0) = \delta(x)$ is given by the Gaussian distribution

$$p(x; \sigma^2) = N(x;\ 0, \sigma).$$

The two crucial points of this derivation are: first, to guarantee that the expansions (3.81) and (3.84) hold, we should consider smooth distributions $p(x; \sigma^2)$ and $q(\Delta x)$; second, to neglect the moments of Δx of order higher than 2, we should at least assume their existence. Thus, discontinuous and heavy-tailed distributions, say, such as the Laplace and Cauchy, are excluded.

Finally, we conclude quoting Jaynes (see Jaynes 2003, p. 206): " ... *this is, in spirit, an incremental version of the central limit theorem; instead of adding up all the small contributions at once, it takes them into account one at a time, requiring that at each step the new probability distribution has the same functional form (to second order in* Δx). ... *This is just the process by which noise is produced in Nature—by addition of many small increments, one at a time (for example, collisions of individual electrons with atoms, each collision radiating another tiny impulse of electromagnetic waves, whose sum is the observed noise). Once a Gaussian form is attained, it is preserved; this process can be stopped at any point, and the resulting final distribution still has the Gaussian form.*"

3.6.3   Properties of the Gaussian distribution

Here we enlist the following properties and characterizations of the Gaussian distribution: (i) the convolution of two Gaussian functions is another Gaussian function; (ii) the Fourier transform of a Gaussian function is another Gaussian function; (iii) the central limit theorem, (iv) maximizing entropy, and (v) minimizing Fisher information for location.

Apparently, the central limit theorem (CLT) and based on it Gaussian approximations of the sums of random variables can be regarded as one of the main reasons for the ubiquitous use of a Gaussian distribution. Nevertheless, we begin from the other ones, which, firstly, also relate to the CLT and, secondly, explain why a Gaussian form, once attained, is further preserved; the remaining properties each play its own role, deserving a separate consideration.

Henceforth, a function $f(x)$ is said to be Gaussian or of a Gaussian form if $f(x) \propto N(x; \mu, \sigma)$.

Convolution of Gaussians

The operation of convolution arises in computing the distribution density $f_Y(y)$ of the sum $Y = X_1 + X_2$ of two independent random variables X_1 and X_2 with densities $f_1(x_1)$ and $f_2(x_2)$, respectively, and is given by

$$f_Y(y) = \int_{-\infty}^{\infty} f_1(x_1) f_2(y - x_1)\, dx_1 = \int_{-\infty}^{\infty} f_1(y - x_2) f_2(x_2)\, dx_2. \qquad (3.85)$$

Let the independent random variables X_1 and X_2 be Gaussian with densities $N(x_1; \mu_1, \sigma_1)$ and $N(x_2; \mu_2, \sigma_2)$. Then it can be shown that the sum of independent Gaussian random variables is distributed according to the Gaussian law

$$f_{X_1 + X_2}(y) = N\left(y; \ \mu_1 + \mu_2, \ \sqrt{\sigma_1^2 + \sigma_2^2} \right).$$

Fourier transform of a Gaussian

The Fourier transform of the Gaussian distribution density is defined as

$$\phi_X(t) = \int_{-\infty}^{\infty} e^{itx} N(x; \mu, \sigma) \, dx \qquad (3.86)$$

and it is well-known as the *characteristic function* of the Gaussian random variable X. Setting $z = x - \mu$, we can rewrite Equation (3.86) as

$$\phi_X(t) = \frac{1}{\sigma\sqrt{2\pi}} \int_{-\infty}^{\infty} e^{-z^2/2\sigma^2} e^{it(\mu+z)} \, dx$$

$$= e^{it\mu} \frac{1}{\sigma\sqrt{2\pi}} \left(\int_{-\infty}^{\infty} e^{-z^2/2\sigma^2} \cos tz \, dz + i \int_{-\infty}^{\infty} e^{-z^2/2\sigma^2} \sin tz \, dz \right).$$

The second integral is zero as the integral of an odd function over a symmetric interval. For computing the first integral, we use the Laplace integral (Abramowitz and Stegun 1972)

$$\int_0^{\infty} e^{-\alpha z^2} \cos \beta z \, dz = \frac{1}{2} \sqrt{\frac{\pi}{\alpha}} \exp \left(-\frac{\beta^2}{4\alpha} \right)$$

and get

$$\phi_X(t) = \exp \left(it\mu - \frac{\sigma^2 t^2}{2} \right).$$

For the standard Gaussian random variable when $\mu = 0$ and $\sigma = 1$, we have that

$$\phi_X(t) = e^{-t^2/2}.$$

The Central Limit Theorem (CLT)

The history of the CLT is long: it begins with the results of de Moivre (1733) and Laplace (1781) who obtained the limit shape of a binomial distribution. Then followed the work of Lyapunov (1901), who invented the method of characteristic functions in probability theory and used it to essentially generalize the de Moivre–Laplace results. Lindeberg (1922), Lévy 1925, 1935, Khintchine (1935), and Feller (1935) formulated general necessary and sufficient conditions of asymptotic normality.

Based on a simple sufficient condition in the case of identical distributions, we formulate and prove the Lindeberg–Lévy CLT (Lindeberg 1922; Lévy, 1925).

Let $X_1, X_2, \ldots, X_n, \ldots$ be independent identically distributed random variables with finite mean μ and variance σ^2. Then the distribution function of the centered and standardized random variable

$$Y_n = \frac{1}{\sigma\sqrt{n}} \sum_{k=1}^{n} (X_k - \mu) \qquad (3.87)$$

tends to the Gaussian distribution function

$$F_{Y_n}(x) = P\{Y_n \le x\} \quad \rightarrow \quad \Phi(x) = \frac{1}{\sqrt{2\pi}} \int_{-\infty}^{x} e^{-t^2/2}\, dt \quad \text{as} \quad n \to \infty$$

for every fixed x.

The proof is based on the asymptotical expansion of the characteristic function of the sum of random variables, so in the sequel, we use some properties of the characteristic functions (Fourier transforms).

Consider the centered and standardized random variables

$$X'_k = \frac{X_k - \mu}{\sigma}, \qquad k = 1, 2, \ldots, n,$$

which are independent and identically distributed; hence they have the same characteristic function $\phi_{X'}(t)$. Next return to Equation (3.87) $Y_n = \sum_{k=1}^{n}(X'_k/\sqrt{n})$ and write out its characteristic function. Since the characteristic function of the random variable $X'_k/\sqrt{n}$ is given by $\phi_{X'}(t/\sqrt{n})$, the characteristic function for Y_n is the product of $\phi_{X'_k}(t/\sqrt{n})$

$$\phi_{Y_n}(t) = \phi^n_{X'}(t/\sqrt{n}).$$

Now expand $\phi_{X'}(t)$ into the Taylor series about the point $t = 0$ with the remainder in the Peano form

$$\phi_{X'}(t) = \phi_{X'}(0) + \phi'_{X'}(0)\, t + [\phi''_{X'}(0)/2 + \alpha(t)]\, t^2.$$

The remainder is $\alpha(t)\, t^2$ where $\alpha(t) \to 0$ as $t \to 0$.

Further, use the properties of the characteristic functions: $\phi_{X'}(0) = 1$ and $\phi^{(k)}_{X'}(0) = i^k\, E[(X')^k], k = 1, 2, \ldots, n$. Since $E[X'] = 0$ and $E[X'^2] = 1$, $\phi'_{X'}(0) = 0$ and $\phi''_{X'}(0) = -1$. Hence,

$$\phi_{X'}(t) = 1 - \frac{t^2}{2} + \alpha(t)\, t^2, \quad \phi_{X'}\left(\frac{t}{\sqrt{n}}\right) = 1 - \frac{t^2}{2n} + \alpha\left(\frac{t}{\sqrt{n}}\right)\frac{t^2}{n},$$

$$\phi_{Y_n}(t) = \left[1 - \frac{t^2}{2n} + \alpha\left(\frac{t}{\sqrt{n}}\right)\frac{t^2}{n}\right]^n.$$

Taking the logarithm of both parts of the last equation and passing to the limit, we get

$$\lim_{n\to\infty} \log \phi_{Y_n}(t) = \lim_{n\to\infty} n\, \log\left[1 - \frac{t^2}{2n} + \alpha\left(\frac{t}{\sqrt{n}}\right)\frac{t^2}{n}\right].$$

Using the relation of equivalency $\log(1 + x) \sim x$ as $x \to 0$, we obtain

$$\lim_{n \to \infty} \log \phi_{Y_n}(t) = \lim_{n \to \infty} \left[n \left(-\frac{t^2}{2n} + \alpha \left(\frac{t}{\sqrt{n}} \right) \frac{t^2}{n} \right) \right]$$

$$= \lim_{n \to \infty} \left(-\frac{t^2}{2} + \alpha \left(\frac{t}{\sqrt{n}} \right) t^2 \right) = -\frac{t^2}{2}.$$

Thus,

$$\lim_{n \to \infty} \log \phi_{Y_n}(t) = e^{-t^2/2}.$$

Since the convergence of the characteristic functions to a certain limit implies the convergence of the distribution functions to the corresponding limit (Feller 1971), the limit law of the random variables Y_n is the standard Gaussian with the parameters $\mu = 0$ and $\sigma = 1$.

If to assume the existence of the third absolute moment of each X_k about its mean $v_{3k} = E[|X_k - \mu_k|^3] < \infty$, then the requirement of distributions identity can be dropped. Asymptotic normality, precisely the Lyapunov CLT (Lyapunov 1901), holds if the X's have different distributions with finite means μ_k and variances σ_k^2, and if $\lim_{n \to \infty} v_3/B_n^3 = 0$, where $v_3 = \sum_1^n v_{3k}$ and $B_n^2 = \sum_1^n \sigma_k^2$. Then the random variable $Y_n = \sum_1^n (X_k - \mu_k)/B_n$ has the limit distribution $\Phi(x)$.

Asymptotic normality may be established under conditions that do not require the existence of third moments. Actually, it is a necessary and sufficient condition that

$$\lim_{n \to \infty} \frac{1}{B_n^2} \sum_{k=1}^n \int_{|x - \mu_k| > \varepsilon B_n} (x - \mu_k)^2 dF_k(x) = 0, \qquad (3.88)$$

where ε is an arbitrary positive number and F_k is the distribution function of X_k, $k = 1, 2, \ldots, n$.

This condition, due to Lindeberg (1922) who proved its sufficiency and Feller (1935) who proved its necessity, implies that the total variance B_n^2 tends to infinity and that every σ_k^2/B_n^2 tends to zero; in fact no random variable dominates the others. The theorem may fail to hold for random variables that do not possess a second moment; for instance, the mean of n variables each distributed according to the Cauchy law

$$dF(x) = \frac{dx}{\pi(1 + x^2)}, \qquad -\infty < x < \infty,$$

is distributed in precisely the same form. This can be easily seen from the characteristic function $\phi(t) = e^{-|t|}$ for the Cauchy distribution (Feller 1971).

The practical applications of the CLT are based on the corresponding Gaussian approximations of the sums of random variables, and their accuracy significantly depends on convergence rate estimates in the CLT.

Return again to the Lindeberg–Lévy formulation of the CLT (3.87) for the sums of i.i.d. random variables $\{X_k\}_{k=1}^n$. In this case, the classical Berry–Esseen convergence rate estimate in the uniform metric is given by

$$\rho(F_{Y_n}, \Phi) = \sup_x |F_{Y_n}(x) - \Phi(x)| \leq C \frac{v_3}{\sigma^3 \sqrt{n}}, \qquad (3.89)$$

where σ^2 and v_3 are respectively the variance and the absolute third moment about the mean of the parent distribution F_X, and C is an absolute constant (this means that there exists such an F_X for which the upper boundary in inequality (3.89) is attained) (Berry 1941; Esseen 1942). The latest improvement of the value of the constant C is given by 0.7655 (Shiganov 1986).

It is noteworthy that the Berry–Esseen boundary with the convergence rate of $1/\sqrt{n}$ is yet pessimistic, fundamentally related to the Gram–Charlier and Edgeworth series (Cramér 1974). The practical implications of the aforementioned results on the CLT probability approximations are usually justified by the relative error of probability approximation of the real-life problem, in many cases, to evaluate a certain tail probability.

The classical Lindeberg–Lévy, Lyapunov and Lindeberg–Feller versions of the CLT establish the convergence of the distribution $F_{Y_n}(x)$ of the standardized sums Y_n to the Gaussian distribution $\Phi(x)$. Evidently, the conditions under which the distribution density of Y_n converges to the Gaussian density should be more strict than for the classical versions of the CLT, since the convergence $F_{Y_n}(x) \to \Phi(x)$ does not imply the convergence $F'_{Y_n}(x) \to \Phi'(x)$. In the case of continuous i.i.d. r.v.'s X_k, the sufficient condition for the uniform convergence of distribution densities is just the existence of mean and variance (Gnedenko and Kolmogorov 1954).

Concluding our remarks on the CLT, we note that the distributions with finite third moments are of a special interest in theory, not only because of the fact that Lyapunov's sufficient condition $v_3 < \infty$ is evidently simpler to verify than the Lindeberg–Feller condition (3.88), but due to the following results. The finiteness of the moment $v_{2+\delta}$, $0 < \delta < 1$, guarantees the decrease rate $n^{-\delta/2}$ for $\rho(F_{Y_n}, \Phi)$ as $n \to \infty$ (Lyapunov 1901); for $\delta \geq 1$, $\rho(F_{Y_n}, \Phi) = O(n^{-1/2})$, that is, just the Berry–Esseen convergence rate, and this rate is unimprovable (Ibragimov 1966).

We have mentioned several classical results on the limit theorems in probability theory dealing with the sums of independent random variables. Further extensions and generalizations of the CLT are concerned (i) with the study of different schemes of dependency between the summands: homogeneous Markov chains with the finite number of states (Renyi 1953), m-dependent random variables (Davidson 1994), martingales (Hall and Heyde 1980); (ii) with the random number of summands, mostly the Poisson and generalized Poisson models being considered (Gnedenko and Korolev 1996), (Bening and Korolev 2002), and (iii) with the properties of various metrics and measures characterizing convergence rates in the CLT (Zolotarev 1997). This topic is still popular among mathematicians: in Zolotarev

(1997), a comprehensive study of the former and recent results in this area is given focusing both on the classical versions of the CLT as well as on the CLT-analogs in the classes of non-Gaussian infinitely divisible and stable distribution laws.

Maximizing entropy

Consider the variational problem of maximizing differential entropy (Cover and Thomas 1991)

$$H(f) = - \int_{-\infty}^{\infty} f(x) \log f(x) \, dx$$

in the class of symmetric distributions with a bounded variance

$$f^*(x) = \arg \max_{f(x)} H(f), \qquad (3.90)$$

$$f(x) \geq 0, \quad f(-x) = f(x),$$

$$\int_{-\infty}^{\infty} f(x) \, dx = 1,$$

$$\sigma^2(f) = \int_{-\infty}^{\infty} x^2 f(x) \, dx \leq \overline{\sigma}^2.$$

Its solution is given by the Gaussian distribution density $f^*(x) = N(x; 0, \overline{\sigma})$ (Cover and Thomas 1991; Shannon 1948).

To show this, first note that the entropy of any distribution increases with an increase in its variance, say, for the Gaussian as $\log \sigma$. Thus, it suffices to solve problem (3.90) under the given variance $\sigma^2(f) = d^2$ assuming $d^2 \leq \overline{\sigma}^2$.

Second, reduce the problem (3.90) of constrained optimization to the problem of unconstrained optimization using the Lagrange functional

$$L(f, \lambda, \mu) = H(f) + \lambda \left(\int_{-\infty}^{\infty} f(x) \, dx - 1 \right) + \mu \left(\int_{-\infty}^{\infty} x^2 f(x) \, dx - d^2 \right),$$

where λ and μ are the Lagrange multipliers corresponding to the conditions of normalization and given variance, respectively.

Third, take the variation of the Lagrange functional and equate it to zero

$$\delta L(f, \lambda, \mu) = \delta \int_{-\infty}^{\infty} (-f(x) \log f(x) + \lambda f(x) + \mu \, x^2 f(x)) \, dx$$

$$= \int_{-\infty}^{\infty} \left(-\delta f(x) \, \log f(x) - f(x) \frac{1}{f(x)} \, \delta f(x) + \lambda \, \delta f(x) + \mu \, x^2 \, \delta f(x) \right) dx$$

$$= \int_{-\infty}^{\infty} (- \log f(x) - 1 + \lambda + \mu \, x^2) \delta f(x) \, dx = 0,$$

from which it follows that the expression in the parentheses should be zero, thus yielding

$$\log f^*(x) = \lambda - 1 + \mu \, x^2.$$

The values of the Lagrange multipliers λ and μ can be obtained using the corresponding side conditions of the variational problem (3.90), namely, the normalization and variance conditions. Finally, we arrive at

$$f^*(x) = N(x; 0, \overline{\sigma}).$$

Therefore, the Gaussian distribution has a higher entropy than any other with the same variance: thus any operation on a distribution, which discards information and preserves variance that is bounded, leads us to a Gaussian. The best example of this is given by the CLT as, evidently, the summation discards information and the appropriate standardizing even conserves variance.

Minimizing Fisher information for location

The notion of Fisher information arises in the Cramér–Rao inequality (Cramér 1974), one of the principal results in statistics, which gives the lower boundary upon the parameter estimate variance

$$var(\widehat{\theta}_n) \geq \frac{1}{n \, I(f)}, \tag{3.91}$$

where $\widehat{\theta}_n$ is an unbiased estimate of a parameter θ of the distribution density $f(x, \theta)$ from a sample of i.i.d. random observations $x_1, \ldots, x_n$ and $I(f)$ is the functional of Fisher information given by

$$I(f) = \int_{-\infty}^{\infty} \left(\frac{\partial \log f(x, \theta)}{\partial \theta} \right)^2 f(x, \theta) \, dx. \tag{3.92}$$

In the case of estimation of a location parameter θ, say, the population mean or the median, when the distribution density depends on θ as $f(x - \theta)$, it can be easily seen that Equation (3.92) takes the following form:

$$I(f) = \int_{-\infty}^{\infty} \left(\frac{f'(x)}{f(x)} \right)^2 f(x) \, dx. \tag{3.93}$$

Now we show that the solution to the variational problem of minimization of Fisher information for location (3.93) over the distributions with a bounded variance is achieved at the Gaussian (Kagan *et al.* 1973), precisely that

$$N(x; 0, \overline{\sigma}) = f^*(x) = \arg \min_{f(x)} I(f)$$

subject to

$$f(x) \geq 0, \quad f(-x) = f(x),$$

$$\int_{-\infty}^{\infty} f(x) \, dx = 1,$$

$$\sigma^2(f) = \int_{-\infty}^{\infty} x^2 f(x) \, dx \leq \overline{\sigma}^2.$$

Similar to the aforementioned derivation for entropy, it suffices to consider the case of a given variance $\sigma^2(f) = d^2 \leq \overline{\sigma}^2$.

Next we use the following version of the Cauchy–Bunyakovskiy inequality

$$\left(\int \phi(x)\psi(x)f(x) \, dx \right)^2 \leq \int \phi^2(x)f(x) \, dx \int \psi^2(x)f(x) \, dx, \tag{3.94}$$

where the functions $\phi(x)$ and $\psi(x)$ should only provide the existence of the integrals in Equation (3.94) and remain arbitrary in all other aspects.

Now choose $\phi(x) = x$ and $\psi(x) = -f'(x)/f(x)$. The integrals in the right-hand part of (3.94) are the distribution variance $\sigma^2(f) = d^2$ and Fisher information (3.93), respectively. Using symmetry and integrating by parts, we compute the integral in the left-hand part of Equation (3.94)

$$-\int_{-\infty}^{\infty} xf'(x) \, dx = -2\int_{0}^{\infty} xf'(x) \, dx = -2\left[xf(x)\big|_0^{\infty} - \int_0^{\infty} f(x) \, dx \right] = 1,$$

assuming that the distribution tails satisfy the natural condition:

$$\lim_{x \to \infty} xf(x) = 0.$$

Collecting the obtained results and substituting them into Equation (3.94), we get the lower boundary upon Fisher information

$$I(f) \geq \frac{1}{d^2}.$$

As this lower boundary just gives the Fisher information value for the Gaussian distribution density

$$\int_{-\infty}^{\infty} \left(\frac{N'(x; 0, d)}{N(x; 0, d)} \right)^2 N(x; 0, d) \, dx = \frac{1}{d^2}$$

and the minimization problem in the class of distributions with a bounded variance allows for the following two-step decomposition:

$$f^* = \arg \min_{f: \, \sigma^2(f) \leq \overline{\sigma}^2} I(f) = \arg \min_{d^2 \leq \overline{\sigma}^2} \left\{ \min_{f: \, \sigma^2(f) = d^2} I(f) \right\},$$

we arrive at the required relation: $f^*(x) = N(x; 0, \overline{\sigma})$.

This important result that the Gaussian distribution is the least informative (favorable) distribution in the class of distributions with a bounded variance gives another reason for the ubiquitous use of the Gaussian distribution in signal processing, and, moreover, it links Huber's results in robustness.

3.6.4   Huber's minimax approach and Gaussianity

In what follows, we show that the least informative (favorable) Gaussian distribution and therefore the most robust in the Huber sense the least squares method naturally arises in the context of robustness, despite the conventional emphasis on the departures from Gaussianity – a kind of a paradigm in robustness.

As has just been shown in the previous subsection, in the class $\mathcal{F}_2$ of distribution densities with a bounded variance

$$\mathcal{F}_2 = \left\{ f : \sigma^2(f) = \int_{-\infty}^{\infty} x^2 f(x) \, dx \leq \overline{\sigma}^2 \right\},$$

the least informative density is the Gaussian

$$f_2^*(x) = N(x; 0, \overline{\sigma}) = \frac{1}{\overline{\sigma}\sqrt{2\pi}} \exp\left(-\frac{x^2}{2\overline{\sigma}^2}\right).$$

The optimal score function is linear $\psi_2^*(x) = x/\overline{\sigma}^2$ and the minimax estimate of location is the sample mean $\overline{x}_n$.

Taking into account the role of a least informative distribution in Huber's minimax approach, we have arrived at a rather strange result: the sample mean $\overline{x}_n$ is robust in the Huber sense in the class of distributions with a bounded variance!

Let us dwell on this phenomenon in more detail. Since Fisher information for the least favorable Gaussian distribution attains its minimum value at $I(f_2^*) = 1/\overline{\sigma}^2$, the sample mean is an estimate of the guaranteed accuracy in $\mathcal{F}_2$, that is,

$$var(\overline{x}_n) \leq \frac{\overline{\sigma}^2}{n} \quad \text{for all} \quad f \in \mathcal{F}_2.$$

Thus, if the bound upon variance $\overline{\sigma}^2$ is small, then the minimax approach yields a reasonable result and the *LS* method can be successfully used with relatively short-tailed distributions in estimation and detection of signals, (e.g., see Shevlyakov and Kim 2006; Shevlyakov and Vilchevski 2011).

On the contrary, if we deal with really heavy-tailed distributions (gross errors, impulse noises) when $\overline{\sigma}^2$ is large or even infinity, like for the Pareto-type distributions, then the minimax solution in the class $\mathcal{F}_2$ is still formally correct as $var(\hat{\theta}_n) \leq \infty$, but practically senseless. In this case, we must use robust versions of the *LS* method, say, such as Huber's *M*-estimates optimal for the class of ε-contaminated Gaussian distributions.

We may also say that the minimax principle gives an unrealistic result in this case. However, as shown in subsection 3.4.3, this disadvantage becomes a significant advantage of the *LS* estimate if one considers the class of nondegenerate distributions with a bounded variance $\mathcal{F}_{12}$, namely, the intersection of the classes $\mathcal{F}_1$ and $\mathcal{F}_2$.

Recall that this class comprises qualitatively different distribution densities, for example, the Gaussian, the heavy-tailed ε-contaminated Gaussian, Laplace, Cauchy-type (with $\bar{\sigma}^2 = \infty$), short-tailed densities, etc. In this class, the least informative density is the Gaussian under relatively short-tailed distributions with the corresponding minimax *LS* estimate of location. This solution is robust and close to Huber's solution for the class $\mathcal{F}_H$ of heavy-tailed distributions due to the presence of the Laplace branch in the least informative density and it is more efficient than Huber's for short-tailed distributions due to the presence of the Gaussian branch (Shevlyakov and Vilchevski 2002, 2011; Shevlyakov and Kim 2006). In other words, *the additional information on the relative weight of distribution tails may significantly improve the efficiency of estimation and detection.*

Finally, we may add that preliminary cleaning of the data in signal processing, say, by sliding median algorithms to make the noise distribution variance bounded (if they were not such), then the subsequent use of the Gaussian-based *LS* techniques becomes justified.

3.6.5   Concluding remarks

Now we return to the question posed in the beginning of this section: "Why the ubiquitous use and success of Gaussian distributions?"

All the arguments *pro* Gaussianity can be classified in the following two groups: the arguments for the gravity and stability of a Gaussian shape, which are the statistical gravity (the CLT and the Landon derivation), the stability (the convolution property), and geometrical invariancy (the Herschel–Maxwell derivation); and the arguments for the optimality of a Gaussian shape (the Gauss derivation, the maximization of entropy, and the minimization of Fisher information). In this list, we skipped various characterization properties of a multivariate Gaussian, especially of a bivariate one, represented in Kagan *et al.* (1973), and the stability aspects related to the Gaussian infinite divisibility analyzed in Zolotarev (1997); some additional reasons *pro* Gaussianity can be found in Cramér (1974), and Jaynes (2003).

On the whole, we may repeat after Jaynes (2003) that, *in Nature, all smooth processes with increasing entropy lead to Gaussianity and once it is reached, it is then preserved.* The fact that a Gaussian is the least favorable distribution minimizing Fisher information is significantly important in signal and data processing.

All the arguments *contra* Gaussianity arise when the aforementioned conditions of smoothness are violated: this refers to the presence of gross errors and outliers in the data, impulse noises in observed signals, etc. Moreover, we may add that most of the formulated properties of a Gaussian, say, the CLT, are of an asymptotic nature, so on finite samples they hold only approximately. For instance, we never know the tails of distributions in real-life data. Generally, these reasons lead to robust methods and algorithms of signal processing, and what is important is that *a Gaussian again naturally emerges in robustness either in the form of various Gaussian ε-neighborhoods or as the least informative (favorable) distribution.*

3.7 Summary

In this chapter, in the framework of the two classical approaches to robustness, namely Huber's and Hampel's, we have briefly described the conventional robust estimates of location designed for neighborhoods of a Gaussian. Further, in the same framework, we have considered new nontraditional robust estimation problem settings including application of the minimax approach to distribution classes of a non-neighborhood nature, stable estimation based on a new global indicator of robustness, namely estimate's stability, stability of the least informative distributions as the solutions to the variational problem of minimizing Fisher information over lattice distributions, and the reasons for the ubiquitous and successful use of Gaussian models in data processing. Finally, we conclude that several new estimates designed within the aforementioned nontraditional problem set-ups, a radical stable estimate in particular, outperform well-known Huber and Hampel *V*-robust estimates of location. In Chapter 11, these methods and algorithms are applied to the problems of robust detection.

References

Abramowitz M and Stegun I 1972 *Handbook of Mathematical Functions*, Dover, New York.

Aczel J 2002 *Functional Equations: History, Applications and Theory*, Kluwer.

Andrews DF, Bickel PJ, Hampel FR, Huber PJ, Rogers WH, and Tukey JW 1972 *Robust Estimates of Location*, Princeton Univ. Press.

Azencott R 1977a Estimation d'un paramètre de translation à partir de tests de rang. *Austérisque* **43–44**, 41–64.

Azencott R 1977b Robustesse des *R*-estimateurs. *Austérisque* **43–44**, 189–202.

Bening VE and Korolev VYu 2002 *Generalized Poisson Models*, VSP, Utrecht.

Ben-Tal A, Gaoui LE and Nemirovski A 1999 *Robust Optimization*, Princeton Univ. Press.

Bernoulli D 1777 Dijudicatio maxime probabilis plurium observationum discrepantium atque verisimillima inductio inde formanda. *Acta Acad. Sci. Petropolit.* **1**, 3–33. (English translation by C.G. Allen 1961. Biometrika 48, 3–13.)

Berry AC 1941 The accuracy of the Gaussian approximation to the sum of independent variables. *Trans. Amer. Math. Soc.* **49**, 122–126.

Bickel PJ 1976 Another look at robustness: a review of reviews and some new developments. *Scand. J. Statist. Theory and Appl.* **3**, 145–168.

Birnbaum A and Laska E 1967 Optimal robustness: a general method with application to linear estimators of location. *J. Amer. Statist. Assoc.* **62**, 1230–1240.

ROBUST ESTIMATION OF LOCATION 103

Bohner M and Guseinov GSH 2003 Improper integrals on time scales. *Dynamic Systems and Applications* **12**, 45–65.

Boltzmann WI 1871 *Wiener Berichte* **63**, 397–418, 679–711, 712–732.

Boscovich RJ 1757 De litteraria expeditione per pontificiam ditionem, et synopsis amplioris operis, ac habenturplura eius ex exemplaria etiam sensorum impressa. *Bononiensi Scientiarum et Artium Instituto Atque Academia Commentarii* **4**, 353–396.

Chapman DG and Robbins HE 1951 Minimum variance estimation without regularity assumptions. *Ann. Math. Statist.* **22**, 581–586.

Cohen L 2005 The history of noise. *IEEE Signal Processing Mag.* **22**, 20–45.

Cover TM and Thomas JM 1991 *Elements of Information Theory*, Wiley.

Cramér H 1974 *Mathematical Methods of Statistics*, Princeton University Press.

Daniel C 1920 Observations weighted according to order. *Amer. J. Math.* **42**, 222–236.

Davidson J 1994 *Stochastic Limit Theory—An Introduction for Econometricians*, Oxford University Press.

De Moivre A 1733 Approximatio ad Summam Terminorum Binomii $(a + b)^n$ in Seriem expansi. *Photographic reproduction in Archibald RC 1926 Isis* **8**, 671–683.

Deniau C, Oppenheim G, and Viano C 1977a M-estimateurs. *Austérisque* **43–44**, 31–40.

Deniau C, Oppenheim G, and Viano C 1977b Robustesse: π-robustesse et minimax-robustesse. *Austérisque* **43–44**, 51–166.

Deniau C, Oppenheim G, and Viano C 1977c Robustesse des M-estimateurs. *Austérisque* **43–44**, 167–187.

Deniau C, Oppenheim G, and Viano C 1977d Courbes d'influence et sensibilité. *Austérisque* **43–44**, 239–252.

Eddington AS 1914 *Stellar Movements and the Structure of the Universe*, Macmillan, London.

Esseen CG 1942 On the Lyapunov limit error in the theory of probability. *Ark. Mat. Astr. Fys.* **28A**, 1–19.

Feller W 1935 Über den zentralen Grenzwertsatz der Wahrscheinlichkeitsrechnung. *Math. Zeitschr.* **40**, 521.

Feller W 1971 *An Introduction to Probability Theory and Its Applications*, Vol. **2**. Wiley.

Fernholz LT 1983 *Von Mises Calculus for Statistical Functionals*. Springer, New York.

Fisher RA 1922 On the mathematical foundations of theoretical statistics. *Philosophical Transactions of the Royal Society* A**222**, 309–368. Reproduced in R.A. Fisher, *Contributions to Mathematical Statistics*, Wiley, 1950.

Gauss CF 1809 *Theoria Motus Corporum Celestium*, Perthes, Hamburg; English translation: *Theory of the Motion of the Heavenly Bodies Moving About the Sun in Conic Sections.* New York: Dover, 1963.

Gauss CF 1823 Theoria Combinationis Observationum Erroribus Minimis Obnoxiae, In *Werke* *4*, Dietirich, Göttingen (1880), pp. 1–108.

Gnedenko BV and Kolmogorov AN 1954 *Limit Distributions for Sums of Independent Random Variables*, Addison–Wesley.

Gnedenko BV and Korolev VYu 1996 *Random Summation: Limit Theorems and Applications*, CRC Press, Boca Raton, FL.

Godambe VP 1960 Optimum property of regular maximum likelihood estimation. *Ann. Math. Statist.* **31**, 1208–1212.

Hall P and Heyde CC 1980 *Martingale Limit Theory and Its Applications*, Academic Press, New York.

Hájek J and Šidák Z 1967 *Theory of Rank Tests*, Academic Press, New York.

Hampel FR 1968 *Contributions to the Theory of Robust Estimation*. PhD thesis, University of California, Berkeley.

Hampel FR 1971 A general qualitative definition of robustness. *Ann. Math. Statist.* **42**, 1887–1896.

Hampel FR 1974 The influence curve and its role in robust estimation. *J. Amer. Statist. Assoc.* **69**, 383–393.

Hampel FR, Ronchetti E, Rousseeuw PJ and Stahel WA 1986 *Robust Statistics. The Approach Based on Influence Functions*, Wiley.

Herschel J 1850 *Edinburgh Rev.* **92**, 14.

Hodges ZL and Lehmann EL 1963 Estimates of location based on rank tests. *Ann. Math. Statist.* **34**, 598–611.

Holland PW and Welsch RE 1977 Robust regression using iteratively reweighted least squares. *Commun. Statist. (Theory and Methods)* **6**, 813–828.

Huber PJ 1964 Robust estimation of a location parameter. *Ann. Math. Statist.* **35**, 73–101.

Huber PJ 1981 *Robust Statistics*, Wiley.

Huber PJ and Ronchetti E (eds) 2009 *Robust Statistics*, 2nd edn, Wiley.

Ibragimov IA 1966 On the accuracy of approximation to the distribution functions of the sums of independent variables by normal distribution. *Theory Probab. Appl.* **11**, 632–655.

Jaeckel LA 1971 Some flexible estimates of location. *Ann. Math. Statist.* **42**, 1540–1552.

Jaynes ET 2003 *Probability Theory. The Logic of Science*, Cambridge University Press.

Jeffreys H 1932 An alternative to the rejection of outliers. *Proc. Royal Soc. London* **A137**, 78–87.

Kagan AM, Linnik YuV and Rao SR 1973 *Characterization Problems in Mathematical Statistics*, Wiley.

Khintchine A 1935 Sul dominio di attrazione della legge di Gauss. *Giorn. Ist. Italiano d. Attuari* **6**, 378.

Kolmogorov AN 1931 On the method of median in the theory of errors. *Math. Sbornik* **38**, 47–50.

Landon VD 1941 The distribution of amplitude with time in fluctuation noise. *Proc. IRE* **29**, 50–54.

Laplace PS 1781 M÷moire sur les probabilit÷s. *Mem. Acad. Roy.* Paris; reprinted in Laplace (1878–1912), **9**, 384–485.

Legendre AM 1805 *Nouvelles méthods pour la détermination des orbits des cométes*, Didot, Paris.

Lindeberg JW 1922 Eine neue Herleitung des Exponentialgesetzes in der Wahrscheinlichkeitsrechnung. *Math. Zeitschr.* **15**, 211–225.

Lévy P 1925 *Calcul des probabilités*, Paris.

Lévy P 1935 *Théorie de l'addition des variables aléatoires*, Paris.

Lyapunov A 1901 Nouvelle forme du théor::me sur la limite de probabilité. *Mém. Acad. Sci. St. Pétersbourg* **12**.

Maxwell JC 1860 Illustration of the dynamical theory of gases. Part I. On the motion and collision of perfectly elastic spheres. *Phil. Mag.*, **56**.

Mendeleyev DI 1895 Course of work on the renewal of prototypes or standard measures of lengths and weights. *Vremennik Glavnoi Palaty Mer i Vesov* **2**, 157–185 (in Russian).

Meshalkin LD 1971 Some mathematical methods for the study of non-communicable diseases. In *Proc. 6th Int. Meeting of Uses of Epidemiol. in Planning Health Services*, Vol. 1, Primosten, Yugoslavia, pp. 250–256.

Mosteller F and Tukey JW 1977 *Data Analysis and Regression*, Addison–Wesley.

Newcomb S 1886 A generalized theory of the combination of observations so as to obtain the best result. *Amer. J. Math.* **8**, 343–366.

Poincaré H 1904 *Science et Hypothesis*, English translation, Dover Publications, Inc., NY.

Polyak BT and Tsypkin YaZ 1976 Robust identification. In: *Identif. Syst. Parameter Estim., Part 1, Proc. 4th IFAC Symp.* Tbilisi, 1976, pp. 203–224.

Prokhorov YuV 1956 Convergence of random processes and limit theorems in probability theory. *Theory Probab. Appl.* **1**, 157–214.

Rényi A 1953 On the theory of order statistics. *Acta Math. Hung.* **4**, 191–232.

Rey WJJ 1978 *Robust Statistical Methods*, Springer, Berlin.

Rousseeuw PJ 1981 A new infinitisemal approach to robust estimation. *Z. Wahrsch. verw. Geb.* **56**, 127–132.

Rychlik T 1987 An asymptotically most bias-stable estimator of location parameter. *Statistics* **18**, 563–571.

Sacks J and Ylvisaker D 1972 A note on Huber's robust estimation of a location parameter. *Ann. Math. Statist.* **43**, 1068–1075.

Shannon CE 1948 A mathematical theory of communication. *Bell Syst. Tech. J.* **27**, 329–423, 623–656.

Shevlyakov GL and Kim K 2006 Robust minimax detection of a weak signal in noise with a bounded variance and density value at the center of symmetry. *IEEE Trans. Inform. Theory* **52**, 1206–1211.

Shevlyakov GL, Morgenthaler S and Shurygin AM 2008 Redescending M-estimators. *J. Statist. Planning and Inference* **138**, 2906–2917.

Shevlyakov GL and Vilchevski NO 2002 *Robustness in Data Analysis: criteria and methods*, VSP, Utrecht.

Shevlyakov GL and Vilchevski NO 2011 *Robustness in Data Analysis*, De Gruyter, Boston.

Shiganov IS 1986 Refinement of the upper bound on the constant in the central limit theorem. *J. Soviet Math.* **35**, 2545–2551.

Shurygin AM 1994a New approach to optimization of stable estimation. In: *Proc. 1 US/Japan Conf. on Frontiers of Statist. Modeling*, Kluwer Academic Publishers, Netherlands, pp. 315–340.

Shurygin AM 1994b Variational optimization of the estimator stability. *Automation and Remote Control* **55**, 1611–1622.

Shurygin AM 2000a Estimator stability in geological applications. *Mathematical Geology* **32**, 19–29.

Shurygin AM 2000b *Applied Stochastics: robustness, estimation and forecasting*, Finances and Statistics, Moscow (in Russian).

Stigler SM 1973 Simon Newcomb, Percy Daniell and the history of robust estimation 1885–1920. *J. Amer. Statist. Assoc.* **68**, 872–879.

Stigler SM 1980 Stigler's law of eponymy. *Trans. NY Acad. Sci.* **39**, 147–159.

Stigler SM 1981 Gauss and the invention of least squares. *Ann. Statist.* **9**, 465–474.

Tsypkin YaZ 1984 *Foundations of Information Theory of Identification*, Nauka, Moscow (in Russian).

Tukey JW 1960 A survey of sampling from contaminated distributions. In: *Contributions to Probability and Statistics*, Olkin I (ed.), pp. 448–485, Stanford Univ. Press.

Tukey JW 1977 *Exploratory Data Analysis*, Addison–Wesley.

Van Eeden C 1963 The relation between Pitman's asymptotic relative efficiency of two tests and the correlation coefficient between their test statistics. *Ann. Math. Stat.* **34**, 1442–1450.

Zieliński R 1987 Robustness of sample mean and sample median under restrictions on outliers. *Applicationae Mathematicae* **19**, 239–240.

Zolotarev VM 1997 *Modern Theory of Summation of Random Variables*, VSP, Utrecht.

4

Robust Estimation of Scale

In this chapter the problem of robust estimation of scale is treated as the problem subordinate to robust estimation of correlation, as some measures of correlation are defined via the measures of scale. Special attention is paid to the former robust Huber minimax variance and bias estimates of scale under ε-contaminated normal distributions as well as to the recently proposed Rousseeuw and Croux (1993) highly robust and efficient estimates of scale with their fast low-complexity modifications (Smirnov and Shevlyakov 2014), for they are applied to the problems of robust estimation of the correlation coefficient. In the first three sections of this chapter, we represent a brief review of different approaches to robust estimation of scale.

4.1 Preliminaries

4.1.1 Introductory remarks

Following Huber (1981), we define the *scale estimate* as a positive statistic S_n that is location-invariant

$$S_n(x_1 + \mu, \ldots, x_n + \mu) = S_n(x_1, \ldots, x_n) \tag{4.1}$$

and scale-equivariant

$$S_n(\lambda x_1, \ldots, \lambda x_n) = |\lambda| S_n(x_1, \ldots, x_n). \tag{4.2}$$

In applied statistics, scale problems, as a rule, do not occur independently of location (or regression) problems, in which the scale usually is a nuisance parameter. Such problems of scale estimation are thoroughly studied in Huber (1981) with the use of M-, L-, and R-estimates. Below, we describe the main representatives of these classes.

Robust Correlation: Theory and Applications, First Edition. Georgy L. Shevlyakov and Hannu Oja.
© 2016 John Wiley & Sons, Ltd. Published 2016 by John Wiley & Sons, Ltd.
Companion Website: www.wiley.com/go/Shevlyakov/Robust

Naturally, we distinguish the data analysis set-up for constructing the measures of data dispersion from the statistical set-up where a scale parameter of the underlying distribution is estimated.

4.1.2   Estimation of scale in data analysis

In data analysis, the role of scale is secondary as compared to location: generally, the problem of estimation of scale is subordinated to the problem of estimation of location.

Consider the optimization approach to designing the estimate $\widehat{\theta}_n$ of location for the data $x_1, \ldots, x_n$ (Shevlyakov and Vilchevski 2002)

$$\widehat{\theta}_n = \arg \min_{\theta} J(\theta), \tag{4.3}$$

where $J(\theta)$ is an objective function.

The estimate S_n of scale (the measure of dispersion of the data about a location estimate) is defined as an appropriately transformed value of the objective function $J(\theta)$ at the optimal point $\widehat{\theta}_n$ (Orlov 1976)

$$S_n \propto J(\widehat{\theta}_n). \tag{4.4}$$

In the case of estimation of location where $J(\theta) = n^{-1} \sum_1^n \rho(x_i - \theta)$,

$$S_n = \propto \rho^{-1}\left(\frac{1}{n} \sum_{i=1}^n \rho(x_i - \widehat{\theta}_n)\right),$$

with

$$\widehat{\theta}_n = \arg \min_{\theta} \sum_{i=1}^n \rho(x_i - \theta).$$

Example 4.1.1. *The standard deviation corresponds to the least squares (LS) or the L_2-norm optimization criterion*

$$S_n = s_x = \sqrt{\frac{1}{n} \sum_{i=1}^n (x_i - \bar{x})^2},$$

with $\rho(u) = u^2$ and $\widehat{\theta}_{LS} = \bar{x} = n^{-1} \sum_{i=1}^n x_i$.

Example 4.1.2. *The mean absolute deviation about median corresponds to the least absolute values (LAVs) or the L_1-norm optimization criterion*

$$S_n = d = \frac{1}{n} \sum_{i=1}^n |x_i - \mathrm{med}\, x|,$$

with $\rho(u) = |u|$ and $\widehat{\theta}_{L_1} = \mathrm{med}\, x$.

Example 4.1.3. *Further, the pth-power deviations (Gentleman 1965) correspond to the L_p-norm optimization criterion*

$$S_n = s_{L_p} = c_n \left(\sum_{i=1}^{n} |x_i - \widehat{\theta}_{L_p}|^p \right)^{1/p},$$

with $\rho(u) = |u|^p$,

$$\widehat{\theta}_{L_p} = \arg \min_{\theta} \sum_{i=1}^{n} |x_i - \theta|^p, \quad p \geq 1,$$

where c_n is a constant chosen, say, from the condition of unbiasedness under the standard normal $E_\Phi(S_n) = \sigma$.

Example 4.1.4. *The semi-sample range*

$$S_n = \frac{R}{2} = \frac{x_{(n)} - x_{(1)}}{2}$$

is defined by the Chebyshev or L_∞-norm criterion with $\rho(u) = \max|u|$ and

$$\widehat{\theta}_{L_\infty} = \arg \min_{\theta} \max_{i} |x_i - \theta| = \frac{x_{(1)} + x_{(n)}}{2}.$$

Example 4.1.5. *The least median squares (LMSs) deviation*

$$S_n = s_{LMS} = \mathrm{med}|x - \widehat{\theta}_{LMS}|$$

for the LMS method with $J(\theta) = \mathrm{med}(x_i - \theta)^2$ and $\widehat{\theta}_{LMS}$ given by

$$\widehat{\theta}_{LMS} = \arg \min_{\theta} \mathrm{med}(x_i - \theta)^2.$$

This estimate is close to the median absolute deviation (MAD)

$$\mathrm{MAD}\, x = \mathrm{med}_i \, |x_i - \mathrm{med}\, x| = \mathrm{med}\, |x - \mathrm{med}\, x|. \tag{4.5}$$

It is noteworthy that the optimization approach to designing scale estimates is close to the scale estimates obtained from S-estimates for location (Hampel *et al.* 1986, p. 115), where the S-estimate of location, defined by

$$\widehat{\theta}_n = \arg \min_{\theta} s(x_1 - \theta, \dots, x_n - \theta),$$

yields the scale estimate $S_n = s(x_1 - \widehat{\theta}_n, \dots, x_n - \widehat{\theta}_n)$.

Summarizing, we may say that any location and, more generally, regression estimate obtained with the optimization approach generates the corresponding estimate of scale.

4.1.3   Measures of scale defined by functionals

Consider a random variable ξ with a a symmetric and absolutely continuous distribution function F. Let ξ be a random variable with a symmetric and absolutely continuous distribution function F. Given the center of symmetry $\theta(F)$, define the measure of dispersion of ξ about θ as a functional $S(F)$ depending, as a rule, on the distance $|\xi - \theta|$.

The general requirement of scale equivariancy should be imposed on the functional $S(F)$ to be a measure of dispersion (Bickel and Lehmann 1973, 1976):

$$S(F_{a\xi+b}) = |a|S(F_\xi)$$

for all a and b.

There are the following three groups of functionals defining qualitatively different measures of scale.

The first group consists of the functionals based on the deviation of each element of a population from a typical (central) element θ, say, the expectation

$$\theta(F) = \mu(F) = \int x \, dF(x),$$

or the median

$$\theta(F) = \text{Med}(F) = F^{-1}(1/2).$$

Denote the distribution function of $|\xi - \theta(F)|$ as F_1. Then this group of functionals of scale can be defined as follows:

$$S(F) = \left\{ \int_0^1 [F_1^{-1}(t)]^p \, dK(t) \right\}^{1/p}, \tag{4.6}$$

where $K(t)$ is a distribution function on the segment $[0, 1]$ and $p > 0$.

If $\theta(F) = \mu(F)$ and $K(t) = t$, $0 < t < 1$, then Equation (4.6) generates such well-known measures of scale as the mean absolute deviation with $p = 1$ and the standard deviation with $p = 2$.

The other functionals of this group are defined by the population median $\text{Med}(F) = F_1^{-1}(1/2)$, yielding, in particular, the median absolute deviation functional $S(F) = \text{MAD}(F) = F_1^{-1}(1/2)$ with $\theta(F) = \text{Med}(F)$.

The second group includes the functionals based on the pair-wise deviations between all the elements of a population $|\xi_1 - \xi_2|$, where the random variables ξ_1 and ξ_2 are independent with common distribution F. Denote the distribution function of the absolute value of the pair-wise differences as F_2. Then related functionals are defined as

$$S(F) = \left\{ \int_0^1 [F_2^{-1}(t)]^p \, dK(t) \right\}^{1/p}. \tag{4.7}$$

For example, if $p = 1$ and $K(t) = t$, $0 \le t \le 1$, then Equation (4.7) yields the Gini mean difference functional.

For $p = 2$, we obtain the standard deviation multiplied by $\sqrt{2}$. The median absolute deviation functional MAD can also be described by Equation (4.7) if $S(F) = \text{MAD}(F) = F_2^{-1}(1/2)$ is set.

The third group comprises the functionals defined by the distances between the quantiles of F of given levels

$$S(F) = \left\{ \int_0^1 |F^{-1}(1 - \alpha) - F^{-1}(\alpha)|^p \, dK(t) \right\}^{1/p}. \tag{4.8}$$

For instance, the inter-α-quantile ranges are in this group

$$S(F) = F^{-1}(1 - \alpha) - F^{-1}(\alpha), \quad 0 < \alpha < 1/2,$$

with the interquartile range functional

$$S(F) = F^{-1}(3/4) - F^{-1}(1/4),$$

yielding the well-known interquartile range IQR x estimate of scale (Kendall and Stuart 1962).

The aforementioned classification of scale functionals can be reduced to the following general scheme of their construction (Shevlyakov and Vilchevski 2002):

- the initial random variable ξ is transformed into $|\xi - \theta(F)|^p$, $|\xi_1 - \xi_2|^p$, etc.;

- the transformed random variables are processed by the operations of averaging, or of "median", or of "Hodges–Lehmann", etc.

Summarizing the above, we may say that a scale functional is defined via a location functional with the transformed variables: the variety of scale measures is determined both by the varieties of transformations and measures of location.

4.2   *M*- and *L*-Estimates of Scale

4.2.1   *M*-estimates of scale

Setting the parameter of location $\theta = 0$, consider the problem of estimating the scale parameter σ for the family of densities, the so-called pure scale problem (Huber and Ronchetti 2009)

$$f(x; \sigma) = \frac{1}{\sigma} f\left(\frac{x}{\sigma}\right), \quad \sigma > 0. \tag{4.9}$$

An *M*-estimate of σ is defined as the solution to the estimating equation

$$\sum_{i=1}^n \chi\left(\frac{x_i}{S_n}\right) = 0, \tag{4.10}$$

where χ is a score function, usually even: $\chi(-x) = \chi(x)$.

The estimate S_n is generated by the functional $S(F)$ defined as follows:

$$\int \chi\left(\frac{x}{S(F)}\right) dF(x) = 0.$$

In this case, the corresponding influence function is of the form Huber (1981, p. 109)

$$IF(x; F, S) = \frac{\chi(x/S(F))S(F)}{\int (x/S(F))\chi'(x/S(F)) \, dF(x)}. \tag{4.11}$$

The breakdown point in the ε-contamination model is given by Huber (1981, p. 110)

$$\varepsilon^* = \min\{\varepsilon_+^*, \varepsilon_-^*\}, \tag{4.12}$$

where

$$\varepsilon_+^* = -\frac{\chi(0)}{\chi(\infty) - \chi(0)}, \quad \varepsilon_-^* = \frac{\chi(\infty)}{\chi(\infty) - \chi(0)}.$$

From Equation (4.12) it follows that the maximal value of the breakdown point of M-estimates of scale is $\varepsilon^* = 0.5$.

Now we enlist the following particular cases of M-estimates of scale defined by Equation (4.10):

- the standard deviation

$$s = \left(\frac{1}{n} \sum x_i^2\right)^{1/2},$$

with $\chi(x) = x^2 - 1$,

- the mean absolute deviation

$$d = \frac{1}{n} \sum |x_i|,$$

with $\chi(x) = |x| - 1$,

- the pth-power deviation

$$S_{L_p} = \left(\frac{1}{n} \sum |x_i|^p\right)^{1/p},$$

with $\chi(x) = |x|^p - 1$,

- the median absolute deviation

$$\text{MAD } x = \text{med}|x_i|,$$

with $\chi(x) = \text{sgn}(|x| - 1)$.

Recall that all these estimates are based on the absolute deviations about zero, since location is assumed given in this setting; their score functions are exhibited in Figs 4.1 to 4.3.

Similarly to estimation of location, M-estimates (4.10) yield the maximum likelihood estimates of the scale parameter σ for the family of densities $\sigma^{-1}f(x/\sigma)$, with

$$\chi(x) = -x\,\frac{f'(x)}{f(x)} - 1. \qquad (4.13)$$

The sufficient conditions of regularity providing Fisher consistency and asymptotic normality of S_n are imposed on the densities f and the score functions χ

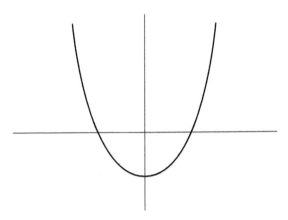

Figure 4.1   Score function for the standard deviation.

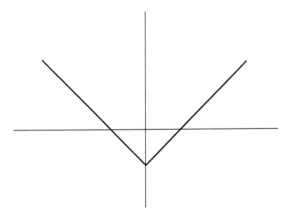

Figure 4.2   Score function for the mean absolute deviation.

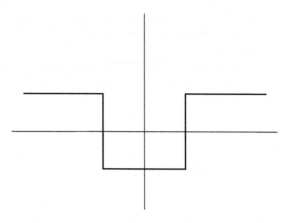

Figure 4.3   Score function for the median absolute deviation.

(Hampel *et al.* 1986, pp. 125, 139):

(F1) f is twice continuously differentiable and satisfies $f(x) > 0$ for all $x \in \mathbb{R}$.

(F2) Fisher information for scale

$$I(f;\sigma) = \frac{1}{\sigma^2} \int \left[-x \frac{f'(x)}{f(x)} - 1 \right]^2 f(x) \, dx \qquad (4.14)$$

satisfies $0 < I(f;\sigma) < \infty$.

(χ1) χ is well-defined and continuous on $\mathbb{R} \setminus C(\chi)$, where $C(\chi)$ is finite. At each point of $C(\chi)$ there exist finite left and right limits of χ, which are different. Moreover, $\chi(-x) = \chi(x)$ if $(-x, x) \subset \mathbb{R} \setminus C(\chi)$, and there exists $d > 0$ such that $\chi(x) \leq 0$ on $(0, d)$ and $\chi(x) \geq 0$ on (d, ∞).

(χ2) The set $D(\chi)$ of points at which χ is continuous but at which χ' is not defined or not continuous is finite.

(χ3) $\int \chi \, dF(x) = 0$ and $\int \chi^2 \, dF(x) < \infty$.

(χ4) $0 < \int x\chi'(x) \, dF(x) < \infty$.

Under the conditions (F1), (F2), (χ1) – (χ4) (Hampel *et al.* 1986), $\sqrt{n}(S_n - \sigma)$ is asymptotically normal with asymptotic variance

$$V(f, \chi) = \frac{\int \chi^2(x) \, dF(x)}{\left(\int x\chi'(x) \, dF(x) \right)^2}. \qquad (4.15)$$

Now we briefly comment on these conditions.

- The condition $\int \chi \, dF(x) = 0$ provides Fisher consistency.

- Using the notation

$$A(\chi) = \int \chi^2(x) \, dF(x), \quad B(\chi) = \int x\chi'(x) \, dF(x),$$

we have for the influence function (4.11)

$$IF(x; F, S) = \frac{\chi(x)}{B(\chi)}.$$

4.2.2   L-estimates of scale

Similarly to the case of location, computationally simpler L-estimates based on order statistics can be proposed for estimation of scale.

Given a sample $x_1, \ldots, x_n$ from a symmetric and absolutely continuous distribution F, the one-sided α-trimmed standard deviation is defined as follows:

$$S_\alpha = \left\{ \frac{1}{n - [\alpha n]} \sum_{i=1}^{n-[\alpha n]} x_{(i)}^2 \right\}^{1/2}, \quad 0 \le \alpha < \frac{1}{2}, \tag{4.16}$$

where $x_{(i)}$ stands for the ith order statistic. The related functional has the form

$$S_\alpha(F) = \left\{ \frac{1}{1 - \alpha} \int_0^{F^{-1}(1-\alpha)} x^2 \, dF(x) \right\}^{1/2}, \quad 0 \le \alpha < \frac{1}{2}.$$

The one-sided α-trimmed mean absolute deviation is defined by

$$d_\alpha = \frac{1}{n - [\alpha n]} \sum_{i=1}^{n-[\alpha n]} |x_{(i)}|, \quad 0 \le \alpha < \frac{1}{2} \tag{4.17}$$

with the functional of the form

$$D_\alpha(F) = \frac{1}{1 - \alpha} \int_0^{F^{-1}(1-\alpha)} |x| \, dF(x), \quad 0 \le \alpha < \frac{1}{2}.$$

Correspondingly, the two-sided α-trimmed standard and mean absolute deviations are defined by

$$S_{2\alpha} = \left\{ \frac{1}{n - [2\alpha n]} \sum_{i=[\alpha n]}^{n-[\alpha n]} x_{(i)}^2 \right\}^{1/2}, \quad 0 \le \alpha < \frac{1}{2} \tag{4.18}$$

and

$$d_{2\alpha} = \frac{1}{n - [\alpha n]} \sum_{i=[\alpha n]}^{n-[\alpha n]} |x_{(i)}|, \quad 0 \le \alpha < \frac{1}{2} \qquad (4.19)$$

with the generating functionals

$$S_{2\alpha}(F) = \left\{ \frac{1}{1 - 2\alpha} \int_{F^{-1}(\alpha)}^{F^{-1}(1-\alpha)} x^2 \, dF(x) \right\}^{1/2}, \quad 0 \le \alpha < \frac{1}{2}$$

and

$$D_{2\alpha}(F) = \frac{1}{1 - 2\alpha} \int_{F^{-1}(\alpha)}^{F^{-1}(1-\alpha)} |x| \, dF(x), \quad 0 \le \alpha < \frac{1}{2}.$$

In the case of $\alpha = 0$, Equations (4.16) to (4.19) give the standard deviation

$$s = \left\{ \frac{1}{n} \sum_{i=1}^{n} x_i^2 \right\}^{1/2}$$

and the mean absolute deviation

$$d = \frac{1}{n} \sum_{i=1}^{n} |x_i| \, ,$$

respectively.

The limiting cases $\alpha \to 1/2$ yield the median absolute deviation $\mathrm{med}|x|$ for both estimates. The expressions for the influence functions and asymptotic variances of L-estimates can be found in Huber (1981, pp. 111–113).

4.3   Huber Minimax Variance Estimates of Scale

4.3.1   Introductory remarks

In this section the Huber minimax variance solution under ε-contaminated normal distributions is given (Huber 1964, 1981), as this solution is essential for constructing the minimax variance estimate of the correlation coefficient in Chapter 5.

Since the problem of estimating the scale parameter for the random variable ξ can be reduced to that of estimating the location parameter $\tau = \log \sigma$ for the random variable $\eta = \log |\xi|$, where σ is a scale parameter for ξ, the minimax solution for scale can be written out with the use of the minimax solution for location (Huber 1964, 1981).

4.3.2   The least informative distribution

For distribution densities depending on the scale parameter

$$p(x; \sigma) = \frac{1}{\sigma} f\left(\frac{x}{\sigma}\right),$$

Fisher information takes the following form (4.14):

$$I(f; \sigma) = \int \left[\frac{\partial \log \, p(x; \sigma)}{\partial \sigma}\right]^2 p(x; \sigma) \, dx$$

$$= \frac{1}{\sigma^2} \int \left[-\frac{f'(x)}{f(x)} x - 1\right]^2 f(x) \, dx.$$

Without any loss of generality set $\sigma = 1$.

The variational problem of minimizing Fisher information for scale (4.14)

$$f^* = \arg \min_{f \in \mathcal{F}_\varepsilon} I(f; 1) \tag{4.20}$$

in the class of ε-contaminated normal densities

$$\mathcal{F}_\varepsilon = \{f : f(x) \geq (1 - \varepsilon) \, \varphi(x), 0 \leq \varepsilon < 1\}, \tag{4.21}$$

where $\varphi(x)$ is the standard normal density, is solved in Huber (1964, 1981) (for details of derivation, see Shevlyakov and Vilchevski 2002, pp. 123–124).

The least informative distribution density has the following three-branch form (Huber 1981, p. 120):

$$f^*(x) = \begin{cases} (1 - \varepsilon) \, \varphi(x_0)\left(\frac{x_0}{|x|}\right)^{(x_0^2)}, & \text{for} \quad |x| < x_0, \\[2mm] (1 - \varepsilon)\varphi(x), & \text{for} \quad x_0 \leq |x| \leq x_1, \\[2mm] (1 - \varepsilon) \, \varphi(x_1)\left(\frac{x_1}{|x|}\right)^{(x_1^2)}, & \text{for} \quad |x| > x_1, \end{cases} \tag{4.22}$$

where the parameters x_0 and x_1 satisfy the following relations:

$$x_0^2 = (1 - k)^+, \qquad x_1^2 = 1 + k,$$

$$2 \int_{x_0}^{x_1} \varphi(x) \, dx + \frac{2x_0\varphi(x_0) + 2x_1\varphi(x_1)}{x_1^2 - 1} = \frac{1}{1 - \varepsilon}. \tag{4.23}$$

For sufficiently small ε ($\varepsilon < 0.205, x_0 = 0, x_1 > \sqrt{2}$), the least informative density f^* is normal in the central zone and is like a t-distribution with $k = x_1^2 - 1 \geq 1$ degrees of freedom in the tails.

For $\varepsilon > 0.205$ and $x_0 > 0$, an additional t-distribution part of the least informative density f^* appears about the center of symmetry $x = 0$: below it is shown that this has an effect in trimming smallest data values along with greatest ones.

4.3.3   Minimax variance M- and L-estimates of scale

For the least informative density (4.22), the efficient maximum likelihood M-estimate of scale defined by the score function

$$\chi^*(x) = -x \, \frac{f^{*\prime}(x)}{f^*(x)} - 1$$

$$= \begin{cases} x_0^2 - 1, & \text{for} \quad |x| < x_0, \\ x^2 - 1, & \text{for} \quad x_0 \le |x| \le x_1, \\ x_1^2 - 1, & \text{for} \quad |x| > x_1. \end{cases} \tag{4.24}$$

yields the minimax variance M-estimate.

The corresponding minimax variance L-estimate is given by the trimmed standard deviation

$$S_n = \left\{ \frac{1}{n - [\alpha_1 n] - [\alpha_2 n]} \sum_{i=[\alpha_1 n]+1}^{n-[\alpha_2 n]} x_i^2 \right\}^{1/2}, \tag{4.25}$$

where

$$\alpha_1 = F^*(x_0) - 1/2, \quad \alpha_2 = 1 - F^*(x_1).$$

Figure 4.4 exhibits the possible forms of the score function. The limiting case $\varepsilon \to 1$ yields the median absolute deviation (MAD) as the limiting M- and L-estimates of the location parameter $\tau = \log \, \sigma$ coincide with the median of $\{\log \, |x_i|\}$; hence the corresponding estimate is the median of $\{|x_i|\}$.

These minimax variance M- and L-estimates are biased at the standard normal distribution Φ. To make them consistent and asymptotically unbiased, they should be divided by an appropriate constant: their values are given in Huber (1981, pp. 125–126).

REMARK It can be expected from the general results of Section 3.2 on robust estimation of a location parameter that the obtained M-estimate of scale is minimax with regard to the asymptotic variance

$$V(f, \chi^*) \le V(f^*, \chi^*) = \frac{1}{I(f^*; 1)}$$

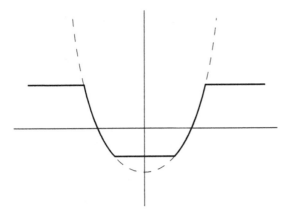

Figure 4.4   Score function for the minimax variance estimate of scale: the trimmed standard deviation.

under ε-contaminated normal distributions satisfying $S(F) = 1$ or, in other words, Fisher's consistency condition

$$\int \chi(x) \, dF(x) = 0. \tag{4.26}$$

It is possible to ignore this rather restrictive condition for sufficiently small ε ($\varepsilon <$ 0.04 and $x_1 > 1.88$) using the standardized variance instead of the asymptotic variance $V(f, \chi)$ (for details, see Huber *1981*, pp. 122–126).

4.4   Highly Efficient Robust Estimates of Scale

4.4.1   Introductory remarks

In this section a parametric family of M-estimates of scale based on the Rousseeuw–Croux Q_n-estimate is introduced, and their estimate bias and efficiency are studied. A low-complexity one-step M-estimate is obtained allowing a considerably faster computation with the greater than 80% efficiency and the highest possible 50% breakdown point. Both the analytical and Monte Carlo modeling results confirm the effectiveness of the proposed approach. Since these novel estimates of scale are essentially connected with the median of absolute deviations (MAD), we begin with it. Recall that this highly robust estimate of scale has already appeared in our book as a particular case of M-estimates of scale and as a limiting case of the Huber minimax variance estimate in subsection 4.3.3.

4.4.2   The median of absolute deviations and its properties

The median of absolute deviations, MAD $x = c\text{Med} |X - \text{Med } x|$, was firstly proposed by Gauss (1816), but it had not been used in practice until Hampel (1974) reopened it. The coefficient c is chosen to provide Fisher consistency of estimation

$$\int \chi(x)\mathrm{d}F(x) = 0 : \quad c = 1/\Phi^{-1}(3/4) = 1.4826.$$

Thus, the median of absolute deviations is of the following form:

$$\text{MAD}_n \, x = 1.4826 \, \text{med } |x - \text{med } x| \, . \tag{4.27}$$

Hampel (1974) introduced MAD as an M-estimate of scale with the minimal gross-error sensitivity $\gamma^* = 2$ at the normal distribution, which means that MAD is the most B-robust estimate of scale as well as the most V-robust estimate (Hampel *et al.* 1986, Theorems 8–10, pp. 142–143).

Its very important property is the maximal value of the breakdown point $\varepsilon^* = 1/2$ (it just inherits the breakdown point $\varepsilon^* = 1/2$ of the sample median).

Hampel (1974) recommends MAD as a useful estimate in the following cases:

- as a rough, but fast, estimate when there is no need for great accuracy;

- as a control value for more detail computations;

- as a basis for outlier detection;

- as an initial value for numerical iterative procedures.

Further, its influence function at the normal is a piece-wise constant function with discontinuities at the symmetric points $x = -1$ and $x = +1$ between which half of the data fall on (see Fig. 4.5)

$$IF(x; \text{MAD}, \Phi) = \frac{\text{sgn}(|x| - \Phi^{-1}(3/4))}{4\Phi^{-1}(3/4)\varphi(\Phi^{-1}(3/4))}. \tag{4.28}$$

Its asymptotic variance is

$$V(\text{MAD}, \Phi) = \int IF^2(x; \text{MAD}, \Phi) \, \mathrm{d}\Phi(x)$$

$$= (4\Phi^{-1}(3/4)\varphi(\Phi^{-1}(3/4)))^{-2} \approx 1.3605, \tag{4.29}$$

yielding the following value of the asymptotic efficiency:

$$\text{eff}(\text{MAD}, \Phi) = \frac{1/I(\Phi)}{V(\text{MAD}, \Phi)} \approx 0.3675, \tag{4.30}$$

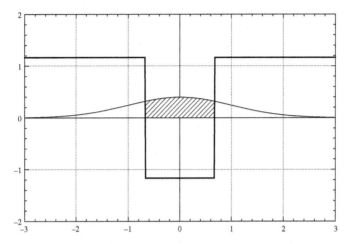

Figure 4.5   The median absolute deviation score function at the standard normal density.

where $I(\Phi) = 2$—Fisher information (4.14) for a scale parameter at the normal distribution.

Besides the low asymptotic efficiency, another drawback of MAD is its symmetry with respect to data dispersion, as, firstly, the center (median) is estimated, and hence the positive and negative deviations from the center have the same weight, which is not natural for asymmetric distributions. Asymmetric weights are provided by the interquartile range (IQR), but at the cost of the lower breakdown point 25% and the low efficiency (4.30) at the normal.

4.4.3   The quartile of pair-wise absolute differences Q_n estimate and its properties

In Rousseeuw and Croux (1993), the authors proposed two new estimates of scale, which can be used instead of the median of absolute deviations. Holding the highest value $1/2$ of the breakdown point, they have considerably greater efficiencies and, moreover, they do not require preliminary estimation of location.

These estimates are based on the pair-wise absolute differences of observations. The first estimate is the Hodges–Lehmann estimate of scale

$$\mathrm{med}_{i<j}|x_i - x_j| \tag{4.31}$$

with the lower breakdown point 0.29 and with the sufficiently high efficiency 0.86.

Rousseeuw and Croux found that it is possible to retain the maximal breakdown point if the median in (4.31) is changed for the other order statistic, simultaneously holding high efficiency. The proposed estimate is defined as the first quartile of the absolute distances between observations

$$Q_n = c \cdot \{|x_i - x_j|\}_{(k)}, \quad k = C_h^2, \tag{4.32}$$

where $h = \lfloor n/2 \rfloor + 1$, that is, approximately half of observations. The constant c providing Fisher consistency of Q_n is equal to

$$c = \frac{1}{\sqrt{2}\Phi^{-1}(5/8)} = 2.2191$$

at the normal.

The influence function of Q_n at symmetric distributions F with density f has the following form:

$$IF(x; Q, \Phi) = c \, \frac{1/4 - F(x + c^{-1}) + F(x - c^{-1})}{\int f(y + c^{-1}) f(y) \, \mathrm{d}y}. \tag{4.33}$$

The gross-error sensitivity $\gamma^*(Q, \Phi) = 2.069$ is slightly greater than that of MAD. By numerical integration, it is possible to get the asymptotic variance

$$V(Q, \Phi) = \int IF^2(x; Q, \Phi) \, \mathrm{d}\Phi(x) \approx 0.6077,$$

from which follows a rather high asymptotic efficiency eff $= 0.823$.

It may seem that the Q_n-estimate does not deserve attention because of its high computational complexity: in general, it requires about $O(n^2)$ of computation time and memory. However, in the particular case of pair-wise distances, Croux and Rousseeuw (1992) proposed a more effective algorithm requiring only $O(n \log n)$ of time and memory.

At present, this estimate is widely used in statistical practice and is available in many statistical packages, for instance, in the R-package.

4.4.4 M-estimate approximations to the Q_n estimate: MQ_n^α, FQ_n^α, and FQ_n estimates of scale

In this subsection we propose an estimate of scale with the influence function similar to $IF(x; Q_n, F)$, thereby ensuring the similarity of derived characteristics, in particular, the asymptotic variance and efficiency.

MQ_n^α estimates of scale

Consider the class of M-estimates of scale S_n defined implicitly by Equation (4.10)

$$\sum_{i=1}^{n} \chi(x_i/S_n) = 0,$$

where χ is a score function even and nondecreasing for $x > 0$. Choosing the shape of the score function χ, one can get a variety of both robust and nonrobust estimates of scale.

It is known that the influence function of such estimates is equal to the chosen score function up to a normalizing factor: $IF(x; S, F) \propto \chi(x)$. We use this fact and construct an M-estimate of scale, which approximates Q_n, setting its score function as

$$\chi_Q(x) = c/4 - c \; (\Phi(x + c^{-1}) - \Phi(x - c^{-1})),$$

where $\Phi(x)$ is the standard normal cumulative distribution function, and the constant $c = 1/(\Phi^{-1}(5/8)\sqrt{2}) = 2.2191$ is chosen from the condition of Fisher consistency: $\int \chi(x) \, d\Phi(x) = 0$ (Rousseeuw and Croux 1993). The influence function of the introduced M-estimate coincides with $IF(x; Q_n, \Phi)$.

For the sake of convenience, we consider a slightly different parametric family with a free parameter $\alpha = c^{-1}$, taking the first terms of the expansion of $\Phi(x)$ in the Taylor series:

$$\Phi(x \pm \alpha) = \Phi(x) \pm \alpha\varphi(x) + \frac{1}{2}\alpha^2\varphi'(x) \pm \frac{1}{6}\alpha^3\varphi''(x) + o(\alpha^3).$$

Hence, we obtain an M-estimate MQ_n^α with the score function of the form

$$\chi_\alpha(x) = c_\alpha - \frac{1}{3}(6 - \alpha^2(x^2 - 1))\varphi(x), \tag{4.34}$$

where $c_\alpha = (12 - \alpha^2)/(12\sqrt{\pi})$ is a Fisher consistency term. Such representation allows the result to be obtained explicitly in terms of elementary functions. Our study shows that it makes no sense to consider the expansion with a greater accuracy –the gain in performance is negligible but the amount of computation increases.

Due to its low complexity, we are also interested in the particular case $\alpha = 0$:

$$\chi_0(x) = \frac{1}{\sqrt{\pi}} - 2\varphi(x). \tag{4.35}$$

Now, we obtain the characteristics of the proposed estimates. The asymptotic variance of the FQ_n^α estimate has the form

$$V(MQ^\alpha, \Phi) = \int IF^2(x; MQ^\alpha, \Phi) \, d\Phi(x) = \frac{\int \chi_\alpha^2(x)\varphi(x) \, dx}{\left(\int x\chi_\alpha'(x)\varphi(x) \, dx\right)^2}$$

$$= \left(\frac{4 - \alpha^2}{8}\right)^{-2} \left(\frac{54 - 12\alpha^2 + \alpha^4}{27\sqrt{3}} - \left(\frac{12 - \alpha^2}{12}\right)^2\right),$$

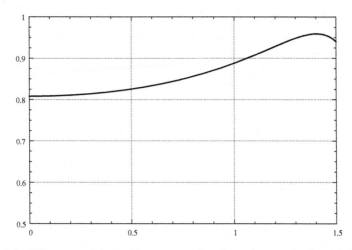

Figure 4.6   Efficiency of the MQ^α estimate of scale at the standard normal. Source: Shevlyakov (2014), Reproduced with permissions from Elsevier.

yielding the Gaussian efficiency

$$\text{eff}_\alpha = \frac{V(SD, \Phi)}{V(MQ^\alpha, \Phi)} = \frac{1}{2\,V(MQ^\alpha, \Phi)}.$$

The maximal achievable efficiency is 0.959 at $\alpha = 1.403$, but even if $\alpha = 0$, it does not fall below the level of 0.808 (see Fig. 4.6).

The breakdown point of this estimate is given by the following formula (Huber 1981, p. 110):

$$\varepsilon_\alpha^* = \min\left\{ \frac{\chi_\alpha(\infty)}{\chi_\alpha(\infty) - \chi_\alpha(0)}, \frac{-\chi_\alpha(0)}{\chi_\alpha(\infty) - \chi_\alpha(0)} \right\}$$

$$= \frac{12(\sqrt{2} - 1) - \alpha^2(2\sqrt{2} - 1)}{(6 - \alpha^2)2\sqrt{2}}$$

in the case of an even $\chi(x)$ monotonically increasing for $x > 0$. These conditions are satisfied with $\alpha \in [0, \sqrt{2}]$, and ε_α^* reaches its maximum at $\alpha = 0$: $\varepsilon_0^* = 0.293$ (see Fig. 4.7).

One-step M-estimates of scale: FQ_n^α and FQ_n

Computation of estimates as the solution of an implicit equation in most cases is difficult, but it is possible to use iterative schemes. In particular, we can confine ourselves to the first iteration of the Newton–Raphson procedure, a so-called one-step

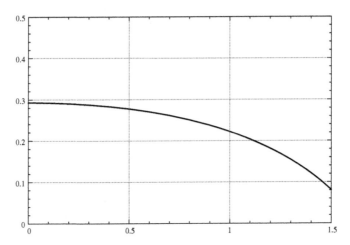

Figure 4.7   Breakdown point of the MQ^α estimate of scale. Source: *Shevlyakov (2014), Reproduced with permissions from Elsevier.*

M-estimate (Huber 1981):

$$S_n^{(1)} = S_n^{(0)} - \left. \frac{\sum\limits_{i=1}^n \chi(x_i/S)}{\frac{\partial}{\partial S}\sum\limits_{i=1}^n \chi(x_i/S)} \right|_{S=S_n^{(0)}}$$

$$= S_n^{(0)}\left(1 + \frac{\sum \chi(x_i/S_n^{(0)})}{\sum (x_i/S_n^{(0)})\chi'(x_i/S_n^{(0)})}\right).$$

The initial approximation $S_n^{(0)}$ should itself be highly robust. Substituting the proposed score function (4.34), we find that the one-step estimate could be defined as the following refinement of the median absolute deviation (4.27):

$$FQ_n^\alpha = \text{MAD}_n\left(1 - \frac{(6-\alpha^2)U_0 + \alpha^2 U_2 - (12-\alpha^2)n/(2\sqrt{2})}{3(2-\alpha^2)U_2 + \alpha^2 U_4}\right), \qquad (4.36)$$

where

$$U_k = \sum_{i=1}^n u_i^k e^{-u_i^2/2}, \quad u_i = x_i/\text{MAD}_n, \quad k = 0, 2, 4.$$

In the particular case $\alpha = 0$, it is reduced to the FQ_n estimate of scale

$$FQ_n = FQ_n^0 = \text{MAD}_n\left(1 - \frac{U_0 - n/\sqrt{2}}{U_2}\right), \qquad (4.37)$$

where $\text{MAD}_n\, x = 1.4826\, \text{med}\, |x - \text{med}\, x|$.

The score function of the FQ_n estimate coincides with the score function $\chi_0(x)$ given by (4.35)

$$\chi_{FQ}(x) = \frac{1}{\sqrt{\pi}} - \sqrt{\frac{2}{\pi}} \, \exp\left(-\frac{x^2}{2}\right). \tag{4.38}$$

The FQ_n estimate efficiency is equal to 0.808 at the normal, and its breakdown point is $1/2$.

REMARK From Fisher's consistency condition $\int \chi(x) \, d\Phi(x) = 0$ it directly follows that the one-step M-estimate converges in probability to unity as $n \to \infty$. Although on finite samples ($n = 20, 60, 1000$) we cannot theoretically guarantee the positiveness of this estimate, our extensive Monte Carlo study ($m = 100,000$ trials) shows that in the normal, Cauchy, and contaminated normal distribution models

$$G(x) = (1 - \varepsilon)\Phi(x/\sigma) + \varepsilon\Phi(x/\sigma_*) \tag{4.39}$$

with $\varepsilon = 0(0.1)0.5$, $\sigma = 1$, $\sigma_* = 0.01, 3, 10$, all the computed one-step M-estimate values are positive.

Anyway, if a nonpositive M-estimate value occurs, one should choose the initial MAD_n estimate.

The important property of such estimates is that they inherit the breakdown point of the initial approximation (Rousseeuw and Croux 1993). When we choose the median of absolute deviations as a basis, the breakdown point of the one-step estimate FQ_n^α is increased up to 0.5.

For the one-step estimate, the following important result holds (P. O. Smirnov, personal communication).

Theorem 4.4.1. *In the scale model $G(x) = F(x/\sigma)$ where $S_n^{(0)}(F) = 1$, the influence function of a Newton–Raphson one-step M-estimate of scale is given by*

$$IF(x; S_n^{(1)}, G) = \frac{\chi(x/S_n^{(0)}(G))S_n^{(0)}(G)}{\int (y/S_n^{(0)}(G))\chi'(y/S_n^{(0)}(G)) \, dG}.$$

Proof. A functional M-estimate $S_*(G)$ of scale is defined as follows:

$$S_*(G) : \int \chi\left(\frac{y}{S_*(G)}\right) \, dG = 0$$

or

$$Q(S_*(G)) = 0,$$

where

$$Q(S) = \int \chi\left(\frac{y}{S}\right) dG .$$

Consider the linear approximation to it at S_0 (Huber 1981, p. 140)

$$Q(S_*(G)) = Q[S_0(G) + (S_*(G) - S_0(G))]$$

$$\approx Q(S_0(G)) + (S_*(G) - S_0(G))\frac{\partial Q}{\partial S}\bigg|_{S=S_0(G)} = 0.$$

Hence,

$$S_*(G) \approx S(G) = S_0(G) - \frac{Q(S_0(G))}{\frac{\partial Q}{\partial S}\big|_{S=S_0(G)}} ,$$

where

$$\frac{\partial Q}{\partial S} = -\int \frac{y}{S^2}\chi'\left(\frac{y}{S}\right) dG.$$

Thus we arrive at the one-step functional M-estimate in the form

$$S(G) = S_0(G)\left(1 + \frac{\int \chi\left(\frac{y}{S_0(G)}\right) dG}{\int \frac{y}{S_0(G)}\chi'\left(\frac{y}{S_0(G)}\right) dG}\right). \qquad (4.40)$$

Introduce the following notations:

$$a_0(x) = \chi\left(\frac{x}{S_0(G)}\right), \quad b_0(x) = \frac{x}{S_0(G)}\chi'\left(\frac{x}{S_0(G)}\right),$$

$$A_0 = \int a_0(y)dG = \int \chi\left(\frac{y}{S_0(G)}\right) dG,$$

$$B_0 = \int a_0(y)dG = \int \frac{y}{S_0(G)}\chi'\left(\frac{y}{S_0(G)}\right) dG,$$

$$c_0(x) = \left(\frac{x}{S_0(G)}\right)^2\chi''\left(\frac{x}{S_0(G)}\right),$$

$$C_0 = \int c_0(y)dG = \int \left(\frac{y}{S_0(G)}\right)^2\chi''\left(\frac{y}{S_0(G)}\right) dG .$$

In these notations, Equation (4.40) takes the form

$$S(G) = S_0(G)\left(1 + \frac{A_0}{B_0}\right) . \qquad (4.41)$$

The influence function of the functional M-estimate is given by Huber (1981, pp. 136–137)

$$IF(x; S_*, G) = \frac{-\chi(x/S_*(G))}{\int \frac{\partial}{\partial S}(\chi(y/S_*))\big|_{S=S_*(G)} dG}$$

$$= S_*(G)\frac{\chi(x/S_*(G))}{\int \frac{y}{S_*(G)}\chi'\left(\frac{y}{S_*(G)}\right) dG} .$$

In our notations, it can be written as

$$IF(x; S_*, G) = S_*(G)\frac{a_*(x)}{B_*} . \tag{4.42}$$

Next, we derive the influence function of the one-step M-estimate using its definition (3.25) (Huber 1981, pp. 136, 140–141):

1. Substitute $G_t = (1 - t)G + t\Delta_x$ into (4.41).

2. Take the derivative of $S(G_t)$ with respect to t at $t = 0$.

Before that, we set several more notations:

$$z_t = \frac{y}{S_0(G_t)} , \qquad \frac{\partial z_t}{\partial t}\bigg|_{t=0} = -\frac{y}{S_0^2(G_t)}\frac{\partial}{\partial t}(S_0(G_t))\bigg|_{t=0}$$

$$= -\frac{y}{S_0(G_t)}\frac{IF(x; S_0, G)}{S_0(G)} .$$

Now substitute G_t into Equation (4.41):

$$S(G_t) = S_0(G_t)\left\{1 + \frac{\int \chi(z_t)d[(1 - t)G + \Delta_x]}{\int z_t\chi'(z_t)d[(1 - t)G + \Delta_x]}\right\}$$

$$= S_0(G_t)\left\{1 + \frac{\int (1 - t)\chi(z_t)dG + t[\chi(z_t)]|_{y=x}}{\int (1 - t)z_t\chi'(z_t)dG + t[z_t\chi'(z_t)]|_{y=x}}\right\}$$

$$= S_0(G_t)(1 + A_t B_t^{-1}),$$

where A_t and B_t are the integrals standing in the nominator and denominator of the last fraction, respectively.

Next compute the derivatives of the terms A_t and B_t:

$$\frac{\partial A_t}{\partial t} = -\int \chi(z_t)dG + \int (1-t)\chi'(z_t)\frac{\partial z_t}{\partial t}dG + [\chi(z_t)]|_{y=x}$$

$$+ t[\chi'(z_t)\frac{\partial z_t}{\partial t}]\Big|_{y=x} \, , \quad \frac{\partial A_t}{\partial t}\Big|_{t=0} = -A_0 - \frac{IF(x; S_0, G)}{S_0(G)}B_0 + a_0(x) \, .$$

$$\frac{\partial B_t}{\partial t} = -\int z_t\chi'(z_t)dG + \int (1-t)\chi'(z_t)\frac{\partial z_t}{\partial t}dG + \int (1-t)z_t\chi''(z_t)\frac{\partial z_t}{\partial t}dG$$

$$+ [z_t\chi'(z_t)]|_{y=x} + t\left[\chi'(z_t)\frac{\partial z_t}{\partial t} + z_t\chi''(z_t)\frac{\partial z_t}{\partial t}\right]\Big|_{y=x} \, ,$$

$$\frac{\partial B_t}{\partial t}\Big|_{t=0} = -B_0 - \frac{IF(x; S_0, G)}{S_0(G)}B_0 - \frac{IF(x; S_0, G)}{S_0(G)}C_0 + b_0(x) \, .$$

Now we use these relations and Equation (4.41) to get the influence function

$$\frac{\partial S(G_t)}{\partial t}\Big|_{t=0} = \frac{\partial S_0(G_t)}{\partial t}\Big|_{t=0}(1 + A_0 B_0^{-1})$$

$$+ S_0(G)\left(\frac{\partial A_t}{\partial t}\Big|_{t=0}B_0^{-1} - A_0 B_0^{-2}\frac{\partial B_t}{\partial t}\Big|_{t=0}\right) \, ,$$

that is,

$$IF(x; S, G) = IF(x; S_0, G)\left(1 + \frac{A_0}{B_0}\right)$$

$$+ \frac{S_0(G)}{B_0}\left(\frac{\partial A_t}{\partial t} - \frac{A_0}{B_0}\frac{\partial B_t}{\partial t}\right)\Big|_{t=0} \, . \tag{4.43}$$

Substitute the derivatives into Equation (4.43)

$$IF(x; S, G) = IF(x; S_0, G)(1 + \frac{A_0}{B_0}) + \frac{S_0(G)}{B_0}\Bigg\{ -A_0 - \frac{IF(x; S_0, G)}{S_0(G)}B_0$$

$$+ a_0(x) + \frac{A_0}{B_0}\left(B_0 + \frac{IF(x; S_0, G)}{S_0(G)}B_0 + \frac{IF(x; S_0, G)}{S_0(G)}C_0 - b_0(x)\right)\Bigg\}$$

$$= IF(x; S_0, G)\left(1 + \frac{A_0}{B_0} - 1 + \frac{A_0}{B_0} + \frac{A_0 C_0}{B_0^2}\right)$$

$$+ \frac{S_0(G)}{B_0}\left(-A_0 + a_0(x) + A_0 - \frac{A_0}{B_0}b_0(x)\right) \, ,$$

from which it follows that

$$IF(x; S, G) = IF(x; S_0, G)\frac{A_0}{B_0^2}(2B_0 + C_0)$$

$$+ \frac{S_0(G)}{B_0^2}(a_0(x)B_0 - b_0(x)A_0) . \tag{4.44}$$

The next step is important: the estimates S_*, S_0 and S are consistent, so at the model distribution F

$$S_*(F) = S_0(F) = S(F) = 1.$$

Thus

$$a_0(x) = a_*(x) = \chi(x), \quad A_0 = A_* = 0,$$

$$b_0(x) = b_*(x) = x\chi'(x), \quad B_0 = B_* = B = \int y\chi'(y)dF.$$

Then we arrive at the claimed result

$$IF(x; S, F) = S_0(F)\frac{a_0(x)}{B_0} = \frac{\chi(x)}{B} = IF(x; S_*, F)$$

or

$$IF(x; S, F) = IF(x; S_*, F) , \tag{4.45}$$

that is, the one-step M-estimate influence function coincides with the full one.

A similar result for location is obtained by (Huber 1981, pp. 140–141). Hence, for the model normal distribution Φ, it follows that the influence function of the one-step approximation is equal to the influence function of the full M-estimate:

$$IF(x; FQ^\alpha, \Phi) = IF(x; MQ^\alpha, \Phi),$$

thus keeping the same asymptotic variance and efficiency (see Fig. 4.8). To the best of our knowledge, these results for M-estimates of scale have not been published.

4.5   Monte Carlo Experiment

In this section we maintain a comparative performance study of several estimates of scale under the standard Gaussian, contaminated Gaussian, and Cauchy underlying distributions.

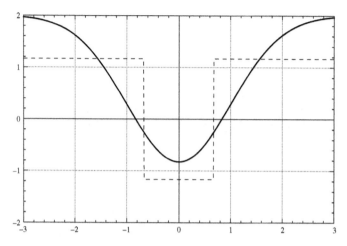

Figure 4.8   Influence function of the FQ_n estimate of scale.

4.5.1   A remark on the Monte Carlo experiment accuracy

We present Monte Carlo results based on $m = 50,000$ trials for each n. Our study shows that 1,000 or 10,000 trials usually used in different papers are not at all sufficient to make reliable conclusions on the bias and variance behavior using observed estimate means and variances.

Moreover, our numerous Monte Carlo experiments (not only in studying the performance of estimates of scale but with many others) generally show that the average achievable accuracy, as a rule, does not exceed 2–3% of the relative error, and this rate is stabilized approximately at $m = 50,000$ trials – it is rather senseless to use considerably more trials to provide a better accuracy of experiment. Certainly, this is our personal experience: apparently, the aforementioned rate of accuracy must depend on the properties of pseudo-random number generating procedures; however, those experimental results deserve a serious justification.

To illustrate our point of view, we present Fig. 4.9, exhibiting a typical dependence of a Monte Carlo experiment accuracy rate (Monte Carlo mean or variance; in this particular case it is the standardized variance; see Bickel and Lehmann 1976) on the number of trials – evidently, the accuracy is stabilized at $m = 50,000 - 60,000$ trials.

4.5.2   Monte Carlo experiment: distribution models

In our experiment, the standard normal distribution is used in two modes: the scale-only model $F(x) = \Phi(x/\sigma)$ and the location-scale model $F(x) = \Phi((x - \mu)/\sigma)$ with $\sigma = 1$, $\mu = 0$. In the latter case, a preliminary estimation of the center of symmetry μ is done by means of the corresponding estimate of location, e.g., the sample median for MAD_n and FQ_n.

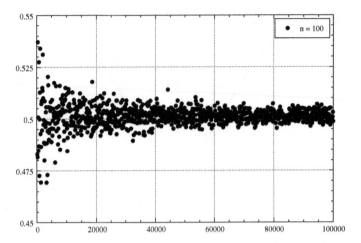

Figure 4.9   Typical dependence of Monte Carlo accuracy on the number of trials.

The contaminated normal distribution is given by the Tukey gross error model:

$$G(x) = (1 - \varepsilon)\Phi(x/\sigma) + \varepsilon\Phi(x/3\sigma)$$

with 10% of contamination ($\varepsilon = 0.1$).
The Cauchy distribution is of the form

$$G(x) = F(x/\sigma) = 1/2 + \arctan(x/\sigma)/\pi.$$

4.5.3   Monte Carlo experiment: estimates of scale

We compare the performance of the following estimates of scale:

- the standard deviation

$$SD_n(\mathbf{x}) = \sqrt{\frac{1}{n}\sum_{i=1}^{n} x_i^2} \ ;$$

- the median absolute deviation

$$\mathrm{MAD}_n(\mathbf{x}) = c \ \mathrm{med}_i \ |x_i|,$$

where $c = 1/\Phi^{-1}(3/4) \approx 1.4826$, and for its computation we use the linear in average *Quickselect* algorithm (Hoare 1961) with the median-of-3 pivot strategy;

- the Q_n-estimate (the lower quartile of absolute pair-wise differences)

$$Q_n(\mathbf{x}) = c \ \{|x_i - x_j|; \ i < j\}_{(k)},$$

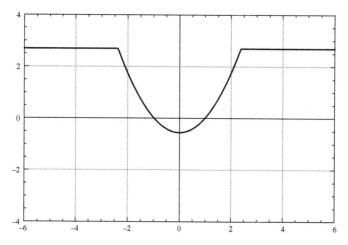

Figure 4.10   Score function $\chi(x)$ for the Huber H95 M-estimate of scale.

$c = 1/(\sqrt{2}\Phi^{-1}(5/8)) \approx 2.2191$, the number k of the order statistic is $k = C_h^2 = h(h-1)/2$, $h = [n/2] + 1$, and for the Q_n-estimate computation the fast $O(n \log n)$ algorithm is used (Croux and Rousseeuw 1992);

• the proposed one-step M-estimate $FQ_n(\mathbf{x})$ (4.37);

• the one-step version of the Huber M-estimate (Huber 1964) with the score function (see Fig. 4.10)

$$\chi(x) = \min(x^2, \alpha^2) - \beta,$$

where α is a tuning parameter and β is chosen to provide Fisher consistency $\int \chi(x)dF(x) = 0$;

$$H95_n(\mathbf{x}) = \text{MAD}_n(\mathbf{x}) \left(1 + \frac{U_- + U_+ - n\beta}{2U_-} \right),$$

$$U_- = \sum_{|u_i| < \alpha} u_i^2, \quad U_+ = \sum_{|u_i| \geq \alpha} \alpha^2,$$

where the parameter $\alpha = 2.38$ yields 95% efficiency, $\beta \approx 0.969$, and $u_i = x_i/\text{MAD}_n(\mathbf{x})$ as before.

4.5.4   Monte Carlo experiment: characteristics of performance

The characteristics of performance are: the estimate means $ave_m(S_n)$ and the standardized variances

$$v_m = n \, var_m(S_n)/ave_m^2(S_n)$$

computed over $m = 50,000$ trials for the aforementioned scale estimates S_n (it was argued by Bickel and Lehmann 1976 that the denominator is needed to obtain a natural measure of accuracy for scale estimates).

We also use the average absolute value of estimate's bias $b_m = ave_m |S_n - \sigma|$.

4.5.5   Monte Carlo experiment: results

The results of Monte Carlo modeling are exhibited in Tables 4.1 to 4.5 and in Figs 4.11 to 4.14.

Table 4.1   Computation time in microseconds.

n	SD_n	MAD_n	Q_n	FQ_n
20	0.5	0.6	2.5	1.6
60	1.6	1.7	20.1	4.7
200	5.1	5.1	84.9	14.8
1000	25.6	24.4	528.9	72.7

Table 4.2   Monte Carlo means and standardized variances in the scale-only standard normal distribution model.

n	SD_n	MAD_n	Q_n	FQ_n	SD_n	MAD_n	Q_n	FQ_n
20	**0.988**	1.016	1.186	0.985	**0.510**	1.219	0.779	0.612
60	**0.996**	1.005	1.061	0.994	**0.504**	1.296	0.667	0.616
200	**0.999**	1.002	1.018	0.998	**0.501**	1.342	0.631	0.629
1000	**1.000**	1.000	1.004	1.000	**0.500**	1.358	0.612	0.610
∞	**1.000**	1.000	1.000	1.000	**0.500**	1.361	0.608	0.619

Table 4.3   Monte Carlo means and standardized variances in the location-scale standard normal distribution model.

n	SD_n	MAD_n	Q_n	FQ_n	SD_n	MAD_n	Q_n	FQ_n
20	**0.986**	0.959	1.188	0.952	**0.537**	1.367	0.787	0.669
60	**0.996**	0.987	1.061	0.983	**0.505**	1.373	0.681	0.636
200	**0.999**	0.996	1.018	0.995	**0.508**	1.352	0.634	0.623
1000	**1.000**	0.999	1.004	0.999	**0.497**	1.378	0.618	0.619
∞	**1.000**	1.000	1.000	1.000	**0.500**	1.361	0.608	0.619

Table 4.4   Monte Carlo means and standardized variances in the Tukey gross error distribution model ($\varepsilon = 0.1$, $\sigma_* = 3\sigma$).

n	SD_n	MAD_n	Q_n	FQ_n	SD_n	MAD_n	Q_n	FQ_n
20	1.291	**1.101**	1.347	1.115	1.596	1.284	0.996	**0.829**
60	1.323	**1.087**	1.195	1.119	1.732	1.363	0.864	**0.807**
200	1.336	**1.084**	1.143	1.121	1.797	1.403	0.829	**0.799**
1000	1.341	**1.082**	1.126	1.121	1.825	1.417	0.804	**0.799**

Table 4.5   Monte Carlo means and standardized variances in the Cauchy distribution model.

n	SD_n	MAD_n	Q_n	FQ_n	SD_n	MAD_n	Q_n	FQ_n
20	49.70	**1.586**	2.362	2.040	$1.8 \cdot 10^5$	2.612	**2.569**	2.658
60	53.01	**1.512**	1.999	1.975	$2.1 \cdot 10^4$	2.488	**2.215**	2.392
200	130.5	**1.491**	1.887	1.958	$3.9 \cdot 10^5$	2.473	**2.081**	2.269
1000	267.6	**1.485**	1.848	1.951	$2.9 \cdot 10^6$	2.475	**2.047**	2.262

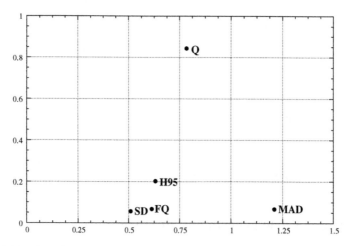

Figure 4.11   Standardized variance v_m (axis x) versus average absolute bias b_m (axis y) performance at the standard normal distribution: $n = 20$.

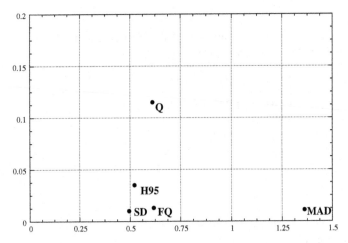

Figure 4.12   Standardized variance v_m (axis x) versus average absolute bias b_m (axis y) performance at the standard normal distribution: $n = 1,000$.

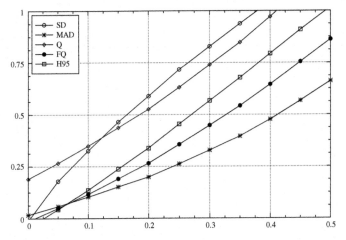

Figure 4.13   Average bias b_m dependence on the contamination fraction $0 \leq \varepsilon \leq 0.5$ at the contaminated normal distribution: $n = 20$.

4.5.6   Monte Carlo experiment: discussion

Computation time comparison

Table 4.1 manifests the superiority of linear algorithms of computation, e.g., of the FQ_n-estimate, over the considerably slower algorithm for the Q_n-estimate based on

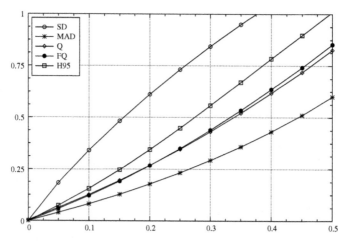

Figure 4.14   Average bias b_m dependence on the contamination fraction $0 \le \varepsilon \le 0.5$ at the contaminated normal distribution: $n = 1,000$.

the pair-wise differences of observations. For the sample size $n = 1,000$, the computation time of Q_n is 9 times greater than that of FQ_n with approximately the same efficiency and an order of lesser bias (see Figs 4.11 and 4.12).

Normal case

Tables 4.2 and 4.3 display Monte Carlo results for the standard normal distribution in the scale-only and location-scale models. The row $n = \infty$ shows theoretical asymptotic values, i.e., mean $E(S; \Phi)$ and variance $V(S; \Phi)$. The results confirm a good asymptotic behavior of the FQ_n-estimate and its better finite sample properties as compared to Q_n, even when estimation of location is required for FQ_n (the best results are in boldface). As it must be, the best in performance is the classical SD_n estimate. The next with respect to bias are the MAD- and the proposed FQ_n-estimates. The bias of the Q_n-estimate is rather large, especially in small samples. The $H95_n$-estimate manifests high efficiency in achieving the claimed 95%. The FQ_n-estimate exhibits uniform n efficiency as compared to the Q_n-estimate, whose efficiency is considerably lesser in small samples.

Contaminated normal case

Table 4.4 exhibits the estimate behavior at the heavy-tailed distributions: the effects of deviations from the normal distribution are examined in the Tukey model of gross errors. The best in bias is the MAD_n-estimate, which is the most B-robust estimate among M-estimates of scale (Hampel *et al.* 1986). The next in bias is FQ_n, at the same time being the first in variance (efficiency). The classical nonrobust estimate

SD_n is the worst with respect to all the performance characteristics with increased bias and variance. The curves in Figs 4.13 and 4.14 confirm all the above-said.

Cauchy case

From Table 4.5 it follows that the performance of the classical SD_n-estimate becomes uncontrolled and, its bias and standardized variance have increased and are being increased with the increasing sample size n. The other estimates being robust are reasonably bounded. The MAD_n-estimate exhibits the minimal bias, the next is FQ_n, but in large samples, the second is Q_n. With respect to efficiency, that estimate is the first with FQ_n the second.

4.5.7   Concluding remarks

A low-complexity FQ_n estimate of scale with the same influence function, asymptotic variance, and efficiency as the highly efficient robust estimate Q_n is proposed. Monte Carlo results show that the one-step approximation algorithm for computing the proposed estimate not only requires much less time but provides a smaller bias than of the Q_n estimate, especially on small samples. With the median absolute deviation as the initial approximation, the breakdown point has the maximum value of 0.5 and an option to enhance efficiency up to 0.95 at the cost of an increased bias in the presence of outliers. This result holds even in the case of an unknown distribution center of symmetry.

Computational algorithms for the FQ_n estimate and its applications are made publicly available as a package for the R-environment for statistical computing and graphics (Smirnov 2013).

4.6   Summary

At present, one of the best robust estimates of scale is given by the Q_n-estimate proposed by Rousseeuw and Croux (1993) – it has the highest efficiency of 82% at the normal and the highest robustness with the breakdown point $1/2$. However, it has the following definite drawbacks: the high computational complexity and the relatively low accuracy (the considerable bias and variance) in small samples. Nearly all those drawbacks are removed with the use of a new fast highly robust and efficient FQ_n-estimate of scale proposed by Smirnov and Shevlyakov (2014).

In the Gaussian case as well as in contaminated Gaussian and Cauchy distribution models, the proposed FQ_n-estimate exhibits a good performance; this allows it to be recommended as

- a faster version of the Q_n-estimate,

- a more efficient specification of the well-known robust MAD-estimate.

We may also add that the MAD-estimate is the best robust estimate with respect to bias, and this is important in applications. Finally, the aforementioned features of the FQ_n-estimate allow it to be applied directly to estimation of a scale parameter of a symmetric distribution or to use it as a basic algorithm in statistical procedures of robust estimation of a correlation coefficient, robust estimation of a power spectrum, and robust exploratory data analysis – this will be shown in the sequel.

References

Bickel PJ and Lehmann EL 1973 Measures of location and scale. In *Proc. Prague Symp. Asymptotic Statist.* **I**, Prague, Charles Univ., pp. 25–36.

Bickel PJ and Lehmann EL 1976 Descriptive statistics for nonparametric models. *Ann. Statist.* **4**, 1139–1158.

Croux C and Rousseeuw PJ 1992 Time-efficient algorithms for two highly robust estimators of scale. *Computational Statistics* **1**, 411–428.

Gauss CF 1816 Bestimmung der Genauigkeit der Beobachtungen. *Z. Astr. verw. Wiss.* In *Werke* **4**, Dietirich, Göttingen (1880), pp. 109–117.

Gentleman WM 1965 Robust Estimation of Multivariate Location by Minimizing pth Power Deviations. Ph.D. Thesis, Princeton University.

Hampel FR 1974 The influence curve and its role in robust estimation. *J. Amer. Statist. Assoc.* **69**, 383–393.

Hampel FR, Ronchetti E, Rousseeuw PJ and Stahel WA 1986 *Robust Statistics. The Approach Based on Influence Functions*, Wiley.

Hoare CAR 1961 Algorithm 65: Find. *Communications of the ACM* **4**, 321–322.

Huber PJ 1964 Robust estimation of a location parameter. *Ann. Math. Statist.* **35**, 73–101.

Huber PJ 1981 *Robust Statistics*, Wiley.

Huber PJ and Ronchetti E (eds) 2009 *Robust Statistics*, 2nd edn, Wiley.

Kendall MG and Stuart A 1962 *The Advanced Theory of Statistics. Distribution Theory*, vol. 1, Griffin, London.

Orlov AI 1976 *Stability in Social and Economic Models*, Nauka, Moscow (in Russian).

Rousseeuw PJ and Croux C 1993 Alternatives to the median absolute deviation. *J. Amer. Statist. Assoc.* **88**, 1273–1283.

Shevlyakov GL and Vilchevski NO 2002 *Robustness in Data Analysis: criteria and methods*, VSP, Utrecht.

Smirnov PO 2013 *robcor: Robust correlations*. R package version 0.1-5. Electronic resource – Vienna, Austria: The Comprehensive R Archive Network, 2013. Available at: http://CRAN.R-project.org/package=robcor (accessed: 06.12.13).

Smirnov PO and Shevlyakov GL 2014 Fast highly efficient and robust one-step M-estimators of scale based on FQ_n. *Computational Statistics and Data Analysis* **78**, 153–158.

5

Robust Estimation of Correlation Coefficients

5.1 Preliminaries

The necessity of robust estimation of correlation was shown in Gnanadesikan and Kettenring (1972) and Huber (1981). In the contamination model of a bivariate normal distribution density

$$f(x, y) = (1 - \epsilon)N(x, y; \mu_1, \mu_2, \sigma_1, \sigma_2, \rho) + \epsilon N(x, y; \mu_1, \mu_2, \sigma_1', \sigma_2', \rho'), \qquad (5.1)$$

$$0 \leq \epsilon < 0.5,$$

the sample correlation coefficient is strongly biased with respect to the estimated parameter ρ, i.e., for any positive $\epsilon > 0$ there exists $k = \sigma_1'/\sigma_1 = \sigma_2'/\sigma_2 \gg 1$ such that $E(r) \approx \rho'$.

The presence of even one or two outliers in the data can completely destroy the sample correlation coefficient up to the change of its sign, as can be seen from Fig. 5.1. Thus, the sample correlation coefficient is extremely sensitive to the presence of outliers in the data, and hence it is necessary to use its robust counterparts.

In this chapter, various groups of robust estimates of the correlation coefficient are studied in the contaminated bivariate normal distribution models and under bivariate distributions with independent components. Conventional and new robust estimates

Robust Correlation: Theory and Applications, First Edition. Georgy L. Shevlyakov and Hannu Oja.
© 2016 John Wiley & Sons, Ltd. Published 2016 by John Wiley & Sons, Ltd.
Companion Website: www.wiley.com/go/Shevlyakov/Robust

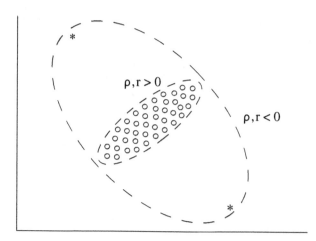

Figure 5.1   Impact of outliers on the Pearson correlation coefficient

are examined on finite samples by the Monte Carlo method and in asymptotics by the influence functions technique.

Comparing the behavior of these estimates, we find the best in each group and show that some of them possess optimal robustness properties. In particular, the asymptotically minimax variance and bias robust estimates of the correlation coefficient are designed for ϵ-contaminated bivariate normal distributions. For the proposed estimates, consistency and asymptotic normality are established, and the explicit expressions for their asymptotic variances and biases are given. The limiting cases of these minimax estimates are the classical sample correlation coefficient with $\epsilon = 0$ and the MAD correlation coefficient as $\epsilon \to 1$.

We also show that the robust estimates of the correlation coefficient based on highly efficient robust estimates of scale outperform many of their competitors.

5.2   Main Groups of Robust Estimates of the Correlation Coefficient

5.2.1   Introductory remarks

In this section we suggest a classification of different groups of robust estimates of the correlation coefficient. It is not new at all: its first version appeared in Pasman and Shevlyakov (1987) and then it was repeated in Shevlyakov and Vilchevsky (2002b, 2011). As it still has sense and may serve as a basis for further considerations, we reproduce it below. Recall that the elements of this classification are used in Chapter 2.

5.2.2   Direct robust counterparts of Pearson's correlation coefficient

A natural approach to construct robust counterparts of the sample correlation coefficient is to replace its linear procedures of averaging by the corresponding nonlinear robust counterparts (Devlin *et al.* 1975, 1981; Gnanadesikan and Kettenring 1972; Huber 1981)

$$r_\alpha(\psi) = \frac{\Sigma_\alpha \psi(x_i - \hat{x})\psi(y_i - \hat{y})}{\left(\Sigma_\alpha \psi^2(x_i - \hat{x})\Sigma_\alpha \psi^2(y_i - \hat{y})\right)^{1/2}}, \qquad (5.2)$$

where

1. $\hat{x}$ and $\hat{y}$ are some robust estimates of location, for example, the sample medians med x and med y;

2. $\psi = \psi(z)$ is a monotone function, for instance, the Huber ψ_H score function;

3. Σ_α is a robust analog of a sum.

The latter transformation is based on trimming the outer terms of the variational series with the subsequent summation of the remaining terms:

$$\Sigma_\alpha z_i = nT_\alpha(z) = n(n - 2r) \sum_{i=r+1}^{n-r} z_i,$$

$$0 \le \alpha \le 0.5, \qquad r = [\alpha\,(n - 1)],$$

where [·] stands for the integer part.

The limiting cases are as follows:

- for $\alpha = 0$, the operations of the common and robust summation coincide for $\Sigma_0 = \Sigma$;

- for $\alpha = 1/2$, the operation of Σ_0 is reduced to the operation of taking the sample median of observations $\Sigma_0 = $ med.

It is easy to see that estimate (5.2) has the following properties:

1. It is invariant under translation and scale transformations of the observations x_i and y_i: $x_i \rightarrow a_1 x_i + b_1, y_i \rightarrow a_2 y_i + b_2$.

2. The condition of boundness $|r_\alpha| \le 1$ definitely holds only for $\alpha = 0$ due to the Cauchy–Bunyakovsky inequality.

3. In the case of linearly dependent observations, $|r_\alpha| = 1$.

Observe that in the experimental study of estimate (5.2), the condition of boundness was never violated under the mixture of normal distributions.

In the literature (Devlin *et al.* 1975; Gnanadesikan and Kettenring 1972; Shevlyakov and Vilchevsky 2002b), the following versions of estimate (5.2) are used:

$$r_\alpha = \frac{\Sigma_\alpha (x_i - \text{med } x)(y_i - \text{med } y)}{(\Sigma_\alpha (x_i - \text{med } x)^2 \Sigma_\alpha (y_i - \text{med } y)^2)^{1/2}},$$

where $\alpha = 0.1, 0.2, 0.5$ and

$$r_0(\psi_H) = \frac{\Sigma \psi_H (x_i - \text{med } x)\psi_H(y_i - \text{med } y)}{(\Sigma \psi_H^2 (x_i - \text{med } x)\Sigma \psi_H^2 (y_i - \text{med } y))^{1/2}},$$

with

$$\psi_H(z) = \max(-c, \min(z, c)), \quad c = 5 \text{ MAD } z.$$

For $\alpha = 0.5, \hat{x} = \text{med } x, \hat{y} = \text{med } y, \psi(z) = z$, Equation (5.2) yields *the correlation median or the comedian estimate* (Falk 1998; Pasman and Shevlyakov 1987)

$$r_{0.5} = r_{COMED} = \frac{\text{med } (x - \text{med } x)(y - \text{med } y)}{\text{MAD } x \text{ MAD } y},$$

where MAD $z = \text{med } (|z - \text{med } z|)$ stands for the median absolute deviation.

5.2.3 Robust correlation via nonparametric measures of correlation

An estimation procedure can be endowed with robustness properties with the use of nonparametric sign and rank statistics. The best known of them are the quadrant (sign) correlation coefficient (2.26) (Blomqvist 1950), the Spearman (2.29) (Spearman 1904), and the Kendall (2.32) (Kendall 1938) rank correlation coefficients defined in Section 2.3.

5.2.4 Robust correlation via robust regression

In this section, first of all, we use the robust versions of the classical formulas (2.6) and (2.7) with various robust estimates of the regression slope and scale.

For instance, using formula (2.7), we suggest the following robust estimate for a correlation coefficient:

$$\hat{\rho} = \text{sgn}(\hat{\beta}_X)\sqrt{\hat{\beta}_X \hat{\beta}_Y}, \tag{5.3}$$

where $\hat{\beta}_X$ and $\hat{\beta}_Y$ are the least absolute values (*LAV*) estimates of regression coefficients

$$\hat{\beta}_X = \arg \min_{\alpha_X, \beta_X} \sum |x_i - \alpha_X - \beta_X y_i|,$$

$$\hat{\beta}_Y = \arg \min_{\alpha_Y, \beta_Y} \sum |y_i - \alpha_Y - \beta_Y x_i|. \tag{5.4}$$

In this case, we denote the obtained estimate of the correlation coefficient as r_{LAV}. It is easy to show that, in contrast to the LS formulas, which yield the parameters of the straight line of the conditional mean, the LAV estimates yield the parameters of the straight line of the conditional median of the normal distribution

$$\text{med}\{X \mid Y = y\} = \text{med } X + \beta_X (y - \text{med } Y),$$

$$\text{med}\{Y \mid X = x\} = \text{med } Y + \beta_Y (x - \text{med } X).$$

Another possibility is given by the least median squares regression

$$\hat{\beta}_X = \arg \min_{\alpha_X, \beta_X} \text{med}(x_i - \alpha_X - \beta_X y_i)^2,$$

$$\hat{\beta}_Y = \arg \min_{\alpha_Y, \beta_Y} \text{med}(y_i - \alpha_Y - \beta_Y x_i)^2. \tag{5.5}$$

The corresponding estimate is referred to as r_{LMS}.

Using Equation (2.3), we arrive at the following robust estimates:

$$r_{m1} = \hat{\beta}_{m1} \frac{\hat{\sigma}_X}{\hat{\sigma}_Y} \tag{5.6}$$

and

$$r_{m2} = \hat{\beta}_{m2} \frac{\hat{\sigma}_X}{\hat{\sigma}_Y}, \tag{5.7}$$

where

$$\hat{\beta}_{m1} = \text{med} \left\{ \frac{y_i - \text{med } y}{x_i - \text{med } x} \right\}$$

and

$$\hat{\beta}_{m2} = \text{med}_{i \neq j} \left\{ \frac{y_i - y_j}{x_i - x_j} \right\};$$

$\hat{\sigma}_X$ and $\hat{\sigma}_Y$ are some robust estimates of scale, for example, the median absolute deviations MAD x and MAD y.

The structure of these estimates can be explained as follows: the distribution density of the ratio of the centered normal random variables is given by the Cauchy formula

$$f(z) = \frac{\sqrt{1 - \rho^2}}{\pi} \left[\frac{\sigma_X}{\sigma_Y} \left(z - \frac{\sigma_Y}{\sigma_X} \rho \right)^2 + \frac{\sigma_Y}{\sigma_X} (1 - \rho^2) \right]^{-1};$$

hence Equations (5.6) and (5.7) yield consistent estimates of the distribution center $\rho \sigma_Y / \sigma_X$.

5.2.5 Robust correlation via robust principal component variances

In this section we develop the interpretation of the correlation coefficient given by the identities (2.14) and (2.19).

We can write the obvious relation for any bivariate random variables (X, Y)

$$\frac{var(X + Y) - var(X - Y)}{var(X + Y) + var(X - Y)} = \frac{2cov(X, Y)}{var(X) + var(Y)};$$

hence we obtain the correlation coefficient ρ provided $var(X) = var(Y)$.

The application of the standardized variables $\widetilde{X}$ and $\widetilde{Y}$ such that $var(\widetilde{X}) = 1$ and $var(\widetilde{Y}) = 1$ lead to the aforementioned Equation (2.14)

$$\rho = \frac{var(\widetilde{X} + \widetilde{Y}) - var(\widetilde{X} - \widetilde{Y})}{var(\widetilde{X} + \widetilde{Y}) + var(\widetilde{X} - \widetilde{Y})}. \tag{5.8}$$

By introducing the robust scale functional

$$S(X) = S(F_X) : S(aX + b) = |a|S(X),$$

we can write a robust analog of variance in the form $S^2(\cdot)$ and the corresponding robust analog of (5.8) in the form

$$\rho^*(X, Y) = \frac{S^2(\widetilde{X} + \widetilde{Y}) - S^2(\widetilde{X} - \widetilde{Y})}{S^2(\widetilde{X} + \widetilde{Y}) + S^2(\widetilde{X} - \widetilde{Y})}, \tag{5.9}$$

where $\widetilde{X}$ and $\widetilde{Y}$ are normalized in the same scale, $\widetilde{X} = X/S(X)$ and $\widetilde{Y} = Y/S(Y)$ (Gnanadesikan and Kettenring 1972; Huber 1981).

The robust "correlation coefficient" $\rho^*(X, Y)$ defined by (5.9) satisfies the principal requirements imposed on the correlation coefficient:

- $|\rho^*(X, Y)| \leq 1$;

- if random variables X and Y are linearly dependent, then $|\rho^*(X, Y)| = 1$;

- in the case of independent random variables X and Y, we generally have $\rho^*(X, Y) = 0$.

However, for the mean and median absolute deviation functionals, the latter property holds for the distributions F_X and F_Y that are symmetric about the center.

Replacing the functionals by their robust estimates in (5.9), we arrive at robust estimates of the correlation coefficient in the form

$$\widehat{\rho}^*(X, Y) = \frac{\widehat{S}^2 \, (\widetilde{X} + \widetilde{Y}) - \widehat{S}^2 \, (\widetilde{X} - \widetilde{Y})}{\widehat{S}^2 \, (\widetilde{X} + \widetilde{Y}) + \widehat{S}^2 \, (\widetilde{X} - \widetilde{Y})}. \tag{5.10}$$

For the median absolute deviation functional $S(X) = \text{MAD } X$, Equation (5.10) takes the form of the *MAD correlation coefficient* (Pasman and Shevlyakov 1987)

$$r_{MAD} = \frac{\text{MAD}^2(\widetilde{X} + \widetilde{Y}) - \text{MAD}^2(\widetilde{X} - \widetilde{Y})}{\text{MAD}^2(\widetilde{X} + \widetilde{Y}) + \text{MAD}^2(\widetilde{X} - \widetilde{Y})}, \qquad (5.11)$$

where

$$\widetilde{x} = \frac{x - \text{med } x}{\text{MAD } x}, \qquad \widetilde{y} = \frac{y - \text{med } y}{\text{MAD } y}.$$

The asymptotically equivalent version of the MAD correlation coefficient is given by the *median correlation coefficient* (Pasman and Shevlyakov 1987; Shevlyakov 1997)

$$r_{med} = \frac{\text{med}^2|u| - \text{med}^2|v|}{\text{med}^2|u| + \text{med}^2|v|}, \qquad (5.12)$$

where u and v are called the *robust principal variables*

$$u = \frac{x - \text{med } x}{\text{MAD } x} + \frac{y - \text{med } y}{\text{MAD } y}, \qquad v = \frac{x - \text{med } x}{\text{MAD } x} - \frac{y - \text{med } y}{\text{MAD } y}. \qquad (5.13)$$

Then the *MAD* correlation coefficient can be rewritten as follows:

$$r_{MAD} = \frac{\text{MAD}^2 u - \text{MAD}^2 v}{\text{MAD}^2 u + \text{MAD}^2 v}. \qquad (5.14)$$

It is natural to use highly efficient robust estimates of scale Q_n and FQ_n of Section 4.4 in Equation (5.10)

$$r_{Qn} = \frac{Q_n^2(u) - Q_n^2(v)}{Q_n^2(u) + Q_n^2(v)} \qquad (5.15)$$

and

$$r_{FQ} = \frac{FQ_n^2(u) - FQ_n^2(v)}{FQ_n^2(u) + FQ_n^2(v)}. \qquad (5.16)$$

Furthermore, we have

- for the mean absolute deviation functional,

$$r_{L1} = \frac{\left(\sum\limits_{i=1}^{n} |u_i|\right)^2 - \left(\sum\limits_{i=1}^{n} |v_i|\right)^2}{\left(\sum\limits_{i=1}^{n} |u_i|\right)^2 + \left(\sum\limits_{i=1}^{n} |v_i|\right)^2},$$

- for the standard deviation functional,

$$r_{L2} = \frac{\sum\limits_{i=1}^{n} u_i^2 - \sum\limits_{i=1}^{n} v_i^2}{\sum\limits_{i=1}^{n} u_i^2 + \sum\limits_{i=1}^{n} v_i^2}, \tag{5.17}$$

- for the trimmed standard deviation functional, we get the *trimmed correlation coefficient*

$$r_{TRIM}(n_1, n_2) = \frac{\sum\limits_{i=n_1+1}^{n-n_2} u_i^2 - \sum\limits_{i=n_1+1}^{n-n_2} v_i^2}{\sum\limits_{i=n_1+1}^{n-n_2} u_i^2 + \sum\limits_{i=n_1+1}^{n-n_2} v_i^2}. \tag{5.18}$$

The particular cases of the latter formula appear in Devlin *et al.* (1975) and Gnanadesikan and Kettenring (1972) with $n_1 = n_2 = [\alpha n]$ and $\alpha = 0.05, 0.1$.

Observe that the general construction (5.18) yields r_{L2} with $n_1 = 0$ and $n_2 = 0$, and, in the case of odd sample sizes, the median correlation coefficient with $n_1 = n_2 = [0.5(n-1)]$. In addition, Equation (5.17) yields Pearson's correlation coefficient r if we use classical estimates in its inner structure: the sample means for location and the mean squared deviations for scale in (5.13).

5.2.6   Robust correlation via two-stage procedures

The preliminary detection of outliers in the data and their rejection with the consequent application of a classical estimate (for example, Pearson's correlation coefficient) to the rest of the observations defines the other group of robust estimates. Their variety mainly depends on the variety of the rules for rejection of outliers. In detail, this approach is considered in Shevlyakov and Vilchevski (2002a, 2011).

5.2.7   Concluding remarks

The aforementioned robust counterparts of the sample correlation coefficient were thoroughly studied in Shevlyakov and Vilchevski (2002a), and the main conclusion, which followed from the extensive Monte Carlo experiment in Gaussian and contaminated Gaussian distribution models, was that the group of robust estimates based on robust principal component variances turned out to be the best among others. Partially in Shevlyakov and Vilchevski (2002a) and most in the works which followed that research (Shevlyakov and Vilchevsky 2002b; Shevlyakov *et al.* 2012), this experimental observation was explained theoretically: in the family of the so-called bivariate independent component distributions, those estimates of the correlation coefficient are minimax bias and variance (in the Huber sense). In what follows, we focus on those estimates and closely related to them robust estimates, as well as on several classical nonparametric estimates of correlation.

5.3   Asymptotic Properties of the Classical Estimates of the Correlation Coefficient

In this section, we present basic results on the asymptotic performance of Pearson's and the maximum likelihood estimates of the correlation coefficient of the standard bivariate normal distribution.

5.3.1   Pearson's sample correlation coefficient

Setting distribution means known and equal to zero, we get for the sample correlation coefficient the scalar product of the standardized samples

$$r = \frac{\sum_{i=1}^{n} x_i y_i}{\sqrt{\sum_{i=1}^{n} x_i^2 \sum_{i=1}^{n} y_i^2}} .$$

At the standard bivariate normal distribution Φ_ρ, the distribution of r is asymptotically normal $N(\rho, V(r; \Phi_\rho))$, where the asymptotic variance is equal to

$$V(r; \Phi_\rho) = (1 - \rho^2)^2.$$

Thus, on large but finite samples, with moderate and low correlation, the sample correlation coefficient r is distributed approximately normal about the exact value ρ with variance

$$var(r) = \frac{(1 - \rho^2)^2}{n} . \tag{5.19}$$

The sample correlation coefficient is a biased estimate such that

$$E(r) - \rho = -\frac{\rho(1 - \rho^2)}{2n} + o\left(\frac{1}{n}\right). \tag{5.20}$$

This bias depends on ρ and is equal to zero only with $\rho \in \{0, \pm 1\}$.

The sample distribution of r converges to ρ rather slowly: it is unreasonable to use the normal approximation with $n < 500$ (Kendall and Stuart 1963). However, there exists a suitable Fisher transformation

$$z = \frac{1}{2} \log \left(\frac{1 + r}{1 - r}\right). \tag{5.21}$$

The variance of z is approximately equal to

$$var(z) \approx \frac{1}{\sqrt{n - 3}}$$

and it practically does not depend on ρ.

The influence function of r is given in Croux and Dehon (2010)

$$IF(x,y;r,\Phi_\rho) = xy - \rho\,\frac{x^2+y^2}{2},$$

and as it is unbounded, the sample correlation coefficient r is not robust. In the Tukey gross-error model (5.28), the influence function of this estimate is

$$IF(x,y;r) = -\frac{E(r)}{2(1-\epsilon+\epsilon k^2)}(x^2+y^2) + \frac{xy}{1-\epsilon+\epsilon k^2},$$

where

$$E(r) = \frac{(1-\epsilon)\rho + \epsilon k^2\rho'}{1-\epsilon+\epsilon k^2}.$$

5.3.2 The maximum likelihood estimate of the correlation coefficient at the normal

Consider the maximum likelihood estimate of the correlation coefficient ρ for the standard bivariate normal distribution with density

$$\varphi_\rho(x,y) = N(x,y;0,0,1,1,\rho).$$

Write down the likelihood equation

$$\sum_{i=1}^{n}\frac{\partial}{\partial\rho}\log\,\varphi_\rho(x_i,y_i) = 0.$$

Since

$$\log\,\varphi_\rho(x,y) = -\log\,(2\pi) - \frac{1}{2}\log\,(1-\rho^2) - \frac{x^2-2\rho xy+y^2}{2(1-\rho^2)},$$

$$\frac{\partial}{\partial\rho}\log\,\varphi_\rho(x,y) = \frac{\rho}{1-\rho^2} - \frac{\rho(x^2-2\rho xy+y^2)}{(1-\rho^2)^2} + \frac{xy}{1-\rho^2},$$

the likelihood equation takes the form

$$\frac{n\rho}{1-\rho^2} - \frac{\rho}{(1-\rho^2)^2}\left(\sum_{i=1}^{n}x_i^2 - 2\rho\sum_{i=1}^{n}x_iy_i + \sum_{i=1}^{n}y_i^2\right) \tag{5.22}$$

$$+\frac{1}{1-\rho^2}\sum_{i=1}^{n}x_iy_i = 0,$$

which leads to the cubic equation

$$\rho(1 - \rho^2) + (1 + \rho^2)\frac{1}{n}\sum_{i=1}^{n} x_i y_i - \rho\left(\frac{1}{n}\sum_{i=1}^{n} x_i + \frac{1}{n}\sum_{i=1}^{n} y_i\right) = 0. \qquad (5.23)$$

For sufficiently large n, the likelihood equation (5.23) has one real root (Kendall and Stuart 1963), which is the sought maximum likelihood estimate $\hat{\rho}_{ML}$.

The asymptotic variance of this estimate is computed via Fisher information for the correlation coefficient. Taking the second derivative of the likelihood equation, we get

$$\frac{\partial^2 \log L}{\partial \rho^2} = \frac{n(1 + \rho^2)}{(1 - \rho^2)^2}$$

$$-\frac{(1 + 3\rho^2)}{(1 - \rho^2)^3}\left(\sum_{i=1}^{n} x_i^2 - 2\rho\sum_{i=1}^{n} x_i y_i + \sum_{i=1}^{n} y_i^2\right) + \frac{4\rho}{(1 - \rho^2)^2}\sum_{i=1}^{n} x_i y_i$$

and substitute it into the formula for Fisher information

$$I(\Phi_\rho) = -E\left(\frac{\partial^2 \log L}{\partial \rho^2}\right) = -\frac{n(1 + \rho^2)}{(1 - \rho^2)^2} + \frac{2n(1 + 3\rho^2)}{(1 - \rho^2)^2} - \frac{4n\rho^2}{(1 - \rho^2)^2} \qquad (5.24)$$

$$= \frac{n(1 + \rho^2)}{(1 - \rho^2)^2}$$

that yields

$$var(\hat{\rho}_{ML}) = \frac{1}{nI(\Phi_\rho)} = \frac{(1 - \rho^2)^2}{n(1 + \rho^2)}. \qquad (5.25)$$

Comparing the obtained variance with the variance of r (5.19), it can be seen that they coincide only at $\rho = 0$; with ρ increasing, the sample correlation coefficient is losing accuracy. The efficiency of r is given by

$$eff(r) = \frac{var(\hat{\rho}_{ML})}{var(r)} = \frac{1}{1 + \rho^2},$$

which is 100% at $\rho = 0$ and 50% as $\rho \to 1$.

Instead of solving the cubic equation (5.23), one can use the iterative Newton–Raphson scheme

$$\hat{\rho}_{k+1} = \hat{\rho}_k - \sum_{i=1}^{n}\frac{\partial \log f(x_i, y_i)}{\partial \rho} \bigg/ \sum_{i=1}^{n}\frac{\partial^2 \log f(x_i, y_i)}{\partial \rho^2} \qquad (5.26)$$

To lessen the computational complexity of each step of iteration, Fisher proposed to use the expectation of the denominator in (5.26), namely $-I(\rho)$ (Kendall and Stuart 1963):

$$\widehat{\rho}_{k+1} = \widehat{\rho}_k + \sum_{i=1}^{n} \frac{\partial \log f(x_i, y_i)}{\partial \rho} \bigg/ I(\widehat{\rho}_k)$$

Substituting the expression for Fisher information (5.24) and for the derivative of the likelihood (5.22), and setting the notations for "scalar products"

$$\mathbf{xx} = \frac{1}{n} \sum_{i=1}^{n} x_i^2, \quad \mathbf{yy} = \frac{1}{n} \sum_{i=1}^{n} y_i^2, \quad \mathbf{xy} = \frac{1}{n} \sum_{i=1}^{n} x_i y_i,$$

we transform the right-hand part as follows:

$$\rho + \frac{1}{I(\rho)} \sum_{i=1}^{n} \frac{\partial \log f(x_i, y_i)}{\partial \rho} = \mathbf{xy} - \frac{\rho}{1 + \rho^2} (\mathbf{xx} + \mathbf{yy} - 2).$$

Finally, we get the iterative algorithm of the form

$$\widehat{\rho}_{k+1} = \frac{1}{n} \sum_{i=1}^{n} x_i y_i - \frac{\widehat{\rho}_k}{1 + \widehat{\rho}_k^2} \left(\frac{1}{n} \sum_{i=1}^{n} x_i^2 + \frac{1}{n} \sum_{i=1}^{n} y_i^2 - 2 \right). \tag{5.27}$$

5.4 Asymptotic Properties of Nonparametric Estimates of Correlation

5.4.1 Introductory remarks

In this section, we present some results on the asymptotic performance of the quadrant, Kendall and Spearman correlation coefficients.

The means and asymptotic variances of some aforementioned robust estimates, for instance, regression estimates, computed in the Tukey gross-error model

$$f(x, y) = (1 - \epsilon)N(x, y; 0, 0, 1, 1, \rho) + \epsilon N(x, y; 0, 0, k, k, \rho'), \tag{5.28}$$

$$0 \le \epsilon < 0.5, \quad k > 1,$$

can be found in Shevlyakov and Vilchevsky (2002a, 2011).

Those and the other results are obtained mostly using the techniques based on the influence functions $IF(x, y; \widehat{\rho})$ (Hampel et al. 1986)

$$E(\widehat{\rho}) \approx \rho + \int IF(x, y; \widehat{\rho}) f(x, y) \, dx \, dy, \tag{5.29}$$

$$var(\hat{\rho}) = \frac{1}{n} \int IF^2(x, y; \hat{\rho}) f(x, y) \, dx \, dy, \qquad (5.30)$$

where the density $f(x, y)$ is given by Equation (5.28).

5.4.2   The quadrant correlation coefficient

The quadrant correlation coefficient q given by (2.26),

$$q = \frac{1}{n} \sum_{i=1}^{n} \operatorname{sgn}(x_i - \operatorname{med} x) \operatorname{sgn}(y_i - \operatorname{med} y),$$

can be defined as the correlation coefficient between the signs of the centered variables.

For the standard bivariate normal distribution $\Phi(x, y; \rho)$ with the correlation coefficient ρ, the statistic q estimates the value $4P(X > 0, Y > 0) - 1$:

$$4 \int_0^\infty \int_0^\infty \frac{1}{2\pi\sqrt{1 - \rho^2}} \exp\left\{ -\frac{x^2 - 2\rho xy + y^2}{2(1 - \rho^2)} \right\} \, dx \, dy - 1 = \frac{2}{\pi} \arcsin \rho,$$

that is, for consistency of this estimate, it is necessary to transform it as follows:

$$r_Q = \sin\left(\frac{\pi q}{2}\right). \qquad (5.31)$$

The influence function and asymptotic variance of this estimate are given by Croux and Dehon (2010) and Shevlyakov and Vilchevski (2002a)

$$IF(x, y; r_Q, \Phi_\rho) = \frac{\pi}{2} \operatorname{sign}(\rho) \sqrt{1 - \rho^2} \left(\operatorname{sgn}(xy) - \frac{2}{\pi} \arcsin \rho \right),$$

$$V(r_Q, \Phi_\rho) = \frac{\pi^2}{4}(1 - \rho^2) \left(1 - \frac{4}{\pi^2} \arcsin^2 \rho \right).$$

The asymptotic relative efficiency of r_Q to r is

$$\operatorname{eff} r_Q = \frac{V(r, \Phi_\rho)}{V(r_Q, \Phi_\rho)} = \frac{1 - \rho^2}{\pi^2/4 - \arcsin^2 \rho},$$

that is, 40.5% at $\rho = 0$ and 15.7% at $\rho = 0.9$.

5.4.3   The Kendall rank correlation coefficient

The Kendall τ-rank correlation coefficient (2.31) (Kendall 1938)

$$\tau = \frac{2}{n(n - 1)} \sum_{i<j} \operatorname{sgn}(x_i - x_j) \operatorname{sgn}(y_i - y_j)$$

estimates the same value $2/\pi \arcsin \rho$ as the quadrant correlation coefficient q, and thus we define the consistent version of τ by

$$r_K = \sin\left(\frac{\pi \tau}{2}\right).$$

The influence function and asymptotic variance of r_K are given by (Croux and Dehon 2010)

$$IF(x, y; r_K, \Phi_\rho) = \pi \operatorname{sgn}(\rho) \sqrt{1 - \rho^2}$$

$$\times \left(4\Phi_\rho(x, y) - 2\Phi(x) - 2\Phi(y) + 1 - \frac{2}{\pi} \arcsin \rho\right),$$

$$V(r_K, \Phi_\rho) = \pi^2(1 - \rho^2)\left(\frac{1}{9} - \frac{4}{\pi^2}\arcsin^2\left(\frac{\rho}{2}\right)\right).$$

The asymptotic relative efficiency of r_K to r is equal to

$$\operatorname{eff} r_K = \frac{V(r, \Phi_\rho)}{V(r_K, \Phi_\rho)} = \frac{1 - \rho^2}{\pi^2/9 - 4\arcsin^2(\rho/2)},$$

that is, 91.2% at $\rho = 0$ and 84.4% at $\rho = 0.9$.

5.4.4   The Spearman rank correlation coefficient

The Spearman rank correlation coefficient given by

$$\widehat{\rho}_S = 1 - \frac{1}{6n(n^2 - 1)} \sum_{i=1}^{n} (R(x_i) - R(y_i))^2$$

estimates the quantity

$$\rho_S = \frac{6}{\pi} \arcsin \frac{\rho}{2}$$

at the standard bivariate normal distribution. Thus, for consistency, we should use the following transform:

$$r_S = 2 \sin\left(\frac{\pi \widehat{\rho}_S}{6}\right).$$

The influence function and asymptotic variance of r_S are given by Croux and Dehon (2010) as

$$IF(x, y; r_S, \Phi_\rho) = \frac{\pi}{3}\operatorname{sign}(\rho) \sqrt{1 - \frac{\rho^2}{4}} \, IF(x, y; \widehat{\rho}_S, \Phi_\rho),$$

where

$$IF(x, y; \widehat{\rho}_S, \Phi_\rho) = 12 \, \Phi(x)\Phi(y) + 12 \, E\left[\Phi(X)\Phi\left(\frac{\rho X - y}{\sqrt{1 - \rho^2}}\right)\right]$$

$$+12E\left[\Phi(Y)\Phi\left(\frac{\rho Y - x}{\sqrt{1-\rho^2}}\right)\right] - 9\left(1 + \frac{2}{\pi}\arcsin\frac{\rho}{2}\right),$$

$$V(\hat{\rho}_S, \Phi_\rho) = \frac{\pi^2}{9}\left(1 - \frac{\rho^2}{4}\right)\left(1 - \frac{9}{4\cdot 144\pi^2}\arcsin^2\frac{\rho}{2} + h(\rho)\right),$$

where

$$h(\rho) = \frac{1}{\pi^2}\int_0^{\arcsin(\rho/2)}\arcsin\left(\frac{\sin x}{1 + 2\cos(2x)}\right)dx$$

$$+\frac{2}{\pi^2}\int_0^{\arcsin(\rho/2)}\arcsin\left(\frac{\sin(2x)}{\sqrt{1 + 2\cos(2x)}}\right)dx$$

$$+\frac{1}{\pi^2}\int_0^{\arcsin(\rho/2)}\arcsin\left(\frac{\sin(2x)}{2\sqrt{\cos(2x)}}\right)dx$$

$$+\frac{1}{2\pi^2}\int_0^{\arcsin(\rho/2)}\arcsin\left(\frac{3\sin x - \sin(3x)}{4\cos(2x)}\right)dx.$$

For the normal distribution, the asymptotic relative efficiency of r_S to r is 91.2% at $\rho = 0$.

In Fig. 5.2, the asymptotic relative efficiencies of nonparametric correlation coefficients are exhibited.

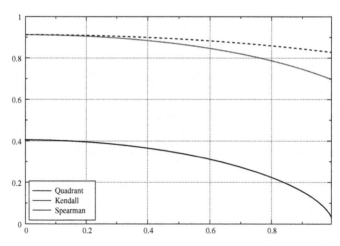

Figure 5.2 Asymptotic relative efficiencies of nonparametric correlation measures (axis y) versus the correlation coefficient ρ (axis x) of the normal distribution

5.5   Bivariate Independent Component Distributions

The family of bivariate independent component distributions (*ICD*) was originally introduced in the Dr. Sc. Thesis of G. Shevlyakov (1991), and since then it has proved its effectiveness in treating the problems of robust correlation (Shevlyakov and Vilchevski 2002a; Shevlyakov *et al.* 2012).

5.5.1   Definition and properties

Consider the family of bivariate distribution densities allowing for factorization in principal components (Shevlyakov and Vilchevski 2002b)

$$f(x,y) = \frac{1}{a}g\left(\frac{u}{a}\right) \cdot \frac{1}{b}g\left(\frac{v}{b}\right) , \tag{5.32}$$

where $g(z)$ is a symmetric distribution density $g(-z) = g(z)$; u and v are the principal components, defined by the orthogonal rotation of the system of coordinates

$$u = \frac{x+y}{\sqrt{2}}, \qquad v = \frac{x-y}{\sqrt{2}} ;$$

a and b are the parameters of scale for density $g(z)$.

Compute the main probabilistic characteristics of this family. The means are

$$E(X) = \frac{1}{\sqrt{2}}(a+b)E_g, \qquad E(Y) = \frac{1}{\sqrt{2}}(a-b)E_g ,$$

$$E(U) = aE_g , \qquad E(V) = bE_g ,$$

where

$$E_g = \int_{-\infty}^{\infty} zg(z)\mathrm{d}\,z = 0$$

for a symmetric function $g(z)$; therefore we get

$$E(X) = E(Y) = 0, \qquad E(U) = E(V) = 0.$$

Analogously, the variances and covariance are

$$var(X) = \frac{1}{2}(a^2 + b^2)D_g, var(Y) = \frac{1}{2}(a^2 + b^2)D_g ,$$

$$cov(X,Y) = \frac{1}{2}(a^2 - b^2)D_g,$$

where

$$D_g = \int_{-\infty}^{\infty} z^2 g(z)\mathrm{d}\,z .$$

For the principal components, we get

$$var(U) = a^2 D_g, \qquad var(V) = b^2 D_g, \qquad cov(U,V) = 0.$$

Then the correlation coefficient $\rho = \rho_{XY}$ takes the form

$$\rho = \frac{cov(X, Y)}{\sqrt{var(X)var(Y)}} = \frac{a^2 - b^2}{a^2 + b^2} = \frac{var(U) - var(V)}{var(U) + var(V)} . \tag{5.33}$$

Thus we arrive at Equation (2.17).

Now assume that the variances $var(X)$ and $var(Y)$ do not depend on the correlation coefficient ρ and are equal to a constant σ^2. Then setting $D_g = 1$, we can express the parameters of scale as follows:

$$a = \sigma\sqrt{1+\rho}, \quad b = \sigma\sqrt{1-\rho},$$

and the distribution density (5.32) can be rewritten as

$$f(x, y; \sigma, \rho) = \frac{1}{\sigma^2\sqrt{1-\rho^2}}g\left(\frac{x+y}{\sigma\sqrt{2(1+\rho)}}\right)g\left(\frac{x-y}{\sigma\sqrt{2(1-\rho)}}\right) . \tag{5.34}$$

Note that in the particular case of the standard normal distribution density $g(z)$

$$g(z) = \varphi(z) = N(z; 0, 1) = \frac{1}{\sqrt{2\pi}}\exp\left(-\frac{z^2}{2}\right) ,$$

Equation (5.34) yields the bivariate normal distribution density

$$f(x, y) = N(x, y; 0, 0, \sigma, \sigma, \rho) = \frac{1}{2\pi\sigma^2\sqrt{1-\rho^2}}\exp\left\{-\frac{x^2 - 2\rho xy + y^2}{2\sigma^2(1-\rho^2)}\right\} .$$

Equation (5.34) also describes bivariate analogs of numerous univariate distribution densities $g(z)$, for instance, the Laplace, Student, and Cauchy. In the case when the second moment of a distribution density $g(z)$ does not exist, the coefficient ρ, certainly, cannot already be interpreted as the classical Pearson correlation coefficient, but it should be considered as a parameter of correlation.

Next we show that the introduced family (5.34) of independent component bivariate distributions represents the widely used in robustness studies Tukey gross-error model.

5.5.2   Independent component and Tukey gross-error distribution models

In what follows, without any loss of generality, we set $\sigma = 1$ in Equation (5.34)

$$f(x, y; \rho) = f_{ICD}(x, y; \rho) = f_{ICD}(u, v; \rho)$$

$$= \frac{1}{\sqrt{1+\rho}}g\left(\frac{u}{\sqrt{1+\rho}}\right)\frac{1}{\sqrt{1-\rho}}g\left(\frac{v}{\sqrt{1-\rho}}\right) . \tag{5.35}$$

The Tukey gross-error model or the contaminated standard bivariate normal distribution density is given by the following equation:

$$f_T(x, y; \rho) = (1 - \epsilon)N(x, y; 0, 0, 1, 1, \rho) + h(x, y), \qquad 0 \le \epsilon < 1, \qquad (5.36)$$

where $h(x, y)$ is an arbitrary density. Then Equation (5.36) can be rewritten in the inequality form

$$f_T(x, y; \rho) \ge (1 - \epsilon)N(x, y; 0, 0, 1, 1, \rho), \qquad 0 \le \epsilon < 1. \qquad (5.37)$$

Now we show that the appropriate choice of the density $g(z)$ implies the inclusion of independent component distribution density into the Tukey gross-error model. Substitute Equation (5.35) into the left-hand part of inequality (5.37):

$$\frac{1}{\sqrt{1+\rho}} g\left(\frac{u}{\sqrt{1+\rho}}\right) \frac{1}{\sqrt{1-\rho}} g\left(\frac{v}{\sqrt{1-\rho}}\right)$$

$$\ge \frac{1-\epsilon}{2\pi\sqrt{1-\rho^2}} \exp\left\{-\frac{u^2}{2(1+\rho)}\right\} \exp\left\{-\frac{v^2}{2(1-\rho)}\right\}$$

or

$$g\left(\frac{u}{\sqrt{1+\rho}}\right) g\left(\frac{v}{\sqrt{1-\rho}}\right) \ge \frac{1-\epsilon}{2\pi} \exp\left\{-\frac{u^2}{2(1+\rho)}\right\} \exp\left\{-\frac{v^2}{2(1-\rho)}\right\}.$$

Setting $g(z) \ge \sqrt{1-\epsilon}\, \varphi(z)$, where $\varphi(z)$ is the standard normal distribution density, we arrive at the required result.

Moreover, from equating $\sqrt{1-\epsilon}$ to $1 - \gamma$ it follows that $\gamma = 1 - \sqrt{1-\epsilon}$, and this means that the density $g(z)$ must belong to the univariate Tukey gross-error model. Finally, the obtained result can be formulated as follows.

Theorem 5.5.1. *A symmetric density $g(z)$ belonging to the γ-contaminated univariate normal densities*

$$g(z) \ge (1 - \gamma)\varphi(z), \qquad 0 \le \gamma < 1,$$

where $\gamma = 1 - \sqrt{1-\epsilon}$ and $0 \le \epsilon < 1$, implies belonging of the independent component distribution density

$$f_{ICD}(u, v; \rho) = \frac{1}{\sqrt{1+\rho}} g\left(\frac{u}{\sqrt{1+\rho}}\right) \frac{1}{\sqrt{1-\rho}} g\left(\frac{v}{\sqrt{1-\rho}}\right)$$

to the ϵ-contaminated bivariate normal densities (Tukey gross-error distribution model)

$$f_{ICD}(x, y; \rho) \ge (1 - \epsilon)N(x, y; 0, 0, 1, 1, \rho) \qquad (5.38)$$

or

$$f_{ICD}(u, v; \rho) \ge (1 - \epsilon)N(u, v; 0, 0, \sqrt{1+\rho}, \sqrt{1-\rho}, 0).$$

5.6   Robust Estimates of the Correlation Coefficient Based on Principal Component Variances

Consider robust estimates of the correlation coefficient $\rho = \rho_{XY}$ based on the previously introduced robust estimates of the principal component variances

$$\widehat{\rho}_n = \frac{\widehat{S}^2\,(U) - \widehat{S}^2\,(V)}{\widehat{S}^2\,(U) + \widehat{S}^2\,(V)}, \tag{5.39}$$

where U and V are the principal component random variables

$$U = \frac{\widetilde{X} + \widetilde{Y}}{\sqrt{2}}, \qquad V = \frac{\widetilde{X} - \widetilde{Y}}{\sqrt{2}},$$

$$\widetilde{X} = \frac{X - \mu_1}{\sigma_1}, \qquad \widetilde{Y} = \frac{Y - \mu_2}{\sigma_2}.$$

In what follows, firstly, we deal with the independent component distribution model (5.34) and, secondly, with robust estimates $\widehat{S}$ defined by M-estimates of scale.

Now we specify the estimation procedure. Given $(x_1, y_1), \dots, (x_n, y_n)$, we propose the following steps for estimating the correlation coefficient:

- transform the initial data

$$u_i = \frac{x_i + y_i}{\sqrt{2}}, \qquad v_i = \frac{x_i - y_i}{\sqrt{2}}, \qquad i = 1, \dots, n;$$

- evaluate the M-estimates of scale $\widehat{S}_u$ and $\widehat{S}_v$ as the solutions to the equations

$$\sum \chi\left(\frac{u_i}{\widehat{S}_u}\right) = 0, \qquad \sum \chi\left(\frac{v_i}{\widehat{S}_v}\right) = 0, \tag{5.40}$$

  where $\chi(\cdot)$ is some score function;

- evaluate the estimate of ρ in the following form:

$$\widehat{\rho}_n = \frac{\widehat{S}_u^2 - \widehat{S}_v^2}{\widehat{S}_u^2 + \widehat{S}_v^2}. \tag{5.41}$$

The asymptotic properties of the proposed estimate (5.41) are completely determined by the asymptotic properties of M-estimates of scale (5.40). The sufficient conditions of regularity providing the desired properties are put on the densities g and score functions χ (Hampel *et al.* (1986), pp. 125, 139):

(g1) g is twice continuously differentiable and satisfies $g(x) > 0$ for all x in $\mathbb{R}$.

(g2) Fisher information for scale $I(g)$ satisfies $0 < I(g) < \infty$.

(χ1) χ is well-defined and continuous on $\mathbb{R}\backslash C(\chi)$, where $C(\chi)$ is finite. In each point of $C(\chi)$ there exist finite left and right limits of χ that are different. Also $\chi(-x) = \chi(x)$ if $(-x, x) \subset \mathbb{R}\backslash C(\chi)$, and there exists $d > 0$ such that $\chi(x) \le 0$ on $(0, d)$ and $\chi(x) \ge 0$ on (d, ∞).

(χ2) The set $D(\chi)$ of points in which χ is continuous but in which χ' is not defined or not continuous is finite.

(χ3) $\int \chi \, dG = 0$ and $\int \chi^2 \, dG < \infty$.

(χ4) $0 < \int x\chi'(x) \, dG(x) < \infty$.

Now we briefly comment on regularity conditions imposed on distribution densities g and score functions χ. In the literature, these conditions take different forms depending on the pursued goals: in general, one may strengthen the conditions on densities and weaken those on scores, and vice versa (various suggestions can be found in Hampel *et al.* (1986) and Huber (1967, 1974).

Here we use a balanced set of conditions given in Hampel *et al.* (1986, pp. 125, 139). The requirement of symmetry is restrictive but necessary for Huber's minimax theory (Huber 1981). Conditions (g2), (χ1), and (χ2) define smooth densities and allow for a finite number of points of discontinuity for scores and their derivatives. Conditions (g1), (χ3), and (χ4) requiring the existence of Fisher information $I(g)$ and other integrals are used for the proofs of consistency (the first equation of (χ3) provides consistency of M-estimates of scale) and asymptotic normality of M-estimates in robust statistics (Hampel *et al.* 1986; Huber 1981).

Now we are in a position to formulate the following result.

Theorem 5.6.1. *In the class of independent component distributions (5.34) under the conditions of regularity (g1)–(χ4), estimate (5.41) is consistent and asymptotically normal*

$$\sqrt{n} \, (\hat{\rho}_n - \rho) \to N(0, AV(\hat{\rho}_n)) \qquad (5.42)$$

with the asymptotic variance

$$AV(\hat{\rho}_n) = 2(1 - \rho^2)^2 \, V(\chi, g), \qquad (5.43)$$

where

$$V(\chi, g) = \frac{\int \chi^2(x)g(x) \, dx}{\left(\int x\chi'(x)g(x) \, dx\right)^2}$$

is the asymptotic variance of the M-estimate of scale defined by (5.40).

Proof. Consistency of estimate (5.39) immediately follows from consistency of M-estimates of scale: as $\hat{\S}_u$ and $\hat{S}_v$ tend to $S_u = \sigma\sqrt{1 + \rho}$ and $S_v = \sigma\sqrt{1 - \rho}$ in probability; hence $\hat{\rho}_n$ tends to ρ in probability.

We get asymptotic normality by the reasoning of Huber (1964, p. 78, Lemma 5): the numerator of the fraction in (5.39) is asymptotically normal, the denominator tends in probability to the positive constant $c = S_u^2 + S_v^2$ and hence $n^{1/2}(\rho_n - \rho)$ is asymptotically normal (Cramér 1946, 20.6).

The precise structure of (5.43) is obtained by a direct routine calculation using the following formula for the variance of a fraction of the random variables ξ and η (Kendall and Stuart 1962):

$$var\left(\frac{\xi}{\eta}\right) = \left(\frac{E(\xi)}{E(\eta)}\right)^2 \left(\frac{var(\xi)}{E^2(\xi)} + \frac{var(\eta)}{E^2(\eta)} - \frac{2cov(\xi,\eta)}{E(\xi)\,E(\eta)}\right) + o\left(\frac{1}{n}\right), \qquad (5.44)$$

where $\xi = \widehat{S}_u^2 - \widehat{S}_v^2$ and $\eta = \widehat{S}_u^2 + \widehat{S}_v^2$.

By the independence of $\widehat{S}_u$ and $\widehat{S}_v$, we get the following components of (5.44):

$$E(\xi) = S_u^2 - S_v^2 + \sigma_u^2 - \sigma_v^2,$$

$$E(\eta) = S_u^2 + S_v^2 + \sigma_u^2 + \sigma_v^2,$$

where

$$S_u^2 = \sigma^2(1 + \rho), \qquad S_v^2 = \sigma^2(1 - \rho),$$

$$\sigma_u^2 = \frac{S_u^2\,V(\chi,g)}{n}, \qquad \sigma_v^2 = \frac{S_v^2\,V(\chi,g)}{n}.$$

Further,

$$var(\xi) = var(\eta) = 4(S_u^2\sigma_u^2 + S_v^2\sigma_v^2) + o\left(\frac{1}{n}\right),$$

$$cov(\xi,\eta) = 4(S_u^2\sigma_u^2 - S_v^2\sigma_v^2) + o\left(\frac{1}{n}\right).$$

By substituting these components into (5.44), we arrive at (5.43).

Theorem 5.6.2. *In the class of independent component distributions (5.34) under the conditions of regularity (g1)–(χ4), the asymptotic bias of estimate (5.41) is given by*

$$E(\widehat{\rho}_n) - \rho = b_n + o\left(\frac{1}{n}\right)$$

with

$$b_n(\chi,g) = -\frac{2\rho(1 - \rho^2)}{n}\,V(\chi,g). \qquad (5.45)$$

Proof. Similarly to the proof of (5.43), we obtain (5.45) from the asymptotic expansion for the expectation of the fraction of asymptotically normal random variables ξ and η (Cramér 1946)

$$E\left(\frac{\xi}{\eta}\right) = \frac{E(\xi)}{E(\eta)} - \frac{1}{E^2(\eta)}\,cov(\xi,\eta) + \frac{E(\xi)}{E^3(\eta)}\,var(\eta) + o\left(\frac{1}{n}\right), \qquad (5.46)$$

where $\xi = \widehat{S}_u^2 - \widehat{S}_v^2$ and $\eta = \widehat{S}_u^2 + \widehat{S}_v^2$ as above. By substituting the required components into (5.46), we obtain (5.45).

REMARK 1   The asymptotic variance and asymptotic bias of estimate *(5.41)* given by *(5.43)* and *(5.45)*, respectively, both have two factors: the first depends only on the correlation coefficient ρ, the second $n^{-1}V(\chi, g)$ is the asymptotic variance of *M*-estimates of scale. Thus, most results on robust estimation of scale *(Huber 1981; Hampel et al. 1986)* can be directly applied to robust estimation of the correlation coefficient of bivariate independent component distributions.

Example 5.6.1.   *From (5.43) it follows that the expression for asymptotic variance of the sample correlation coefficient r of a bivariate normal distribution with $\chi(x) = x^2 - 1$ and $g(x) = \phi(x)$ has the well-known form:*

$$var(r) = \frac{(1 - \rho^2)^2}{n}.$$

Example 5.6.2.   *The choice $\chi(x) = \text{sign}(|x| - 1)$ and $g(x) = \phi(x)$ yields the asymptotic variance of the MAD correlation coefficient*

$$var(r_{MAD}) = \frac{(1 - \rho^2)^2}{8n\phi^2(\zeta_{3/4})\zeta_{3/4}^2}, \zeta_{3/4} = \Phi^{-1}(3/4),$$

where $\Phi(x)$ is a standard normal cumulative, $\phi(x) = \Phi'(x)$.

Example 5.6.3.   *The asymptotic bias given by (5.45) has the same structure as the well-known result on the asymptotic bias of the sample correlation coefficient at the normal distribution (Fisher 1915):*

$$E(r) = \rho - \frac{\rho(1 - \rho^2)}{n} + o\left(\frac{1}{n}\right).$$

5.7   Robust Minimax Bias and Variance Estimates of the Correlation Coefficient

5.7.1   Introductory remarks

The first result on minimax robust estimation of the correlation coefficient belongs to Huber (1981): the quadrant correlation coefficient

$$q = \frac{1}{n} \sum_{i=1}^{n} \text{sgn}(x_i - \text{med } x) \, \text{sgn}(y_i - \text{med } y), \qquad (5.47)$$

is asymptotically minimax with respect to bias at the mixture $F = (1 - \epsilon)G + \epsilon H$ with G and H centrosymmetric in $\mathbb{R}^2$.

As already mentioned, since the asymptotic variance and bias of the robust estimates of correlation based on principal component variances linearly depend of the asymptotic variance of a scale parameter in the family of independent component distributions (Theorem 5.6.1 and Theorem 5.6.2), this allows Huber's results on the robust minimax estimation of scale (Section 4.3) to be applied to robust minimax estimation of the correlation coefficient (Shevlyakov and Vilchevsky 2002b; Shevlyakov *et al.* 2012).

5.7.2   Minimax property

To design minimax bias and variance estimates of the correlation coefficient, here we briefly recall Huber's result on minimax variance M-estimates of scale (Huber 1981): under the conditions of regularity $(g1)$–$(\chi4)$ of Theorem 5.6.1, M-estimates $\widehat{S}$ of scale defined by the estimating equation $\sum \chi(x_i/\widehat{S}) = 0$ are consistent, asymptotically normal and possess the minimax (saddle-point) property with regard to the asymptotic variance $V(\chi, g)$:

$$V(\chi^*, g) \leq V(\chi^*, g^*) \leq V(\chi, g^*). \tag{5.48}$$

Here g^* is the least informative density minimizing Fisher information $I(g)$ for scale

$$g^* = \arg\min_{g \in \mathcal{G}} I(g), \qquad I(g) = \int \left(-x\, \frac{g'(x)}{g(x)} - 1\right)^2 g(x)\, dx$$

in the class of ϵ-contaminated normal distribution densities

$$\mathcal{G} = \{g :\ g(x) \geq (1 - \epsilon)\varphi(x)\}, \qquad 0 \leq \epsilon < 1.$$

In this case, the optimal score function $\chi^*(x)$ is given by

$$\chi^*(x) = \begin{cases} x_0^2 - 1, & \text{for}\quad |x| < x_0, \\[2mm] x^2 - 1, & \text{for}\quad x_0 \leq |x| \leq x_1, \\[2mm] x_1^2 - 1, & \text{for}\quad |x| > x_1, \end{cases} \tag{5.49}$$

where the parameters $x_0(\epsilon)$ and $x_1(\epsilon)$ depend on the contamination parameter ϵ being tabulated in Huber (1981, p. 121).

Note that the median absolute deviation $\widehat{S} = \text{MAD}\, x$ is a limit case of the minimax variance estimator as $\epsilon \to 1$ with $\chi^*(x) = \chi_{MAD}(x) = \text{sign}(|x| - 1)$.

The saddle-point inequality (5.48) shows that the estimate $\widehat{S}$ determined by the score function χ^* provides the guaranteed level of the accuracy of estimation for all g in $\mathcal{G}$

$$V(\chi^*, g) \leq V(\chi^*, g^*) = \frac{1}{n\,I(g^*)}.$$

The following result is obtained by direct application of the above solution.

Theorem 5.7.1. *In the subclass (5.38) (Theorem 5.5.1) of ϵ-contaminated bivariate normal distributions*

$$f(x, y) \geq (1 - \epsilon)\,N(x, y|0, 0, 1, 1, \rho), \qquad 0 \leq \epsilon < 1, \tag{5.50}$$

the minimax bias and variance estimate of ρ is given by the the trimmed correlation coefficient (5.18), where the numbers $n_1 = n_1(\epsilon)$ and $n_2 = n_2(\epsilon)$ of the trimmed smallest and greatest order statistics $u_{(i)}$ and $v_{(i)}$ depend on the value of the contamination parameter ϵ through the auxiliary parameter $\gamma = 1 - \sqrt{1 - \epsilon}$. The precise character of the dependencies $n_1 = n_1(\gamma)$ and $n_2 = n_2(\gamma)$ can be found in Huber (1981, 5.6).

The proof of Theorem 5.7.1 immediately follows from Theorem 5.5.1.

If $\epsilon = 0$, then the minimax bias and variance estimate of the correlation coefficient is given by the modified analog of the sample correlation coefficient r_{L2} (5.17). In the limit case as $\epsilon \to 1$, the minimax bias and variance estimate tends to the *MAD* correlation coefficient r_{MAD} (5.11).

5.7.3   Concluding remarks

The obtained result represents the correlation analogs of Huber's results on the minimax variance robust estimation of location and scale (Huber 1964, 1981) with the limit case in the form of the *MAD* correlation coefficient as the correlation analog of the sample median estimate for location.

Moreover, under the aforementioned regularity conditions $(g1)$–$(\chi4)$ imposed on score functions χ and densities g, the most B and V-robust estimates of the correlation coefficient in the sense of Theorem 10 (Hampel *et al.* 1986, pp. 142–143) are also given by the *MAD* correlation coefficient.

5.8   Robust Correlation via Highly Efficient Robust Estimates of Scale

5.8.1   Introductory remarks

In this section, we extend formulas (5.45) and (5.43) for the asymptotic bias and variance of the robust estimates of the correlation coefficient based on M-estimates of principal component variances onto a generalized class of robust estimates of correlation.

5.8.2   Asymptotic bias and variance of generalized robust estimates of the correlation coefficient

Consider again robust estimates of the correlation coefficient ρ based on the robust estimates of the principal component variances (5.39)

$$\hat{\rho}_n = \frac{\hat{S}_n^2 (U) - \hat{S}_n^2 (V)}{\hat{S}_n^2 (U) + \hat{S}_n^2 (V)} , \qquad (5.51)$$

where S is a robust functional, U and V are the principal component random variables

$$U = \frac{\tilde{X} + \tilde{Y}}{\sqrt{2}}, \qquad V = \frac{\tilde{X} - \tilde{Y}}{\sqrt{2}}.$$

$\tilde{X}$ and $\tilde{Y}$ are standardized random variables

$$\tilde{X} = \frac{X - \mu_1}{\sigma_1}, \qquad \tilde{Y} = \frac{Y - \mu_2}{\sigma_2}$$

belonging to the family of independent component distributions (5.34).

Next, assume that $\hat{S}_n$ is a consistent and asymptotically normal estimate of a principal variable scale at a distribution density g with the expectation $E_g(\hat{S})$ and variance $var_g(\hat{S})$, respectively.

Then the following result holds.

Theorem 5.8.1. *Under regularity conditions imposed on the scale functionals S and densities g, estimates $\hat{\rho}_n$ are consistent and asymptotically normal with the variance*

$$var(\hat{\rho}_n) = 2(1 - \rho^2)^2 \frac{var_g(\hat{S})}{E_g(\hat{S})^2} \qquad (5.52)$$

and the asymptotic bias

$$E(\hat{\rho}_n) - \rho = b_n + o(1/n)$$

where

$$b_n = -\frac{2\rho(1 - \rho^2)}{n} \frac{var_g(\hat{S})}{E_g(\hat{S})^2} . \qquad (5.53)$$

The proof is performed by applying the techniques used in the proofs of Theorems 5.6.1 and 5.6.2.

From Equations (5.52) and (5.53) it can be seen that the obtained results are similar to those of Theorems 5.6.1 and 5.6.2 with only one difference: instead of the expression for the asymptotic variance $V(\chi, g)$ of Huber's M-estimates of scale, the ratio

$$\frac{var_g(\hat{S})}{E_g(\hat{S})^2}$$

is used, which is just the standardized variance of the estimate $\widehat{S}_n$ (Bickel and Lehmann 1973).

Note that we arrive at the statements of Theorems 5.6.1 and 5.6.2 naturally by setting $var_g(\widehat{S}) = V(\chi, g)$ and $E_g(\widehat{S}) = 1$.

5.8.3 Concluding remarks

The obtained result gives another way to enhance robustness and efficiency of robust estimates (5.51) of the correlation coefficient: namely, to use in their structure highly robust and efficient estimates of scale, not necessarily M-estimates, for instance, the Q_n estimate of scale proposed by Rousseeuw and Croux (1993). However, in our comparative study we use a low-complexity one-step M-estimate of scale approximating the Q_n estimate with the greater than 80% efficiency at a Gaussian and the highest possible 50% breakdown point. We denote this M-estimate as FQ_n (see Section 4.4): it has the score function (4.38)

$$\chi_{FQ}(x) = 1/\sqrt{\pi} - 2\varphi(x).$$

The corresponding robust estimate of the correlation coefficient is denoted as r_{FQ}.

5.9 Robust M-Estimates of the Correlation Coefficient in Independent Component Distribution Models

5.9.1 Introductory remarks

In this section, we implement the classical Huber approach (Huber 1964) to designing robust estimates of the correlation coefficient.

5.9.2 The maximum likelihood estimate of the correlation coefficient in independent component distribution models

Consider the maximum likelihood estimate of of the correlation coefficient ρ for independent component distributions $f(u, v)$ (5.34), where we without any loss of generality set $\sigma = 1$:

$$\sum_{i=1}^{n} \psi_{ML}(u_i, v_i; \widehat{\rho}_{ML}) = 0 \tag{5.54}$$

with the score function

$$\psi_{ML}(u, v; \rho) = \frac{\partial \log\ f(u, v; \rho)}{\partial \rho}\ .$$

It has the following form:

$$\psi_{ML}(u, v; \rho) = \frac{1}{1+\rho}\chi_{ML}\left(\frac{u}{\sqrt{1+\rho}}\right) - \frac{1}{1-\rho}\chi_{ML}\left(\frac{v}{\sqrt{1-\rho}}\right), \tag{5.55}$$

where χ_{ML} is the maximum likelihood score function for M-estimates of scale (Huber 1981)

$$\chi_{ML}(t) = -1 - \frac{tg'(t)}{g(t)} \ . \tag{5.56}$$

In the particular case of the bivariate normal distribution when $g(z) = \varphi(z)$, the routine calculation of the Fisher information $I(\rho)$ for the correlation coefficient yields

$$I(\rho) = \frac{1 + \rho^2}{(1 - \rho^2)^2},$$

which in its turn gives the classical value of the asymptotic variance (Kendall and Stuart 1963)

$$AV(\hat{\rho}_{ML}) = \frac{1}{I(\rho)} = \frac{(1 - \rho^2)^2}{(1 + \rho^2)} \ . \tag{5.57}$$

From the comparison of this result with the asymptotic variance $AV(r) = (1 - \rho^2)^2$ of the sample correlation coefficient r at the bivariate normal distribution it follows that the maximum likelihood estimator $\hat{\rho}_{ML}$ is asymptotically more accurate than r, being equal to it only at $\rho = 0$.

5.9.3 M-estimates of the correlation coefficient

Taking into account the structure of the maximum likelihood estimating equation (5.54) and the corresponding score function (5.55) for estimation of the correlation coefficient, we implement Huber's program realized for robust estimation of location and scale (Huber 1981) and define a class of M-estimates of ρ for independent component distributions with an arbitrary score function χ:

$$\sum_{i=1}^{n} \psi_M(u_i, v_i; r_M) = 0 \ , \tag{5.58}$$

where

$$\psi_M(u, v; \rho) = \frac{1}{1 + \rho} \chi \left(\frac{u}{\sqrt{1 + \rho}} \right) - \frac{1}{1 - \rho} \chi \left(\frac{v}{\sqrt{1 - \rho}} \right) \ .$$

The particular case of an M-estimate of the correlation coefficient is given by $\chi_{MAD}(z) = \text{sign}(|z| - 1)$, the score function for the MAD estimate of scale in (5.58), with the corresponding $MMAD$-estimate r_{MMAD}.

5.9.4 Asymptotic variance of M-estimators

Now we show that the asymptotic variance of M-estimates has a structure similar to that of formula (5.43).

Theorem 5.9.1. *Under the regularity conditions (g1)–(χ4) of Theorem 5.6.1 imposed on symmetric densities g and scores χ and in the family of bivariate independent component distribution densities (5.34), the asymptotic variance of M-estimates has the form*

$$AV(r_M) = \frac{2(1 - \rho^2)^2}{1 + \rho^2} V(\chi, g), \tag{5.59}$$

where

$$V(\chi, g) = \frac{\int \chi^2(x)g(x)\, dx}{\left(\int x\chi'(x)g(x)\, dx\right)^2}$$

is the asymptotic variance of M-estimates of scale.

Here we give a sketch of the proof. The asymptotic variance of r_M is obtained by direct computation from the following formula of Hampel *et al.* (1986, pp. 125, 139):

$$AV(r_m) = V(\psi_M, f) = \frac{A}{B^2},$$

where

$$A = E(\psi_M^2) = \int \int \psi_M^2(u, v; \rho) f(u, v; \rho)\, du\, dv,$$

$$B = E(\partial \psi_M / \partial \rho) = \int \int \frac{\partial \psi_M(u, v; \rho)}{\partial \rho} f(u, v; \rho)\, du\, dv.$$

Formula (5.59) for the asymptotic variance of r_M is similar to formula (5.43), also having two factors: the first depends only on ρ, the second $V(\chi, g)$ is the asymptotic variance of M-estimates of scale. The difference is only in the factor $1 + \rho^2$ in the denominator of (5.59), which reduces the variance value and, therefore, enhances the estimate efficiency.

Here as before, most results on robust estimation of scale (Hampel *et al.* 1986; Huber 1981), can be directly applied to robust estimation of the correlation coefficient of bivariate independent component distributions.

5.9.5   Minimax variance M-estimates of the correlation coefficient

The further result based on Theorem 5.9.1 and Huber's results on minimax variance estimation of scale (Huber 1981, pp. 120–121) is partially analogous to Theorem 5.7.1.

Theorem 5.9.2. *In the subclass (5.38) (Theorem 5.5.1) of ε-contaminated bivariate normal distributions*

$$f(x, y) \geq (1 - \epsilon)\, N(x, y | 0, 0, 1, 1, \rho), \qquad 0 \leq \epsilon < 1, \tag{5.60}$$

the minimax variance M-estimate of ρ is given by the M-estimate (5.58) with the score function (5.49), where $x_1 = x_1(\gamma)$ and $x_2 = x_2(\gamma)$ depend on the value of the contamination parameter ϵ through the auxiliary parameter $\gamma = 1 - \sqrt{1 - \epsilon}$.

The proof of Theorem 5.9.2 follows from Theorem 5.5.1.

If $\epsilon = 0$, then the minimax variance M-estimate of the correlation coefficient coincides with the maximum likelihood estimate defined by (5.55) and (5.56). In the limit case as $\epsilon \to 1$, the minimax variance M-estimate tends to the *MMAD*-estimate (5.58) with the score function $\chi_{MAD}(x) = \text{sign}(|x| - 1)$.

5.9.6   Concluding remarks

Under the regularity conditions $(g1)$–$(\chi4)$ of Theorem 5.6.1 imposed on score functions χ and densities g, the most V-robust estimate of the correlation coefficient in the sense of Theorem 10 (Hampel *et al.* 1986, pp. 142–143) is given by the *MMAD*-estimate (5.58) with the optimal score $\chi_{MAD}(x) = \text{sign}(|x| - 1)$.

5.10   Monte Carlo Performance Evaluation

5.10.1   Introductory remarks

The asymptotic behavior of robust estimates of the correlation coefficient given above evidently reflects their performance on large samples, but in no way does it represent the peculiarities of their real-life performance on medium, and especially on small samples. Beginning with the first works on robust estimation of correlation, namely, in Devlin *et al.* (1975, 1981) and Gnanadesikan and Kettenring (1972), great attention is paid to Monte Carlo examination of robust estimates, but at the certain expense of their theoretical study. On the contrary, in Huber (1981) and Huber and Ronchetti (2009), the Monte Carlo experiment is rather an exception; however, Hampel *et al.* (1986) is more balanced in this aspect. In our study, we apply a balanced approach combining both ways of research and partially following our former works (Shevlyakov and Vilchevski 2002a, 2011).

5.10.2   Monte Carlo experiment set-up

Distribution models

In Tables 5.1 to 5.8, we exhibit experimental results (50,000 trials) on the comparative performance of the proposed and classical estimates on small ($n = 20$) and large ($n = 1000$) samples at the bivariate normal and ϵ-contaminated normal distributions with the density

$$f(x, y) = (1 - \epsilon)N(x, y; 0, 0, 1, 1, \rho) + \epsilon N(x, y; 0, 0, k, k, \rho') ,$$

Table 5.1   Normal distribution $n = 20$: $\rho = 0.9$

	r	r_Q	r_S	r_K	r_{MAD}	r_{FQ}	r_{MFQ}	r_{MMAD}	r_{MCD}
Mean	0.895*	0.858	0.875	0.892	0.873	0.889	**0.897**	0.906	0.874
MSE	0.050	0.139	0.069	0.063	0.093	0.056*	**0.037**	0.074	0.154
RE	0.729*	0.103	0.431	0.467	0.224	0.596	**1.323**	0.334	0.078

Table 5.2   Normal distribution $n = 1000$: $\rho = 0.9$

	r	r_Q	r_S	r_K	r_{MAD}	r_{FQ}	r_{MFQ}	r_{MMAD}	r_{MCD}
Mean	0.900	0.899	0.900	0.900	0.899	0.900	0.900	0.900	0.900
MSE	0.006*	0.015	0.007	0.007	0.010	0.007	**0.005**	0.007	0.010
RE	0.987*	0.154	0.739	0.820	0.363	0.804	**1.453**	0.662	0.380

Table 5.3   Contaminated normal distribution $n = 20$: $\rho = 0.9$, $\epsilon = 0.1$, $\rho' = 0$,
$k = 3$

	r	r_Q	r_S	r_K	r_{MAD}	r_{FQ}	r_{MFQ}	r_{MMAD}	r_{MCD}
Mean	0.641	0.789	0.708	0.777	0.855	0.849	0.831	0.875*	**0.876**
MSE	0.504	0.205	0.269	0.195	0.103	**0.088**	0.101*	**0.088**	0.143
RE	0.014	0.061	0.051	0.079	0.194	**0.295**	0.273*	0.256	0.091

Table 5.4   Contaminated normal $n = 1000$: $\rho = 0.9$, $\epsilon = 0.1$, $\rho' = 0$, $k = 3$

	r	r_Q	r_S	r_K	r_{MAD}	r_{FQ}	r_{MFQ}	r_{MMAD}	r_{MCD}
Mean	0.553	0.835	0.730	0.787	0.890*	0.868	0.845	0.876	**0.899**
MSE	0.452	0.059	0.172	0.105	0.015*	0.024	0.046	0.026	**0.009**
RE	0.010	0.087	0.052	0.101	0.304	**0.465**	0.418	0.414	0.438*

Table 5.5   ICD Cauchy distribution $n = 20$: $\rho = 0.9$

	r	r_Q	r_S	r_K	r_{MAD}	r_{FQ}	r_{MFQ}	r_{MMAD}	r_{MCD}
Mean	0.624	0.681	0.624	0.716	0.856*	**0.857**	0.667	0.768	0.823
MSE	0.625	0.312	0.352	0.272	0.148*	**0.143**	0.281	0.202	0.258
RE	0.006	0.037	0.038	0.045	0.090*	**0.097**	0.073	0.077	0.030

Table 5.6   ICD Cauchy distribution $n = 1000$: $\rho = 0.9$

	r	r_Q	r_S	r_K	r_{MAD}	r_{FQ}	r_{MFQ}	r_{MMAD}	r_{MCD}
Mean	0.628	0.743	0.743	0.639	**0.899**	**0.899**	0.688	0.799	**0.899**
MSE	0.613	0.160	0.159	0.263	0.014*	**0.013**	0.213	0.103	0.018
RE	0.000	0.046	0.051	0.038	0.198*	**0.212**	0.074	0.111	0.107

Table 5.7   Bivariate Cauchy t-distribution $n = 20$: $\rho = 0.9$

	r	r_Q	r_S	r_K	r_{MAD}	r_{FQ}	r_{MFQ}	r_{MMAD}	r_{MCD}
Mean	0.844	0.852	0.836	**0.884**	0.872	0.880*	0.679	0.777	0.876
MSE	0.295	0.145	0.140	0.096*	0.105	**0.082**	0.258	0.184	0.148
RE	0.022	0.096	0.117	0.203*	0.177	**0.288**	0.103	0.097	0.085

Table 5.8   Bivariate Cauchy t-distribution $n = 1000$: $\rho = 0.9$

	r	r_Q	r_S	r_K	r_{MAD}	r_{FQ}	r_{MFQ}	r_{MMAD}	r_{MCD}
Mean	0.846	0.899	0.856	**0.900**	0.899	**0.900**	0.689	0.799	**0.900**
MSE	0.289	0.015	0.047	0.011*	0.011*	**0.009**	0.212	0.103	0.012
RE	0.000	0.154	0.132	0.306*	0.297	**0.471**	0.092	0.118	0.256

the bivariate ICD Cauchy density (5.34) with $g(z) = \pi^{-1}(1 + z^2)^{-1}$, and the bivariate Cauchy t-distribution (a particular case of the bivariate t-distribution) with the density

$$f(x, y) = \frac{1}{2\pi\sqrt{1 - \rho^2}} \left(1 + \frac{x^2 + y^2 - 2\rho xy}{1 - \rho^2} \right)^{-3/2}.$$

Robust estimates of the correlation coefficient

To provide unbiasedness of estimation, the quadrant r_Q and Kendall correlation coefficients r_K are transformed by taking $\sin(\pi/2)$ of their initial values, whereas the Spearman correlation coefficient r_S is transformed by $2\sin(\pi/6)$ (Kendall and Stuart 1963).

All of the aforementioned estimates were studied, but in what follows, we exhibit results of the best competitors:

- the most robust minimax variance and bias MAD correlation coefficient defined by (5.11);

- the FQ estimate computed by formula (5.39) with the M-estimate $\widehat{S}$ of scale defined by the score function (4.38);

- a highly robust and efficient M-estimate of the correlation coefficient denoted as r_{MFQ} obtained by substituting the score function (4.38) into (5.58);

- the limit case of the minimax variance M-estimate r_{MMAD} defined in Section 5.9 (the most V-robust estimate);

- the minimum covariance determinant (MCD) estimate (Rousseeuw 1985, p. 877) of correlation computed by the means of the package R.

Evaluation criteria

The Monte Carlo estimate mean squared error (MSE) is computed as follows:

$$MSE(\widehat{\rho}) = \frac{1}{M} \sum_{k=1}^{M} (\widehat{\rho} - \rho)^2,$$

where M is a number of trials ($M = 50{,}000$).

The relative estimate efficiency (RE) is defined as the ratio of the asymptotic variance of the sample correlation coefficient r and the experimental estimate of the $\widehat{\rho}$-variance:

$$RE(\widehat{\rho}) = \frac{(1 - \rho^2)^2}{n \ var(\widehat{\rho})} .$$

The best performances in table rows are in boldface, the next values to them are starred.

5.10.3   Discussion

Normal distribution

From Tables 5.1 and 5.2 it follows that

- on small and large samples, the best estimate among the chosen set of robust alternatives to the sample correlation coefficient is the r_{MFQ} M-estimate; the next to it in performance is the sample correlation coefficient r;

- the Kendall correlation coefficient r_K is the best in performance among the nonparametric measures, especially on small samples;

- on large samples, estimate biases can be neglected, but not their variances.

Contaminated normal distributions

From Tables 5.3 and 5.4 it follows that

- the sample correlation coefficient r is catastrophically bad under contamination;

- on small and large samples, the r_{MCD} correlation coefficient is the best with respect to bias; the next to it is the most B- and V-robust MAD correlation coefficient, and as the computational complexity of r_{MCD} is much higher than that of r_{MAD}, the latter is preferable.

- the set of performed experiments does not allow sorting of the other estimates, namely r_{FQ} and r_{MFQ}.

The quadrant correlation coefficient r_Q is also an asymptotically minimax bias estimate of the correlation coefficient (Huber 1981) as r_{MAD}. Nevertheless, as it follows from Tables 5.1 to 5.8, its overall performance is inferior to the performance of r_{MAD}. This can be explained by the choice of the class of direct robust counterparts of the sample correlation coefficient (Huber 1981; Shevlyakov and Vilchevsky 2002b) at which the minimax property of r_Q is established – the class of estimates based on principal variable variances is richer and more advantageous than the competing class. However, we may cautiously recommend the quadrant coefficient r_Q as a moderate robust alternative to the sample correlation coefficient r both due to its low complexity and to its finite sample binomial distribution (Blomqvist 1950).

Furthermore, the minimax variance and bias r_{TRIM} and r_{MTRIM} estimates computed at $\epsilon = 0.1$ were inferior in performance not only to their limit cases r_{MAD} and r_{MMAD} estimates, respectively, but also to the r_{FQ} and r_{MFQ} estimates. This can be explained by the fact that the highly robust and efficient FQ_n estimate of scale dominates over Huber's robust minimax variance trimmed standard deviation estimate (Smirnov and Shevlyakov 2014).

Bivariate Cauchy distributions

From Tables 5.6 and 5.8 it follows that

- the sample correlation coefficient r is again catastrophically bad at both heavy-tailed distributions, especially at the ICD Cauchy density; however, we expected a much worse performance with respect to the estimate's mean, as the population means do not exist in this case;

- on small and large samples, the r_{FQ} correlation coefficient is the best with respect to all performance characteristics; the next to it are the most B- and V-robust MAD correlation coefficients at the ICD Cauchy density and the non-parametric Kendall correlation at the bivariate Cauchy t-distribution;

- the aforementioned advantages of the MCD correlation coefficient in the contaminated normal case disappear at these heavy-tailed distributions;

- it seems that the bivariate independent component Cauchy distribution poses more challenges for estimation of correlation as compared to the bivariate Cauchy t-distribution.

5.10.4   Concluding remarks

In this section, the comparative performance of various robust estimates of the correlation coefficient is studied at the conventional contaminated normal and new heavy-tailed bivariate ICD Cauchy densities.

In the Monte Carlo experiment, the proposed r_{MFQ} and r_{MMAD} M-estimates proved to be the best on small and large samples at slightly contaminated normal distributions.

At heavy-tailed bivariate distributions, our earlier proposed r_{FQ} estimate of the correlation coefficient based on the fast highly robust and efficient Rousseeuw–Croux FQ_n estimate of scale can be recommended.

5.11   Robust Stable Radical M-Estimate of the Correlation Coefficient of the Bivariate Normal Distribution

5.11.1   Introductory remarks

In this section, we apply the stable estimation approach in robustness developed in Section 3.5 to robust stable estimation of the correlation coefficient of the standard bivariate normal distribution. Now we reformulate the aforementioned notions, adapting them to the case of stable estimation of the correlation coefficient.

For the data $(x_1, y_1), \ldots , (x_n, y_n)$ from a bivariate distribution with density $f(x, y; \theta)$ defined on the support $\mathbb{R}^2$, consider the M-estimates $\widehat{\theta}$ of an unknown scalar parameter θ in the conventional form

$$\sum_{i=1}^{n} \psi(x_i, y_i; \widehat{\theta}) = 0 \, , \qquad (5.61)$$

where $\psi(x, y; \theta)$ is a score function belonging to a certain class Ψ.

Within the proposed approach (for details, see Section 3.5), a new global measure of the estimate's sensitivity, the Lagrange functional derivative of the asymptotic variance

$$VS(\psi, f) = \frac{\partial V(\psi, f)}{\partial f}$$

is introduced and, and in this case, it takes the following form:

$$VS(\psi, f) = \frac{\displaystyle\int_{-\infty}^{\infty} \int_{-\infty}^{\infty} \psi^2(x, y; \theta) \, \mathrm{d}x \, \mathrm{d}y}{\left[\displaystyle\int_{-\infty}^{\infty} \int_{-\infty}^{\infty} \frac{\partial \psi(x, y; \theta)}{\partial \theta} f(x, y; \theta) \, \mathrm{d}x \, \mathrm{d}y\right]^2} \, . \qquad (5.62)$$

An M-estimate and the corresponding score function are called *stable* if there exists the integral $\int_{\chi} \psi^2(x, y; \theta) \, dx \, dy < \infty$ and unstable otherwise. Recall that for M-estimates of location (see Section 3.5), the latter requirement leads to highly robust redescending M-estimates.

An optimal score function minimizing the estimate's sensitivity $VS(\psi, f)$ is given by

$$\psi_*(x, y; \theta) = \arg \min_{\psi \in \Psi} VS(\psi, f) = \frac{\partial f(x, y; \theta)}{\partial \theta} + \beta f(x, y; \theta), \qquad (5.63)$$

where the constant β is determined from the condition of Fisher consistency for M-estimates.

An estimate with the score function (5.63) is called the estimate of maximal stability when $VS_{\min} = VS(\psi_*, f)$, and a new global indicator of robustness called the *stability* of an M-estimate is introduced as the following ratio:

$$\text{stb } \widehat{\theta} = \frac{VS_{\min}}{VS(\psi, f)},$$

naturally lying in the $[0, 1]$ range.

Setting different weights for efficiency and stability, various criteria of optimization of estimation can be proposed (see Shurygin 1994a,b). In particular, a reasonable choice is associated with the equal weights of the efficiency and stability functionals, i.e., when eff $\widehat{\theta} = \text{stb } \widehat{\theta}$: this estimate is called *radical*, and in this case, the score function of the radical M-estimator is given by

$$\psi_{rad}(x, y; \theta) = \left[\frac{\partial \log f(x, y; \theta)}{\partial \theta} + \beta \right] \sqrt{f(x, y; \theta)}, \qquad (5.64)$$

where the constant β is obtained from the condition of Fisher consistency.

Note that the factor $\sqrt{f(x, y; \theta)}$ in Equation (5.64) yields relatively lesser weights to the relatively greater observations in the equation

$$\sum_{i=1}^{n} \psi_{rad}(x_i, y_i; \widehat{\theta}) = 0$$

for M-estimates and therefore provides robustness of the estimation procedure to gross errors.

5.11.2   Asymptotic characteristics of the stable radical estimate of the correlation coefficient

Now we apply the aforementioned results to stable estimation of the correlation coefficient of the standard bivariate normal distribution density (implicitly assuming that the parameters of location and scale are given)

$$f(x, y; \rho) = N(x, y; 0, 0, 1, 1, \rho) = \frac{1}{2\pi\sqrt{1 - \rho^2}} \exp\left\{ -\frac{x^2 - 2\rho xy + y^2}{2(1 - \rho^2)} \right\}.$$

From Equation (5.64) it follows that the radical M-estimate of ρ is the solution to the equation

$$\sum_{i=1}^{n} \left(\frac{\partial \log f(x_i, y_i; r_{rad})}{\partial \rho} + \beta \right) \sqrt{f(x_i, y_i; r_{rad})} = 0,$$

where

$$\beta = -\frac{\rho}{3(1 - \rho^2)}.$$

Tedious calculations yield the asymptotic variance of the radical estimate

$$var(r_{rad}) = \frac{81(9 + 10\rho^2)(1 - \rho^2)^2}{512(1 + \rho^2)^2 \, n}$$

with the corresponding values of efficiency and stability

$$\text{eff}(r_{rad}) = \text{stb}(r_{rad})$$

$$= g(\rho^2) = \frac{512(1 + \rho^2)}{81(9 + 10\rho^2)}$$

varying in a rather narrow range from

$$g_{\min} = g(1) = 0.6654$$

to

$$g_{\max} = g(0) = 0.7023.$$

Thus, the radical estimate r_{rad} possesses reasonable levels of efficiency and stability, but its asymptotic efficiency at the normal is considerably lower than that of the best robust estimates of the correlation coefficient based on robust principal component variances, say r_{Qn} or r_{FQ}.

5.11.3   Concluding remarks

The radical stable estimate has revealed its high robustness and efficiency in the problems of robust estimation of a location parameter (see Section 3.5), but our study of its application to the problems of robust estimation of scale and correlation shows that the best robust estimates obtained by the conventional methods, for instance, the Q_n estimate of a scale parameter and the FQ estimate of the correlation coefficient, dominate over the radical estimate.

5.12   Summary

In this chapter, several groups of robust estimates of the correlation coefficient are examined under contaminated bivariate normal, independent component bivariate, and heavy-tailed Cauchy bivariate distributions with the use of influence function techniques in asymptotics and by the Monte Carlo method on finite samples.

As a result of the comparative study of various robust estimates, we exposed the best in each group and showed that some of them possess optimal robustness properties. In particular, the asymptotically minimax variance and bias robust estimate of the correlation coefficient is given by the MAD correlation coefficient, a correlation analog of the sample median, and median absolute deviation estimates of location and scale.

We also show that the robust estimates of the correlation coefficient based on highly efficient robust estimates of scale proposed in Chapter 4 outperform many of their competitors.

The family of independent component bivariate distributions proved to be most prospective in the theoretical study of robust estimation of the correlation coefficient. The generalization of those results for the case of robust estimation of correlation matrices of independent component multivariate distributions is an open problem. However, conventional approaches to robust estimation of correlation matrices are treated in Chapters 7 and 8.

References

Bickel PJ and Lehmann EL 1973 Measures of location and scale. In *Proc. Prague Symp. Asymptotic Statist.* I, Prague, Charles Univ., pp. 25–36.

Blomqvist N 1950 On a measure of dependence between two random variables. *Ann. Math. Statist.* **21**, 593–600.

Cramér H 1946 *Mathematical Methods of Statistics*. Princeton Univ. Press, Princeton.

Croux C and Dehon C 2010 Influence functions of the Spearman and Kendall correlation measures. *Statistical Methods and Applications* **19**, 497–515.

Devlin SJ, Gnanadesikan R and Kettenring JR 1975 Robust estimation and outlier detection with correlation coefficients. *Biometrika* **62**, 531–545.

Devlin SJ, Gnanadesikan R and Kettenring JR 1981 Robust estimation of dispersion matrices and principal components. *J. Amer. Statist. Assoc.* **76**, 354–362.

Falk M 1998 A note on the correlation median for elliptical distributions. *J. Mult. Anal.* **67**, 306–317.

Fisher RA 1915 Frequency distribution of the values of the correlation coefficient in samples from indefinitely large population. *Biometrika* **10**, 507–521.

Gnanadesikan R and Kettenring JR 1972 Robust estimates, residuals, and outlier detection with multiresponse data. *Biometrics* **28**, 81–124.

Hampel FR, Ronchetti E, Rousseeuw PJand Stahel WA 1986 *Robust Statistics. The Approach Based on Influence Functions*, Wiley.

Huber PJ 1964 Robust estimation of a location parameter. *Ann. Math. Statist.* **35**, 73–101.

Huber PJ 1967 The behaviour of maximum likelihood estimates under nonstandard conditions. In *Proc. 5th Berkeley Symp. on Math. Statist. Prob.*, Vol. 1, Berkeley Univ. California Press, pp. 221–223.

Huber PJ 1974 Fisher information and spline interpolation. *Ann. Statist.* **2**, 1029–1033.

Huber PJ 1981 *Robust Statistics*, Wiley.

Huber PJ and Ronchetti E (eds) 2009 *Robust Statistics*, 2nd edn, Wiley.

Kendall MG 1938 A new measure of rank correlation. *Biometrika* **30**, 81–89.

Kendall MG and Stuart A 1962 *The Advanced Theory of Statistics. Distribution Theory*, Vol. 1, Griffin, London.

Kendall MG and Stuart A 1963 *The Advanced Theory of Statistics. Inference and Relationship*, Vol. 2, Griffin, London.

Pasman VR and Shevlyakov GL 1987 Robust methods of estimation of correlation coefficients. *Automation and Remote Control* **48**, 332–340.

Rousseeuw PJ 1985 Multivariate estimation with high breakdown point. In *Mathematical Statistics and Applications*, Grossman W. Pflug G. Vincze I. and Wertz W. (eds), Reidel, Dodrecht, pp. 283–297.

Rousseeuw PJ and Croux C 1993 Alternatives to the median absolute deviation. *J. Amer. Statist. Assoc.* **88**, 1273–1283.

Shevlyakov GL 1997 On robust estimation of a correlation coefficient. *J. Math. Sci.* **83**, 434–438.

Shevlyakov GL and Vilchevski NO 2002a *Robustness in Data Analysis: criteria and methods*, VSP, Utrecht.

Shevlyakov GL and Vilchevsky NO 2002b Minimax variance estimation of a correlation coefficient for epsilon-contaminated bivariate normal distributions. *Statistics and Probability Letters* **57**, 91–100.

Shevlyakov GL and Vilchevski NO 2011 *Robustness in Data Analysis*, De Gruyter, Boston.

Shevlyakov GL, Smirnov PO, Shin VI, and Kim K 2012 Asymptotically minimax bias estimation of the correlation coefficient for bivariate independent component distributions. *J. Mult. Anal.* **111**, 59–65.

Shurygin AM 1994a New approach to optimization of stable estimation. In *Proc. 1 US/Japan Conf. on Frontiers of Statist. Modeling* Kluwer Academic Publishers, Netherlands, pp. 315–340.

Shurygin AM 1994b Variational optimization of the estimator stability. *Automation and Remote Control* **55**, 1611–1622.

Spearman C 1904 The proof and measurement of association between two things. *Amer. J. Psychol.* **15**, 88–93.

Smirnov PO and Shevlyakov GL 2014 Fast highly efficient and robust one-step M-estimators of scale based on FQ_n. *Computational Statistics and Data Analysis* **78**, 153–158.

6

Classical Measures of Multivariate Correlation

This chapter provides an overview of classical multivariate correlation measures and inference tools based on the covariance and correlation matrix.

6.1 Preliminaries

So far, we have treated the correlation or linear dependence between two random variables only. Considering bivariate correlations may not be satisfactory in a more general case of $p > 2$ variables where one is often interested in the dependence between groups of variables, or one hopes to find latent uncorrelated or independent components, or one wishes to know whether the dependence between two variables can be fully explained by a third variable, and so on.

In this chapter we adopt the following matrix notation. The *"vec" operation* is used to vectorize a matrix. If $\mathbf{A} = (\mathbf{a}_1, \dots, \mathbf{a}_s)$ is an $r \times s$ matrix with columns $\mathbf{a}_1, \dots, \mathbf{a}_s$, then

$$\mathrm{vec}(\mathbf{A}) = \begin{pmatrix} \mathbf{a}_1 \\ \dots \\ \mathbf{a}_s \end{pmatrix}$$

is an $s \cdot r$ vector that stacks the columns of $\mathbf{A}$ on top of each other. A useful result for vectorizing the matrix product is

$$\mathrm{vec}(\mathbf{BCD}) = (\mathbf{D}' \otimes \mathbf{B}) \; \mathrm{vec}(\mathbf{C}).$$

Robust Correlation: Theory and Applications, First Edition. Georgy L. Shevlyakov and Hannu Oja.
© 2016 John Wiley & Sons, Ltd. Published 2016 by John Wiley & Sons, Ltd.
Companion Website: www.wiley.com/go/Shevlyakov/Robust

Here $\mathbf{D}' \otimes \mathbf{B}$ is the so-called Kronecker product of matrices $\mathbf{D}'$ and $\mathbf{B}$. Clearly then, with two matrices $\mathbf{B}$ and $\mathbf{C}$,

$$\text{vec}(\mathbf{BC}) = (\mathbf{I} \otimes \mathbf{B}) \ \text{vec}(\mathbf{C}) = (\mathbf{C}' \otimes \mathbf{I}) \ \text{vec}(\mathbf{B}).$$

Next, let $\mathbf{e}_i$ be a p-vector with ith element one and other elements zero. We further write $\mathbf{E}^{ij} := \mathbf{e}_i \mathbf{e}_j^T$, $i,j = 1, \ldots, p$, and

$$\mathbf{J}_{p,p} := \sum_{i=1}^{p} \sum_{j=1}^{p} \mathbf{E}^{ij} \otimes \mathbf{E}^{ij} = \text{vec}(\mathbf{I}_p)\text{vec}(\mathbf{I}_p)^T, \quad \mathbf{K}_{p,p} := \sum_{i=1}^{p} \sum_{j=1}^{p} \mathbf{E}^{ij} \otimes \mathbf{E}^{ji},$$

$$\mathbf{I}_{p,p} := \sum_{i=1}^{p} \sum_{j=1}^{p} \mathbf{E}^{ii} \otimes \mathbf{E}^{ij} = \mathbf{I}_{p^2} \quad \text{and} \quad \mathbf{D}_{p,p} := \sum_{i=1}^{p} \mathbf{E}^{ii} \otimes \mathbf{E}^{ii}.$$

Then, for any $p \times p$ matrix $\mathbf{A}$,

$$\mathbf{J}_{p,p}\text{vec}(\mathbf{A}) = \text{tr}(\mathbf{A})\text{vec}(\mathbf{I}_p), \quad \mathbf{K}_{p,p}\text{vec}(\mathbf{A}) = \text{vec}(\mathbf{A}^T),$$

and $\mathbf{D}_{p,p}\text{vec}(\mathbf{A}) = \text{vec}(\text{diag}(\mathbf{A}))$. The matrix $\mathbf{K}_{p,p}$ is called a commutation matrix.

A $p \times p$ matrix $\mathbf{U}$ is *orthogonal* if its rows and columns are orthonormal, that is, $\mathbf{U}^T\mathbf{U} = \mathbf{U}\mathbf{U}^T = \mathbf{I}_p$. A *sign-change matrix* is a square matrix with diagonal elements ± 1. A $p \times p$ *permutation matrix* is a square matrix that is obtained by permuting the rows and/or the columns of $\mathbf{I}_p$. For a symmetric non-negative definite matrix $\mathbf{S}$, the matrix $\mathbf{S}^{-1/2}$ is taken to be symmetric and satisfy $\mathbf{S}^{-1/2}\mathbf{S}\mathbf{S}^{-1/2} = \mathbf{I}_p$. The matrix norm $\| \cdot \|$ is defined by $\|\mathbf{A}\| = \text{tr}(\mathbf{A}'\mathbf{A})^{1/2}$.

6.2   Covariance Matrix and Correlation Matrix

Assume that $\mathbf{X} = (X_1, \ldots, X_p)^T$ is a continuous *multivariate (p-variate) random variable*. Its random variation is often described by its *multivariate density function* $f(x_1, \ldots, x_p)$ or *multivariate cumulative density function* $F(x_1, \ldots, x_p)$. Distributions of the univariate $X_1, X_2, \ldots, X_p$ are called *marginal distributions*. Joint distribution naturally uniquely determines the distributions of all univariate random variables that are obtained using $X_1, X_2, \ldots, X_p$, such as X_1^2, $X_1 X_2$ or *linear combinations*

$$\mathbf{a}^T\mathbf{X} = a_1 X_1 + a_2 X_2 + \ldots + a_p X_p.$$

On the other hand, *if one knows the distributions of all linear combinations $\mathbf{a}^T\mathbf{X}$, then one knows also the (multivariate) distribution of X.*

Marginal distributions can naturally be described, for example, using their expectations

$$E(X_1), E(X_2), \ldots, E(X_p)$$

and variances

$$var(X_1), var(X_2), ..., var(X_p)$$

but marginal distributions are not sufficient to describe *co variation*, e.g., linear dependence. Linear dependence is usually described using *covariances*. The covariance between X_i and X_j is

$$cov(X_i, X_j) = E[(X_i - E(X_i))(X_j - E(X_j))], \quad i,j = 1, ..., p.$$

A standardized measure of linear correlation between two variables is the *correlation coefficient*: the correlation coefficient between variables X_i and X_j discussed throughout the book is thus

$$cor(X_i, X_j) = \frac{cov(X_i, X_j)}{\sqrt{var(X_i)var(X_j)}}, \quad i,j = 1, ..., p.$$

The location center of a multivariate distribution of **X** is often defined as the *mean vector*

$$E(\mathbf{X}) = \begin{pmatrix} E(X_1) \\ E(X_2) \\ ... \\ E(X_p) \end{pmatrix}$$

that minimizes $E(\|\mathbf{X} - \boldsymbol{\mu}\|^2)$, and the co variation around the mean vector is described by the *covariance matrix*

$$cov(\mathbf{X}) = \begin{pmatrix} var(x_1) & cov(x_1, x_2) & ... & cov(x_1, x_p) \\ cov(x_2, x_1) & var(x_2) & ... & cov(x_2, x_p) \\ ... & ... & ... & ... \\ cov(x_p, x_1) & cov(x_p, x_2) & ... & var(x_p) \end{pmatrix}.$$

Note that the covariance matrix is a symmetric non-negatively definite matrix and its diagonal elements are the variances of the marginal distributions: $var(x_i) = cov(x_i, x_i), i = 1, .., p$.

Using the mean vector and the covariance matrix, one can easily compute the mean and variance of any linear combination, namely,

$$E(\mathbf{a}^T \mathbf{X}) = \mathbf{a}^T E(\mathbf{X}) = a_1 E(X_1) + a_2 E(X_2) + \cdots + a_p E(X_p)$$

and

$$var(\mathbf{a}'\mathbf{X}) = \mathbf{a}' cov(\mathbf{X}) \mathbf{a} = \sum_{i=1}^{p} \sum_{j=1}^{p} a_i a_j cov(X_i, X_j).$$

One is also often interested in the correlations between the variables, that is, on the *correlation matrix*

$$cor(\mathbf{X}) = \begin{pmatrix} 1 & cor(x_1,x_2) & ... & cor(x_1,x_p) \\ cor(x_2,x_1) & 1 & ... & cor(x_2,x_p) \\ ... & ... & ... & ... \\ cor(x_p,x_1) & cor(x_p,x_2) & ... & 1 \end{pmatrix}.$$

Assume now that

$cov(\mathbf{X})$ is positive definite, that is, $var(\mathbf{a}^T\mathbf{X}) > 0$ for all $\mathbf{a} \neq \mathbf{0}$.

Then the inverse $cov(\mathbf{X})^{-1}$ exists and, for any possible value $\mathbf{x} \in \mathbf{R}^p$, one can calculate the number

$$D(\mathbf{x}) = \sqrt{(\mathbf{x} - E(\mathbf{X}))^T cov(\mathbf{X})^{-1}(\mathbf{x} - E(\mathbf{X}))},$$

which is the so-called *Mahalanobis distance* between $\mathbf{x}$ and the mean value $E(\mathbf{X})$. The Mahalanobis distance between two possible values $\mathbf{x}$ and $\mathbf{y}$ is defined in a similar way:

$$D(\mathbf{x}, \mathbf{y}) = \sqrt{(\mathbf{x} - \mathbf{y})^T cov(\mathbf{X})^{-1}(\mathbf{x} - \mathbf{y})}.$$

Interpretation: Mahalanobis distance is invariant under affine transformations. One can thus think that the covariance matrix $cov(\mathbf{X})$ gives a *multivariate measurement device* for which the distances do not depend on the used coordinate system (e.g., units of measurements or rotations of coordinate axes) at all. The distance in a certain direction then depends on the variation (standard deviation) in that direction.

6.3 Sample Mean Vector and Sample Covariance Matrix

Let $\mathbf{X}_1, ..., \mathbf{X}_n$ be a random sample from a p-variate distribution with mean vector $E(\mathbf{X}) = \mu$ and covariance matrix $cov(\mathbf{X}) = \Sigma$. For the observed sample we write $\mathbf{x}_1, ..., \mathbf{x}_n$ and then

x_{ij} = the measured value of the jth variables for the ith observation.

Natural estimates of the mean vector and the covariance matrix are then their sample counterparts, the *sample mean vector*

$$\bar{\mathbf{X}} = \frac{1}{n} \sum_{i=1}^{n} \mathbf{X}_i$$

and *sample covariance matrix*

$$\mathbf{S} = \frac{1}{n-1} \sum_{i=1}^{n} (\mathbf{X}_i - \bar{\mathbf{X}})(\mathbf{X}_i - \bar{\mathbf{X}})^T = \frac{1}{n-1} \left[\sum_{i=1}^{n} \mathbf{X}_i \mathbf{X}_i^T - n\bar{\mathbf{X}}\bar{\mathbf{X}}^T \right].$$

Note that using $n - 1$ as a divisor instead of n makes $\mathbf{S}$ an unbiased estimate of $\mathbf{\Sigma}$. For the observation vectors $\mathbf{x}_1, ..., \mathbf{x}_n$, we write

$$\bar{\mathbf{x}} = \begin{pmatrix} \bar{x}_1 \\ \bar{x}_2 \\ ... \\ \bar{x}_p \end{pmatrix} \quad \text{and} \quad \mathbf{S} = \begin{pmatrix} s_{11} & s_{12} & ... & s_{1p} \\ s_{21} & s_{22} & ... & s_{2p} \\ ... & ... & ... & ... \\ s_{p1} & s_{p2} & ... & s_{pp} \end{pmatrix},$$

respectively, where

$$\bar{x}_j = \frac{1}{n} \sum_{i=1}^{n} x_{ij} \quad \text{and} \quad s_{jk} = \frac{1}{n-1} \sum_{i=1}^{n} (x_{ij} - \bar{x}_j)(x_{ik} - \bar{x}_k), \quad j, k = 1, ..., p.$$

Further, the population correlation matrix can be estimated with the *sample correlation matrix*

$$\mathbf{R} = \begin{pmatrix} 1 & r_{12} & ... & r_{1p} \\ r_{21} & 1 & ... & r_{2p} \\ ... & ... & ... & ... \\ r_{p1} & r_{p2} & ... & 1 \end{pmatrix},$$

where

$$r_{jk} = \frac{s_{jk}}{\sqrt{s_{jj} s_{kk}}}, \quad j, k = 1, ..., p.$$

With matrix notation, if $\mathbf{X} = (\mathbf{X}_1, ..., \mathbf{X}_n)^T$, then

$$\bar{\mathbf{X}} = \frac{1}{n} \mathbf{X}^T \mathbf{1}_n \quad \text{and} \quad \mathbf{S} = \frac{1}{n-1} \mathbf{X}^T \left(\mathbf{I}_n - \frac{1}{n} \mathbf{J}_n \right) \mathbf{X}$$

with $\mathbf{J}_n = \mathbf{1}_n \mathbf{1}_n^T$ (an $n \times n$ matrix with all elements 1) and

$$\mathbf{R} = \text{diag}(\mathbf{S})^{-1/2} \, \mathbf{S} \, \text{diag}(\mathbf{S})^{-1/2},$$

where $\text{diag}(\mathbf{S})$ is a diagonal matrix with the same diagonal elements as $\mathbf{S}$.

6.4   Families of Multivariate Distributions

6.4.1   Construction of multivariate location-scatter models

In this section we consider different *parametric and semiparametric models* for multivariate continuous observations. Consider a data matrix consisting of n observed values of a p-variate response variable,

$$\mathbf{X} = (\mathbf{X}_1, ..., \mathbf{X}_n)^T.$$

The *multivariate location-scatter model* is obtained if the p-variate observations $\mathbf{X}_i$, $i = 1, ..., n$, are assumed to be independent and to be generated by

$$\mathbf{X}_i \ = \ \mu \ + \ \Omega \, \mathbf{Z}_i, \quad i = 1, ..., n,$$

where the p-vectors $\mathbf{Z}_i$ are called *standardized variables*, μ is a *location p-vector*, Ω *is a full-rank $p \times p$ transformation matrix*, and $\Sigma = \Omega\Omega^T > 0$ is called a *scatter matrix*. Notation $\Sigma > 0$ means that Σ is positive definite (with rank p).

We will be explicit regarding what is meant by a standardized random vector $\mathbf{Z}_i$. It is usual to say that $\mathbf{Z}_i$ is standardized if $E(\mathbf{Z}_i) = \mathbf{0}$ and $cov(\mathbf{Z}_i) = \mathbf{I}_p$. Then $E(\mathbf{X}_i) = \mu$ and $cov(\mathbf{X}_i) = \Sigma$. Other ways to standardize are of course possible if one does not wish to assume the existence of the second moments, for example.

Different parametric and semiparametric (or nonparametric) models are obtained by making different assumptions on the distribution of $\mathbf{Z}_i$. In the following we write $\mathbf{X} \sim \mathbf{Y}$, if $\mathbf{X}$ and $\mathbf{Y}$ possess the same probability distribution.

6.4.2   Multivariate symmetrical distributions

Symmetry of a distribution of the standardized p-variate random variable $\mathbf{Z}$ may be seen as an invariance property of the distribution under orthogonal, sign-change, and/or permutation transformations. The random p-vector $\mathbf{Z}$ is said to be (i) *spherically symmetrical* if $\mathbf{UZ} \sim \mathbf{Z}$ for all orthogonal matrices $\mathbf{U}$, (ii) *marginally symmetrical* if $\mathbf{SZ} \sim \mathbf{Z}$ for all sign-change matrices $\mathbf{S}$, (iii) *symmetrical (or centrally symmetrical)* if $-\mathbf{Z} \sim \mathbf{Z}$, and (iv) *exchangeable* if $\mathbf{PZ} \sim \mathbf{Z}$ for all permutation matrices $\mathbf{P}$. Note that a spherically symmetrical random variable is also marginally symmetrical, centrally symmetrical and exchangeable as $\mathbf{J}$, $-\mathbf{I}_p$, and $\mathbf{P}$ are all orthogonal. A marginally symmetrical random variable is naturally also centrally symmetrical. As in Oja (2010), a hierarchy of symmetrical models is then obtained if we assume that

(A0) $\mathbf{Z}_i \sim N_p(\mathbf{0}, \mathbf{I}_p)$, which is discussed soon,

(A1) $\mathbf{Z}_i$ spherically symmetrical,

(A2) $\mathbf{Z}_i$ marginally symmetrical and exchangeable,

(A3) $\mathbf{Z}_i$ marginally symmetrical, and

(A4) $\mathbf{Z}_i$ symmetrical.

The first assumption (A0) gives the regular multivariate normal model with $\mathbf{X}_i \sim N_p(\mu, \mathbf{\Sigma})$. This is a *fully parametric model*. The classical multivariate inference methods (Hotelling's T^2, multivariate analysis of variance (MANOVA), multivariate regression analysis, principal component analysis (PCA), canonical correlation analysis (CCA), factor analysis (FA), etc.) assume multivariate normality.

The models (A1) to (A4) are *semiparametric models* with parameters μ and $\mathbf{\Sigma}$ and provide extensions of the multivariate normal model (A0). In all these models, the location parameter μ is the well-defined symmetry center of the distribution of $\mathbf{X}_i$.

For models (A0) to (A2), the scatter matrix $\mathbf{\Sigma}$ is proportional to the covariance matrix if it exists. We next consider some models in more detail.

6.4.3   Multivariate normal distribution

Assume now that $\mathbf{Z}_i$ is spherically symmetric and that $R_i^2 = \|\mathbf{Z}_i\|^2 = Z_{i1}^2 + \cdots + Z_{ip}^2 \sim \chi_p^2$. This gives the model (A0). It follows that the distribution of $\mathbf{Z}_i$ is the so-called *standard multivariate normal* distribution $N_p(\mathbf{0}, \mathbf{I}_p)$ with the (probability) density function

$$f(\mathbf{z}) = (2\pi)^{-p/2} \exp\left\{-\frac{1}{2}\mathbf{z}^T\mathbf{z}\right\}.$$

We then write $\mathbf{Z}_i \sim N_p(\mathbf{0}, \mathbf{I}_p)$. The p components of $\mathbf{Z}_i$ are independent and distributed as $N(0, 1)$. In fact, $\mathbf{Z}_i$ *is spherically symmetric and has independent components if and only if $\mathbf{z}_i$ has a multivariate normal distribution.*

In the location-scatter model (A0), the distribution of $\mathbf{X}_i$ is then a p-variate normal distribution $N_p(\mu, \mathbf{\Sigma})$ with the density function

$$f_{\mathbf{X}}(\mathbf{x}) = (2\pi)^{-p/2}\det(\mathbf{\Sigma})^{-1/2} \exp\left\{-\frac{1}{2}(\mathbf{x} - \mu)'\mathbf{\Sigma}^{-1}(\mathbf{x} - \mu)\right\}.$$

Now $E(\mathbf{X}_i) = \mu$ and $cov(\mathbf{X}_i) = \mathbf{\Sigma}$. A random vector $\mathbf{X}$ has a p-variate normal distribution if and only if $\mathbf{a}'\mathbf{X}$ has a univariate normal distribution for all $\mathbf{a} \in \mathbf{R}^p$.

If $\mathbf{X}_i \sim N_p(\mu, \mathbf{\Sigma})$ then

$$\mathbf{A}\mathbf{X}_i + \mathbf{b} \sim N_q(\mathbf{A}\mu + \mathbf{b}, \mathbf{A}\mathbf{\Sigma}\mathbf{A}')$$

for any $q \times p$ matrix $\mathbf{A}$ and any $\mathbf{b} \in \mathbf{R}^q$, $q \leq p$. For $q > p$, $\mathbf{X}_i$ has a so-called singular multivariate normal distribution.

6.4.4   Multivariate elliptical distributions

Consider then the model (A1): now

$$\mathbf{X}_i \;=\; \mu \;+\; \mathbf{\Omega}\,\mathbf{Z}_i, \quad i = 1, ..., n,$$

where $\mathbf{Z}_1, ..., \mathbf{Z}_n$ are independent and identically distributed (i.i.d.) random vectors from a *spherically symmetrical and continuous distribution*. As the distribution of $\mathbf{Z}$

is spherically symmetrical around the origin, then the (multivariate) density function $f(\mathbf{z})$ depends on $\mathbf{z}$ only through the modulus $\|\mathbf{z}\| = (z_1^2 + \cdots + z_p^2)^{1/2}$ and we can write

$$f(\mathbf{z}) = \exp\{-\rho(\|\mathbf{z}\|)\}$$

for some function $\rho(r)$. Note that *the equal density contours are spheres.*

It can be shown that the modulus $R_i = \|\mathbf{Z}_i\|$ and direction $\mathbf{U}_i = \|\mathbf{U}_i\|^{-1}\mathbf{U}_i$ are independent in this model. The direction vector $\mathbf{U}_i$ is uniformly distributed on the p-dimensional unit sphere S_p and the density of the modulus is

$$g(r) = c_p r^{p-1} \exp\{-\rho(r)\}, \quad r > 0,$$

where $c_p = 2\pi^{p/2}/\Gamma(p/2)$. Note that Ω is well-defined only up to a rotation, and also the scatter matrix Σ is confounded with ρ. For the uniqueness of Σ, ρ (and $\mathbf{z}_i$) are often chosen so that $E(r_i^2) = p$ or $\mathrm{Med}(r_i^2) = \chi_{p,0.5}^2$ (the constant $\chi_{p,0.5}^2$ is the median of a chi-square distribution with p degrees of freedom); with these choices, Σ is the regular covariance matrix in the multivariate normal case.

Under these assumptions, the random sample $\mathbf{X} = (\mathbf{X}_1, ..., \mathbf{X}_n)^T$ comes from a p-variate elliptical distribution (we write $\mathbf{X}_i \sim E_p(\mu, \Sigma, \rho)$) with the probability density function

$$f_\mathbf{X}(\mathbf{x}) = \det(\Sigma)^{-1/2} f\left(\Sigma^{-1/2}(\mathbf{x} - \mu)\right)$$

$$= \det(\Sigma)^{-1/2} \exp\left\{-\rho(\sqrt{(\mathbf{x} - \mu)'\Sigma^{-1}(\mathbf{x} - \mu)})\right\},$$

where μ is the symmetry center and $\Sigma > 0$ is the scatter matrix (parameter). The location parameter μ is the mean vector (if it exists) and the scatter matrix Σ is proportional to the regular covariance matrix (if it exists).

The equal density contours $\{\mathbf{x} : (\mathbf{x} - \mu)'\Sigma^{-1}(\mathbf{x} - \mu) = c\}$ *are now ellipses (ellipsoids) with the center* μ *and the size and shape determined by* Σ. With a fixed ρ, the model is *parametric* and, with an unspecified ρ, we have a *semiparametric* model.

Multivariate t distribution: assume that $\mathbf{Z}_i$ is spherically symmetric and that $R_i^2/p \sim F(p, v)$ (F distribution with p and v degrees of freedom). Then the distribution of $\mathbf{Z}_i$ is a p-variate spherical t distribution with v degrees of freedom; write $\mathbf{Z}_i \sim t_{v,p}$. The density function of $\mathbf{Z}_i$ is then

$$f(\mathbf{z}) = \frac{\Gamma((p + v)/2)}{\Gamma(v/2)(\pi v)^{p/2}}\left(1 + \frac{\mathbf{z}^T \mathbf{z}}{v}\right)^{-(p+v)/2}.$$

The components of $\mathbf{z}_i$ *are uncorrelated, but not independent,* and their marginal distribution is a univariate t_v distribution. The smaller v is, the heavier are the tails of the distribution. The expected value exists if $v \geq 2$, and the covariance matrix exists for degrees of freedom $v \geq 3$. The very heavy-tailed distribution with $v = 1$ is called the *multivariate Cauchy distribution.* The multivariate normal distribution is obtained as a limit case as $v \to \infty$.

Multivariate power exponential family: in the so-called multivariate power exponential family, the density of the distribution of $\mathbf{Z}_i$ is

$$f(\mathbf{z}) = k_{p,v} \exp\left\{ -\frac{(\mathbf{z}^T \mathbf{z})^v}{2} \right\},$$

where

$$k_{p,v} = \frac{p\Gamma(p/2)}{\pi^{p/2}\Gamma\left((2v+p)/(2v)\right)2^{(2v+p)/(2v)}}$$

is determined so that the density integrates up to 1. Now $R_i^{2v} \sim \Gamma(1/2, p/(2v))$, which can be used to simulate the observations from this flexible parametric model. If $v = 1$ then $\mathbf{Z}_i \sim N_p(\mathbf{0}, \mathbf{I}_p)$. The model includes both heavy-tailed ($v < 1$) and light-tailed ($v > 1$) elliptical distributions. The (heavy-tailed) multivariate double exponential (Laplace) distribution is given by $v = 1/2$ and a (light-tailed) multivariate uniform elliptical distribution is obtained as a limit when $v \to \infty$.

6.4.5   Independent component model

In the independent component model the independent and identically distributed random variables $\mathbf{X}_i$ are generated by

$$\mathbf{X}_i = \mu + \mathbf{\Omega}\mathbf{Z}_i, \quad i = 1, ..., n,$$

where μ is the location center, the $p \times p$ matrix $\mathbf{\Omega}$ is *the mixing matrix*, and the standardized p-vector $\mathbf{z}_i$ ($E(\mathbf{Z}_i) = \mathbf{0}$ and $cov(\mathbf{Z}_i) = \mathbf{I}_p$) has latent (mutually) *independent components*. The distribution of $\mathbf{Z}_i$ is then fully determined by the marginal densities $f_1, ..., f_p$ of $Z_{i1}, ..., Z_{ip}$. Note that this model provides another natural extension of the multivariate normal model that is often more realistic than the elliptic model.

In the so-called independent component analysis (ICA), the aim is, based on the data $\mathbf{X} = (\mathbf{X}_1, ..., \mathbf{X}_n)^T$, to find an estimate of an *unmixing matrix* Γ such that $\Gamma\mathbf{X}_i$ has independent components (see, e.g., Hyvärinen *et al.* 2001). Note that the model and the problem are not well-posed as the standardized independent components are defined only up to their signs and order, but this does not cause any problem in practical data analysis.

6.4.6   Copula models

Copula models are semiparametric models where (i) there are no assumptions on the marginal distributions f_j of $X_{ij}, j = 1, ..., p$ (nonparametric part) but (ii) the dependence is described with a parametric model for $F_1(X_{i1}), ..., F_p(X_{ip})$. The Gaussian copula is then constructed using a multivariate normal distribution with an unknown correlation matrix $\mathbf{P}$ (see, e.g., Nelsen 1999). The correlation matrix estimates are then based on the marginal ranks $\mathbf{X}_1, ..., \mathbf{X}_n$. This is discussed in Chapter 8.

6.5 Asymptotic Behavior of Sample Covariance Matrix and Sample Correlation Matrix

Let first $\mathbf{Z}_1, ..., \mathbf{Z}_n$ come from a multivariate normal distribution $N_p(\mathbf{0}, \mathbf{I}_p)$. Then

$$\bar{\mathbf{Z}} \sim N_p\left(\mathbf{0}, \frac{1}{n}\mathbf{I}_p\right) \quad \text{and} \quad (n-1)\mathbf{S}_Z \sim W_p(n-1),$$

where $W_p(n)$ is the so-called Wishart distribution with n degrees of freedom. For Wishart distribution, see, for example, Bilodeau and Brenner (1999).

Next, $\mathbf{X}_1 = \mathbf{\Omega}\mathbf{Z}_1 + \mu, ..., \mathbf{X}_n = \mathbf{\Omega}\mathbf{Z}_n + \mu$ is a random sample is from $N_p(\mu, \mathbf{\Sigma})$ with $\mathbf{\Sigma} = \mathbf{\Omega}\mathbf{\Omega}^T$. As

$$\bar{\mathbf{X}} = \mathbf{\Omega}\bar{\mathbf{Z}} + \mu \quad \text{and} \quad \mathbf{S}_X = \mathbf{\Omega}\mathbf{S}_Z\mathbf{\Omega}^T,$$

we say that *the mean vector and covariance matrix are affine equivariant*. This further implies that

$$\bar{\mathbf{X}} \sim N_p\left(\mu, \frac{1}{n}\mathbf{\Sigma}\right) \quad \text{and} \quad (n-1)\mathbf{\Sigma}^{-1/2}\mathbf{S}_X\mathbf{\Sigma}^{-1/2} \sim W_p(n-1).$$

Moreover, $\bar{\mathbf{X}}$ and $\mathbf{S}$ are independent. This is one of the properties that characterize the (multivariate) normal distribution.

Consider next the limiting behavior of $\bar{\mathbf{X}}$ and $\mathbf{S}_X$. Assume that $\mathbf{X}_1, ..., \mathbf{X}_n$ is a random sample from *any* p-variate distribution of $\mathbf{X}$ with mean vector μ and covariance matrix $\mathbf{\Sigma}$. The multivariate version of the central limit theorem then says that

$$\sqrt{n}(\bar{\mathbf{X}} - \mu) \to_d N_p(\mathbf{0}, \mathbf{\Sigma})$$

and, if the fourth moments exist,

$$\sqrt{n}(\mathbf{S}_X - \mathbf{\Sigma}) \to_d S,$$

where S is a multivariate normal distribution with zero mean matrix. The variances and covariances of S in some interesting cases are listed below.

1. If $\mathbf{X} \sim N_p(\mathbf{0}, \mathbf{I}_p)$ then

$$cov(vec(S)) = (\mathbf{I}_{p,p} + \mathbf{K}_{p,p}).$$

Thus $S_{ij}, i \leq j$, are independent and

$$var(S_{ii}) = 2 \quad \text{and} \quad var(S_{ij}) = 1, i \neq j.$$

2. if $\mathbf{X} \sim N_p(\mu, \mathbf{\Sigma})$ then, due to affine equivariance of the covariance matrix,

$$cov(vec(S)) = \mathbf{\Sigma}^{1/2} \otimes \mathbf{\Sigma}^{1/2}(\mathbf{I}_{p,p} + \mathbf{K}_{p,p})\mathbf{\Sigma}^{1/2} \otimes \mathbf{\Sigma}^{1/2}$$
$$= (\mathbf{I}_{p,p} + \mathbf{K}_{p,p})\mathbf{\Sigma} \otimes \mathbf{\Sigma}.$$

3. If $\mathbf{X}_1, ..., \mathbf{X}_n$ come from a spherical distribution of $\mathbf{Z}$ with $E(\mathbf{Z}) = \mathbf{0}$ and $cov(\mathbf{Z}) = \mathbf{I}_p$, then

$$cov(vec(S)) = E(Z_1^2 Z_2^2)(\mathbf{I}_{p,p} + \mathbf{K}_{p,p}) + (E(Z_1^2 Z_2^2) - 1)\mathbf{J}_{p,p}.$$

The limiting distribution of the covariance matrix thus depends only on

$$E(Z_1^2 Z_2^2) = Var(Z_1 Z_2) = E(\|\mathbf{Z}\|^4)/(p(p+2)),$$

a measure of multivariate kurtosis. Note also that $E(Z_1^2 Z_2^2) - 1 = cov(Z_1^2, Z_2^2)$. In the multivariate normal case, $E(Z_1^2 Z_2^2) = 1$. The variables $(S_{11}, ..., S_{pp})$ and $S_{ij}, i < j$, are independent; $S_{11}, ..., S_{pp}$ are independent only in the multivariate normal case.

4. If $\mathbf{X}_1, ..., \mathbf{X}_n$ come from an elliptical distribution of $\mathbf{X} = \boldsymbol{\Sigma}^{1/2}\mathbf{Z} + \mu$, $\mathbf{Z}$ as above, then

$$cov(vec(S)) = E(Z_1^2 Z_2^2)(\mathbf{I}_{p,p} + \mathbf{K}_{p,p})\boldsymbol{\Sigma} \otimes \boldsymbol{\Sigma}$$
$$+ (E(Z_1^2 Z_2^2) - 1)vec(\boldsymbol{\Sigma})vec(\boldsymbol{\Sigma})^T.$$

Therefore,

$$var(S_{ij}) = (\mathbf{e}_i \otimes \mathbf{e}_j)^T cov(vec(S))(\mathbf{e}_i \otimes \mathbf{e}_j)$$
$$= E(Z_1^2 Z_2^2)(\sigma_{ii}\sigma_{jj} + \sigma_{ij}^2) + (E(Z_1^2 Z_2^2) - 1)\sigma_{ij}^2$$

and

$$cov(S_{ij}, S_{rs}) = (\mathbf{e}_i \otimes \mathbf{e}_j)^T cov(vec(S))(\mathbf{e}_r \otimes \mathbf{e}_s)$$
$$= E(Z_1^2 Z_2^2)(\sigma_{ir}\sigma_{js} + \sigma_{jr}\sigma_{is}) + (E(Z_1^2 Z_2^2) - 1)\sigma_{ij}\sigma_{rs}.$$

For these and similar results, see Tyler (1982) and Theorem 3.2 in Oja (2010).

Consider next the limiting behavior of

$$\mathbf{R}_X = diag(\mathbf{S}_X)^{-1/2} \mathbf{S}_X \ diag(\mathbf{S}_X)^{-1/2}.$$

If $\mathbf{S}_X \to_P \boldsymbol{\Sigma}$ then naturally

$$\mathbf{R}_X \to_P \mathbf{P} = (diag\boldsymbol{\Sigma})^{-1/2} \boldsymbol{\Sigma} \ (diag\boldsymbol{\Sigma})^{-1/2}$$

where $\mathbf{P} = (\rho_{ij})$ is the population correlation matrix. Assume that $\mathbf{X}_1, ..., \mathbf{X}_n$ is a random sample from a p-variate distribution of $\mathbf{X}$ with mean vector μ and covariance matrix $\boldsymbol{\Sigma}$. As we are interested in the statistical behavior of the correlation matrix and the correlation matrix is invariant under rescaling of the components of $\mathbf{X}$, it is not a restriction to assume that $\mathbf{P} = \boldsymbol{\Sigma}$, and then $\boldsymbol{\Sigma}_{ij} = \rho_{ij}$, the correlation between X_i and X_j, $i,j = 1, ..., p$. As before, if the fourth moments exist,

$$\sqrt{n}(\mathbf{S}_X - \mathbf{P}) \to_d S,$$

where S is a multivariate normal matrix with zero expectation. Using Slutsky's theorem, one then sees that

$$\sqrt{n}(\mathbf{R}_X - \mathbf{P}) \to_d \mathcal{R} \sim S - \frac{1}{2}[\mathrm{diag}(S)\boldsymbol{\Sigma} + \boldsymbol{\Sigma}\mathrm{diag}(S)].$$

It then follows that, for all $i \neq j$,

$$\sqrt{n}(r_{ij} - \rho_{ij}) \to_d \mathcal{R}_{ij} \sim S_{ij} - \frac{\rho_{ij}}{2}(S_{ii} + S_{jj}).$$

In the multivariate normal case, using the results above,

$$\begin{pmatrix} S_{ii} \\ S_{jj} \\ S_{ij} \end{pmatrix} \sim N_3 \left(\begin{pmatrix} 0 \\ 0 \\ 0 \end{pmatrix}, \begin{pmatrix} 2 & 2\rho_{ij}^2 & 2\rho_{ij} \\ 2\rho_{ij}^2 & 2 & 2\rho_{ij} \\ 2\rho_{ij} & 2\rho_{ij} & 1+\rho_{ij}^2 \end{pmatrix} \right)$$

and one then sees, for example, that

$$\sqrt{n} \frac{\mathcal{R}_{ij} - \rho_{ij}}{1 - \rho_{ij}^2} \sim N(0, 1).$$

For more general results under the assumption of ellipticity, note first that

$$vec(\mathcal{R}) = \left[\mathbf{I}_{p,p} - \frac{1}{2}(\boldsymbol{\Sigma} \otimes \mathbf{I}_p + \mathbf{I}_p \otimes \boldsymbol{\Sigma})\mathbf{D}_{p,p} \right] vec(S)$$

and write $\mathbf{h}_{ij} = (\mathbf{e}_i \otimes \mathbf{e}_j - \frac{1}{2}\sigma_{ij}(\mathbf{e}_i \otimes \mathbf{e}_i + \mathbf{e}_j \otimes \mathbf{e}_j)^T$. Then

$$var(\mathcal{R}_{ij}) = \mathbf{h}_{ij}^T cov(S)\mathbf{h}_{ij} = E(Z_1^2 Z_2^2)(1 - \rho_{ij}^2)^2$$

and

$$cov(\mathcal{R}_{ij}, \mathcal{R}_{rs}) = \mathbf{h}_{ij}^T cov(S)\mathbf{h}_{rs}.$$

6.6   First Uses of Covariance and Correlation Matrices

Assume next that $\mathbf{X} \sim N_p(\mu, \boldsymbol{\Sigma})$ with $E(\mathbf{X}) = \mu$ and $cov(\mathbf{X}) = \boldsymbol{\Sigma}$. The set

$$\{\mathbf{x} \in \mathbb{R}^p \ : \ (\mathbf{x} - \mu)^T \boldsymbol{\Sigma}^{-1}(\mathbf{x} - \mu) \leq \chi_p^2(\alpha)\}$$

gives a $100(1 - \alpha)\%$ *tolerance region* (interval/ellipse/ellipsoid) for the random variable $\mathbf{X}$. Here $\chi_p^2(\alpha)$ is the $100(1 - \alpha)$-quantile of the χ^2-distribution with p degrees of freedom. Tolerance ellipses describe the properties of a multivariate normal distribution (location, dispersion, and shape) in a sufficient way. (In the elliptic case, an approximate tolerance ellipsoid is obtained in the same way just by replacing $\chi_p^2(\alpha)$ by the $(1 - \alpha)$-quantile of the distribution of $\|\mathbf{Z}\|^2$.) If $\mathbf{x}_1, ..., \mathbf{x}_n$ is the observed random sample from $N_p(\mu, \mathbf{\Sigma})$, the corresponding estimated tolerance ellipsoids are given by

$$\{\mathbf{x} \in \mathbf{R}^p \ : \ (\mathbf{x} - \bar{\mathbf{x}})^T \mathbf{S}^{-1}(\mathbf{x} - \bar{\mathbf{x}}) \le \chi_p^2(\alpha)\}.$$

With these ellipsoids, when plotted together with the data points, one can graphically check the multinormality assumption for the data and, for example, nicely display the (location, dispersion, and shape) differences between the distributions in groups.

Standardized observations can be used to find *Mahalanobis distances* and multivariate *skewness and kurtosis statistic*s in the following way: let $\mathbf{X} = (\mathbf{x}_1, ..., \mathbf{x}_n)^T$ be an observed random sample (now given as an $n \times p$ matrix) from a p-variate distribution and let

$$\bar{\mathbf{x}} \quad \text{and} \quad \mathbf{S}$$

be the sample mean vector and sample covariance matrix, respectively. Then the data matrix of standardized observations is

$$\mathbf{Z} = (\mathbf{z}_1, ..., \mathbf{z}_n)^T = \left(\mathbf{I}_n - \frac{1}{n}\mathbf{J}_n\right)\mathbf{X}\mathbf{S}^{-1/2}.$$

Consider next the $n \times n$ matrix

$$\mathbf{D} = \mathbf{Z}\mathbf{Z}^T = (\mathbf{z}_i^T \mathbf{z}_j)_{i,j=1,...,n}.$$

Then

1. The matrix $\mathbf{D} = \mathbf{D}(\mathbf{X})$ is invariant under affine transformations, that is,

$$\mathbf{D}(\mathbf{X}\mathbf{A}^T + \mathbf{1}_n\mathbf{b}^T) = \mathbf{D}(\mathbf{X}).$$

   In fact, it is even *maximal invariant*.

2. Diagonal elements of D_{ii} give the squared Mahalanobis distances between $\mathbf{x}_i$ and $\bar{\mathbf{x}}$, $i = 1, ..., n$.

3. Squared Mahalanobis distances between $\mathbf{x}_i$ and $\mathbf{x}_j$ are given by $d_{ii} + d_{jj} - 2d_{ij}$, $i, j = 1, ..., n$.

4. Mahalanobis angles between $\mathbf{x}_i$ and $\mathbf{x}_j$ are given by $d_{ij}/\sqrt{d_{ii}d_{jj}}$, $i, j, = 1, ..., n$.

5. Mardia's (Mardia 1970) multivariate measures of skewness and kurtosis are

$$\frac{1}{n^2} \sum_{i=1}^{n} \sum_{j=1}^{n} d_{ij}^3 \quad \text{and} \quad \frac{1}{n} \sum_{i=1}^{n} d_{ii}^2,$$

respectively. Being functions of $\mathbf{D}$ only, these measures are invariant under affine transformations.

6.7 Working with the Covariance Matrix–Principal Component Analysis

6.7.1 Principal variables

In the principal component analysis one considers linear combinations

$$\mathbf{v}^T \mathbf{X} = v_1 X_1 + v_2 X_2 + \cdots + v_p X_p,$$

with the constraint $\mathbf{v}'\mathbf{v} = 1$. The general idea then is to replace the original variables by a (much) smaller number of variables (linear combinations $\mathbf{v}_1^T \mathbf{X}, \ldots, \mathbf{v}_k^T \mathbf{X}$) without losing too much information. In this procedure, it is then assumed that *"the more variation the more information"*. Assume that $cov(\mathbf{X}) = \mathbf{\Sigma}$. The principal variables are found one-by-one as follows.

1. Find $\mathbf{v}_1$, satisfying $\mathbf{v}_1^T \mathbf{v}_1 = 1$, such that the variance of $Z_1 = \mathbf{v}_1^T \mathbf{X}$ (i.e., $\mathbf{v}_1^T \mathbf{\Sigma} \mathbf{v}_1$) is as large as possible.

2. Find a vector $\mathbf{v}_2$, satisfying $\mathbf{v}_2^T \mathbf{v}_2 = 1$ and $\mathbf{v}_2^T \mathbf{v}_1 = 0$, such that the variance of $Z_2 = \mathbf{v}_2^T \mathbf{X}$ (i.e., $\mathbf{v}_2^T \mathbf{\Sigma} \mathbf{v}_2$) is as large as possible.

3. Find a vector $\mathbf{v}_3$, satisfying $\mathbf{v}_3^T \mathbf{v}_3 = 1$, $\mathbf{v}_3^T \mathbf{v}_1 = 0$, and $\mathbf{v}_3^T \mathbf{v}_2 = 0$, such that the variance of $Z_3 = \mathbf{v}_3^T \mathbf{X}$ (i.e., $\mathbf{v}_3^T \mathbf{\Sigma} \mathbf{v}_3$) is as large as possible.

4. Continue in the same way until the number of variables is p.

New variables $Z_1, Z_2, \ldots, Z_p$ are called *principal variables (components)* and, typically, only a few first variables are used in the future analysis (dimension reduction). As $\mathbf{v}_k$ maximizes $\mathbf{v}^T \mathbf{\Sigma} \mathbf{v}$ under the constraints $\mathbf{v}^T \mathbf{\Sigma} \mathbf{v}_j = \delta_{kj}, j = 1, \ldots, k$, the solution can be found via optimizing the Lagrangian function

$$D(\mathbf{v}, \lambda_k) = \mathbf{v}^T \mathbf{\Sigma} \mathbf{v} - \sum_{j=1}^{k} \lambda_{kj} (\mathbf{v}^T \mathbf{\Sigma} \mathbf{v}_j - \delta_{kj})$$

with $\lambda_k = (\lambda_{k1}, \ldots, \lambda_{kk})$. Here $\delta_{kj} = 1$ (0) if $k = j$ ($k \neq j$) is the so-called Kronecker delta.

Next write $\mathbf{V} = (\mathbf{v}_1, ..., \mathbf{v}_p)$ and let

$$\boldsymbol{\Sigma} = \mathbf{UDU}^T = \sum_{i=1}^{p} d_i \mathbf{u}_i \mathbf{u}_i^T,$$

$d_1 \geq \cdots \geq d_p > 0$ be the eigenvector–eigenvalue decomposition of $\boldsymbol{\Sigma}$. For the unique-ness of this decomposition, assume that the eigenvalues are distinct so that $d_1 > \cdots > d_p$. Then $\mathbf{V} = \mathbf{U}$ and $cov(\mathbf{Z}) = \mathbf{D}$. This is seen as follows:

1. $\mathbf{v}_1$ maximizes

$$\mathbf{v}_1^T \boldsymbol{\Sigma} \mathbf{v}_1 = \mathbf{v}_1^T \left(\sum_{i=1}^{p} d_i \mathbf{u}_i \mathbf{u}_i^T \right) \mathbf{v}_1 = \sum_{i=1}^{p} d_i (\mathbf{v}_1^T \mathbf{u}_i)^2.$$

Write $w_i = (\mathbf{v}_1^T \mathbf{u}_i)^2, i = 1, ..., p$. Then $\sum_{i=1}^{p} w_i = 1$ and $\sum_{i=1}^{p} d_i w_i$ is maximized if $w_1 = 1$, that is, if $\mathbf{v}_1 = \mathbf{u}_1$.

2. $\mathbf{v}_2$ maximizes

$$\mathbf{v}_2^T \boldsymbol{\Sigma} \mathbf{v}_2 = \sum_{i=1}^{p} d_i (\mathbf{v}_2^T \mathbf{u}_i)^2$$

under the constraint $\mathbf{v}_2^T \mathbf{u}_1 = 0$. Write now $w_i = (\mathbf{v}_2^T \mathbf{u}_i)^2$, $i = 1, ..., p$. Then again $\sum_{i=1}^{p} w_i = 1$ and $\sum_{i=1}^{p} d_i w_i$ is maximized under the constraint $w_1 = 0$ if $w_2 = 1$, that is, if $\mathbf{v}_2 = \mathbf{u}_2$.

3. Continue in the same way for $\mathbf{v}_3, ..., \mathbf{v}_{p-1}$.

Write

$$\mathbf{Z} = \begin{pmatrix} Z_1 \\ Z_2 \\ ... \\ Z_p \end{pmatrix} = \begin{pmatrix} \mathbf{u}_1^T \mathbf{X} \\ \mathbf{u}_2^T \mathbf{X} \\ ... \\ \mathbf{u}_p^T \mathbf{X} \end{pmatrix} = \mathbf{U}^T \mathbf{X}.$$

The covariance matrix of $\mathbf{Z}$ is a diagonal matrix

$$cov(\mathbf{Z}) = \mathbf{U}^T \boldsymbol{\Sigma} \mathbf{U} = \mathbf{U}^T \mathbf{UDU}^T \mathbf{U} = \mathbf{D},$$

where the diagonal elements of $\mathbf{D}$, also called the *principal values*, satisfy $d_1 \geq \cdots \geq d_p > 0$. The principal variables are thus uncorrelated (and independent in the multi-variate normal case).

The total variation of $\mathbf{X}$ (and $\mathbf{Z}$) is measured by

$$E \left(\|\mathbf{X} - E(\mathbf{X})\|^2 \right) = \sum_{i=1}^{p} E[(X_i - E(X_i))^2] = \text{tr}(\boldsymbol{\Sigma}) = \text{tr}(\mathbf{D}) = d_1 + \cdots + d_p.$$

The ith principal variable explains

$$100 \times \frac{d_i}{d_1 + \cdots + d_p} \; \%$$

of the variation. The proportions

$$\frac{d_1}{d_1 + \cdots + d_p}, \quad \frac{d_1 + d_2}{d_1 + \cdots + d_p}, \quad \frac{d_1 + d_2 + d_3}{d_1 + \cdots + d_p}, \quad \ldots$$

yield the proportion explained by the first, the two first, the three first, etc., principal components. These proportions are often graphically described with so-called *scree plots*.

6.7.2   Interpretation of principal components

New variables are naturally hoped to be easily interpretable. In principal component analysis (PCA), the matrix

$$\mathbf{U} = (\mathbf{u}_1, \ldots, \mathbf{u}_p)$$

is also called a *matrix of loadings*, and

$$\mathbf{X} = \begin{pmatrix} X_1 \\ X_2 \\ \ldots \\ X_p \end{pmatrix} = Z_1 \mathbf{u}_1 + Z_2 \mathbf{u}_2 + \cdots + Z_p \mathbf{u}_p.$$

$Z_1, \ldots, Z_p$ give the values of $\mathbf{X}$ in a new (rotated) coordinate system with coordinate axes $\mathbf{u}_1, \mathbf{u}_2, \ldots, \mathbf{u}_p$. In other words, the rotation produces new uncorrelated random variables $Z_1, \ldots, Z_p$.

The interpretation of the $\mathbf{Z}$ variables can be made by using the matrix of loadings or by considering the correlations between the $\mathbf{Z}$-variables and $\mathbf{X}$-variables, that is, by looking at the correlation matrix

$$corr(\mathbf{Z}, \mathbf{X}) = (corr(Z_i, X_j))_{i,j=1,\ldots,p} = \mathbf{D}\mathbf{U}^T.$$

Naturally, the estimated eigenvectors and eigenvalues of Σ (and estimated principal components and principal values) are obtained simply by replacing in the above derivations the population covariance matrix Σ by the sample covariance matrix $\mathbf{S}_X$.

Warning: the solution (principal variables) depends on the measurement units used for original variables $X_1, \ldots, X_p$. To circumvent this problem, the covariance matrix is often replaced by the correlation matrix in the analysis. Then the measurement unit for each variable is its own standard deviation. The eigenvectors and eigenvalues based on the covariance matrix are rotation equivariant and invariant, respectively, but this is no longer true for the estimates based on the correlation matrix. On the other hand, the estimates based on the correlation matrix are invariant under the heterogeneous rescaling of the components. The use of the correlation matrix is recommended if the measurement units for the marginal variables are chosen in an arbitrary way.

6.7.3 Asymptotic behavior of the eigenvectors and eigenvalues

As the eigenvectors and eigenvalues of S_X are rotation equivariant and invariant, respectively, it is not a restriction to assume that $\Sigma = \Lambda$ and

$$\sqrt{n}(S_X - \Lambda) \to_d S,$$

where $\Lambda = \text{diag}(\lambda_1, ..., \lambda_p)$ is a diagonal matrix with distinct diagonal elements in a decreasing order. If $X_1, ..., X_n$ comes from an elliptical distribution of $X = \Sigma^{1/2} Z + \mu$, Z as above, then

$$cov(vec(S)) = E(Z_1^2 Z_2^2)(I_{p,p} + K_{p,p})\Lambda \otimes \Lambda$$
$$+ (E(Z_1^2 Z_2^2) - 1)vec(\Lambda)vec(\Lambda)^T$$

and the variables S_{ij}, $i < j$, are independent and

$$var(S_{ij}) = E(Z_1^2 Z_2^2)\lambda_i \lambda_j.$$

Let $\hat{U}$ and $\hat{D}$ be the matrix of eigenvectors and the (diagonal) matrix of eigenvalues of S_X, respectively. Then

$$\sqrt{n}(\hat{U} - I_p) \to_d U \quad \text{and} \quad \sqrt{n}(\hat{D} - \Lambda) \to_d D,$$

where the distributions of U and D are obtained via

$$S \sim U\Lambda - \Lambda U + D.$$

Then

$$U_{ij} \sim \frac{1}{\lambda_i - \lambda_j} S_{ij}, \quad i \neq j,$$

and

$$U_{ii} \sim 0 \quad \text{and} \quad D_{ii} \sim S_{ii}, \qquad i = 1, ..., p.$$

Thus U_{ij}, $i < j$, are independent and

$$var(U_{ij}) = E(Z_1^2 Z_2^2)\frac{\lambda_i \lambda_j}{(\lambda_i - \lambda_j)^2}.$$

Thus the limiting distribution of the eigenvectors and eigenvalues can be easily found from that of the covariance matrix. In the general case $\Sigma = \mathbf{U}\Lambda\mathbf{U}^T$, the above formulas still hold with

$$\sqrt{n}\mathbf{U}^T(\hat{\mathbf{U}} - \mathbf{U}) \to_d \mathcal{U} \quad \text{and} \quad \sqrt{n}(\hat{\mathbf{D}} - \Lambda) \to_d \mathcal{D}.$$

For the eigenvectors and eigenvalues of the correlation matrix, we refer to Proposition 10.3 in Bilodeau and Brenner (1999).

6.8   Working with Correlations–Canonical Correlation Analysis

6.8.1   Canonical variates and canonical correlations

The purpose of canonical correlation analysis (CCA) is to describe the linear interrelations between two random multivariate vectors, $\mathbf{X}_1$ and $\mathbf{X}_2$. New coordinate systems are found for both vectors in such a way that, in both systems, the random vectors are standardized and that the covariance matrix between the two random vectors is a diagonal matrix with descending positive diagonal elements. The new variables and their correlations are called *canonical variates* and *canonical correlations*, respectively. Moreover, the rows of the transformation matrix are called canonical vectors. Canonical analysis is one of the fundamental contributions to multivariate inference by Harold Hotelling (Hotelling 1936).

The aim of canonical correlation analysis is to consider linear dependence between the sets of variables, $\mathbf{X}_1 = (X_1, ..., X_r)^T$ and $\mathbf{X}_2 = (X_{r+1}, ..., X_{r+s})^T$. One then considers linear combinations

$$\mathbf{a}^T\mathbf{X}_1 = a_1 X_1 + ... + a_r X_r \quad \text{and} \quad \mathbf{b}^T\mathbf{X}_2 = b_1 X_{r+1} + ... + b_s X_{r+s}$$

such that $var(\mathbf{a}^T\mathbf{X}_1) = 1$ and $var(\mathbf{b}^T\mathbf{X}_2) = 1$. The general idea is again to replace the original variables $X_1, ..., X_r$ and $X_{r+1}, ..., X_{r+s}$ by a (much) smaller number of variables (linear combinations $\mathbf{a}^T\mathbf{X}_1$ and $\mathbf{b}^T\mathbf{X}_2$) without losing any information about the dependence between the two sets of variables.

One then proceeds as follows.

1. Find vectors $\mathbf{a}_1$ and $\mathbf{b}_1$ such that $var(\mathbf{a}_1^T\mathbf{X}_1) = var(\mathbf{b}_1^T\mathbf{X}_2) = 1$ and $cov(\mathbf{a}_1^T\mathbf{X}_1, \mathbf{b}_1^T\mathbf{X}_2)$ is as large as possible.

2. Find vectors $\mathbf{a}_2$ and $\mathbf{b}_2$ such that $var(\mathbf{a}_2^T\mathbf{X}_1) = var(\mathbf{b}_2^T\mathbf{X}_2) = 1$, $cov(\mathbf{a}_2^T\mathbf{X}_1, \mathbf{a}_2^T\mathbf{X}_1) = cov(\mathbf{b}_1^T\mathbf{X}_2, \mathbf{b}_2^T\mathbf{X}_2) = 0$, and $cov(\mathbf{a}_2^T\mathbf{X}_1, \mathbf{b}_2^T\mathbf{X}_2)$ is as large as possible.

3. Find vectors $\mathbf{a}_3$ and $\mathbf{b}_3$ such that $var(\mathbf{a}_3^T\mathbf{X}_1) = var(\mathbf{b}_3^T\mathbf{X}_2) = 1$, $\mathbf{a}_1^T\mathbf{X}_1$, $\mathbf{a}_2^T\mathbf{X}_1$, and $\mathbf{a}_3^T\mathbf{X}_1$ are uncorrelated, $\mathbf{b}_1^T\mathbf{X}_2$, $\mathbf{b}_2^T\mathbf{X}_2$, and $\mathbf{b}_3^T\mathbf{X}_2$ are uncorrelated, and $cov(\mathbf{a}_3^T\mathbf{X}_1, \mathbf{b}_3^T\mathbf{X}_2)$ is as large as possible.

4. Continue in the same way until the number of pairs of variables is $\min\{r, s\}$.

New variables $\mathbf{a}_1^T\mathbf{X}_1, \mathbf{a}_2^T\mathbf{X}_1, \ldots$ and $\mathbf{b}_1^T\mathbf{X}_2, \mathbf{b}_2^T\mathbf{X}_2, \ldots$ are called *canonical variables (components)* and, typically, only the few first pairs of variables $(\mathbf{a}^T\mathbf{X}_1, \mathbf{b}^T\mathbf{X}_2)$ are used in the future analysis (dimension reduction). The interpretations of new variables can again be found by considering their loadings or correlations between the original variables and the canonical variables.

The limiting distribution of the canonical correlations in the multivariate normal case was derived in Hsu (1941). His result is valid under very general assumptions on the population canonical correlations. The limiting distributions of the canonical vectors have been considered in several papers. It was only quite recently that Anderson (1999) completed the asymptotic theory for canonical correlation analysis based on the sample covariance matrix in the multivariate normal case.

Anderson (1999) gave the complete limiting distributions of the canonical correlations and vectors assuming that the population correlations are distinct, that is, $\rho_1 > \cdots > \rho_p > 0$. He showed among other things that if $p = q$, then for $i = 1, \ldots, p$, the marginal distributions of $\sqrt{n}(\hat{r}_i - \rho_i)$, $\sqrt{n}(\hat{\mathbf{a}}_i - \mathbf{a}_i)$, and $\sqrt{n}(\hat{\mathbf{b}}_i - \mathbf{b}_i)$ are asymptotically normal with zero mean and asymptotic variances

$$\text{ASV}(\hat{r}_i) = (1 - \rho_i)^2,$$

$$\text{ASV}(\hat{\mathbf{a}}_i) = \frac{1}{2}\mathbf{a}_i\mathbf{a}_i^T + \sum_{\substack{k=1 \\ k\neq i}}^{p} \frac{(\rho_k^2 + \rho_i^2 - 2\rho_k^2\rho_i^2)(1 - \rho_i^2)}{(\rho_i^2 - \rho_k^2)^2}\mathbf{a}_k\mathbf{a}_k^T,$$

and

$$\text{ASV}(\hat{\mathbf{b}}_i) = \frac{1}{2}\mathbf{b}_i\mathbf{b}_i^T + \sum_{\substack{k=1 \\ k\neq i}}^{p} \frac{(\rho_k^2 + \rho_i^2 - 2\rho_k^2\rho_i^2)(1 - \rho_i^2)}{(\rho_i^2 - \rho_k^2)^2}\mathbf{b}_k\mathbf{b}_k^T.$$

Write now

$$E\left(\begin{pmatrix}\mathbf{X}_1 \\ \mathbf{X}_2\end{pmatrix}\right) = \begin{pmatrix}\mu_1 \\ \mu_2\end{pmatrix} \quad \text{and} \quad cov\left(\begin{pmatrix}\mathbf{X}_1 \\ \mathbf{X}_2\end{pmatrix}\right) = \begin{pmatrix}\mathbf{\Sigma}_{11} & \mathbf{\Sigma}_{12} \\ \mathbf{\Sigma}_{21} & \mathbf{\Sigma}_{22}\end{pmatrix}.$$

The canonical variables and correlations can also be found as follows.

1. Standardize $\mathbf{X}_1$ and $\mathbf{X}_2$:

$$\mathbf{X}_1 \to \mathbf{\Sigma}_{11}^{-1/2}(\mathbf{X}_1 - \mu_1) \quad \text{and} \quad \mathbf{X}_2 \to \mathbf{\Sigma}_{22}^{-1/2}(\mathbf{X}_2 - \mu_2).$$

2. Let $\mathbf{K} = \boldsymbol{\Sigma}_{11}^{-1/2} \boldsymbol{\Sigma}_{12} \boldsymbol{\Sigma}_{22}^{-1/2}$ be the covariance matrix between the standardized vectors, that is, the correlation matrix between the original vectors. Find the singular value decomposition (SVD)

$$\mathbf{K} = \mathbf{UDV}^T.$$

3. Rotate standardized vectors

$$\boldsymbol{\Sigma}_{11}^{-1/2}(\mathbf{X}_1 - \boldsymbol{\mu}_1) \to \mathbf{U}^T \boldsymbol{\Sigma}_{11}^{-1/2}(\mathbf{X}_1 - \boldsymbol{\mu}_1)$$

and

$$\boldsymbol{\Sigma}_{22}^{-1/2}(\mathbf{X}_2 - \boldsymbol{\mu}_2) \to \mathbf{V}^T \boldsymbol{\Sigma}_{22}^{-1/2}(\mathbf{X}_2 - \boldsymbol{\mu}_2)$$

to get canonical variables. The canonical correlations are in $\mathbf{D}$.

Assume now, without loss of generality, that $\mathbf{X}_1, ..., \mathbf{X}_n$ is a random sample from a $p + q$-variate distribution with zero mean vector and covariance matrix

$$\boldsymbol{\Sigma} = \begin{pmatrix} \boldsymbol{\Sigma}_{11} & \boldsymbol{\Sigma}_{12} \\ \boldsymbol{\Sigma}_{21} & \boldsymbol{\Sigma}_{22} \end{pmatrix} = \begin{pmatrix} \mathbf{I}_p & \boldsymbol{\Lambda} \\ \boldsymbol{\Lambda} & \mathbf{I}_q \end{pmatrix},$$

where $\boldsymbol{\Lambda} = (\boldsymbol{\Lambda}_0, \mathbf{0})$ with a $p \times p$ diagonal matrix $\boldsymbol{\Lambda}_0$. Assume also that

$$\sqrt{n}(\mathbf{S} - \boldsymbol{\Sigma}) \to_d S = \begin{pmatrix} S_{11} & S_{12} \\ S_{21} & S_{22} \end{pmatrix},$$

where $\text{vec}(S)$ has a multivariate normal distribution with zero mean vector and covariance matrix as above. If

$$\widehat{\mathbf{K}} = \mathbf{S}_{11}^{-1/2} \mathbf{S}_{12} \mathbf{S}_{22}^{-1/2},$$

then, using Slutsky's theorem,

$$\sqrt{n}(\widehat{\mathbf{K}} - \boldsymbol{\Lambda}) \to_d \mathcal{K} \sim S_{12} - \frac{1}{2} S_{11} \boldsymbol{\Lambda} - \frac{1}{2} \boldsymbol{\Lambda} S_{22}.$$

This can further be used to find the limiting distributions of the eigenvalues of $\widehat{\mathbf{K}}\widehat{\mathbf{K}}^T$ and $\widehat{\mathbf{K}}^T\widehat{\mathbf{K}}$, those of $(\mathcal{U}_1, \mathcal{U}_2)$, say. Finally, the limiting distributions of $\sqrt{n}(\widehat{\mathbf{A}} - \mathbf{A})$ and $\sqrt{n}(\widehat{\mathbf{B}} - \mathbf{B})$ are those by

$$\mathcal{A} \sim \mathcal{U}_1^T - \frac{1}{2} S_{11} \quad \text{and} \quad \mathcal{B} \sim \mathcal{U}_2^T - \frac{1}{2} S_{22}.$$

6.8.2   Testing for independence between subvectors

In the bivariate normal case ($p = q = 1$) with the population and sample covariance matrices

$$\boldsymbol{\Sigma} = \begin{pmatrix} \sigma_{11} & \sigma_{12} \\ \sigma_{21} & \sigma_{22} \end{pmatrix} \quad \text{and} \quad \mathbf{S} = \begin{pmatrix} s_{11} & s_{12} \\ s_{21} & s_{22} \end{pmatrix},$$

respectively, a natural test statistic for testing H_0: $\rho = 0$ (independence) is the Pearson product moment correlation coefficient

$$r = \frac{s_{12}}{\sqrt{s_{11}s_{22}}},$$

that is, the maximum likelihood estimator of

$$\rho = \frac{\sigma_{12}}{\sqrt{\sigma_{11}\sigma_{22}}}.$$

Under H_0 and under ellipticity, $\sqrt{n}\ r \to_d N(0,1)$.

Assume next that $p \geq 2$ and $q \geq 2$ and the population and sample covariance matrices are

$$\Sigma = \begin{pmatrix} \Sigma_{11} & \Sigma_{12} \\ \Sigma_{21} & \Sigma_{22} \end{pmatrix} \quad \text{and} \quad S = \begin{pmatrix} S_{11} & S_{12} \\ S_{21} & S_{22} \end{pmatrix},$$

respectively. Wilks (1935) showed that the likelihood ratio test statistic for testing H_0: $\Sigma_{12} = 0$ (independence) is

$$W = \frac{\det(S)}{det(S_{11})\, det(S_{22})}.$$

Under H_0,

$$-n \log\ W \to_d \chi^2_{pq}.$$

Note that when W is used in testing independence in non-normal populations, one needs to assume that the fourth moments of underlying distribution are finite. Another classical test of independence is the Pillai trace (Pillai 1955)

$$P = tr(S_{11}^{-1} S_{12} S_{22}^{-1} S_{21}),$$

which is asymptotically equivalent with Wilks' test, that is, $n(P + \log\ W) \to_p 0$.

Muirhead (1982) showed that Wilks' test W is affine invariant, that is, its value is not changed under the group of transformations

$$\mathcal{G} = \{X_i \to AX_i + b\},$$

where b is a $p + q$ vector and

$$A = \begin{pmatrix} A_1 & 0 \\ 0 & A_2 \end{pmatrix}$$

is a block diagonal matrix with nonsingular $p \times p$ and $q \times q$ matrices A_1 and A_2. Invariance implies that the performance of W does not depend on the underlying covariance structures of X_1 and X_2. Also Pillai's trace is naturally affine invariant under $\mathcal{G}$.

Assumptions are as above with $p = q$. We considered tests for the null hypothesis H_0: $\Sigma_{12} = 0$. Since, in the canonical correlation analysis, the matrices $\mathbf{A}$, $\mathbf{B}$, and $\mathbf{R}$ satisfy

$$\mathbf{A}^T \Sigma_{12} \mathbf{B} = (\mathbf{R}, \mathbf{0}),$$

the hypothesis can also be written as

$$H_0: \rho_1 = \cdots = \rho_p = 0,$$

and the test statistics given in Wilks and Pillai can be used for testing the canonical correlations also. Note that Wilks' test statistic can be expressed in terms of sample canonical correlations as

$$W = \frac{det(\mathbf{S})}{det(\mathbf{S}_{11}) \, det(\mathbf{S}_{22})} = \prod_{i=1}^{p} (1 - \hat{r}_i^{\,2})$$

and Pillai's trace becomes

$$P = tr(\mathbf{S}_{11}^{-1} \mathbf{S}_{12} \mathbf{S}_{22}^{-1} \mathbf{S}_{21}) = \sum_{i=1}^{p} \hat{r}_i^{\,2}.$$

Muirhead (1982) showed that under the group of transformations $\mathcal{G}$, any invariant test is a function of squared sample canonical correlations. The invariance of the test statistics thus follows.

If H_0: $\rho_1 = \cdots = \rho_p = 0$ is rejected, it may be of interest to study how many population canonical correlations differ from 0, that is, how many canonical correlations are needed to describe the relationships between $\mathbf{X}_1$ and $\mathbf{X}_2$. It can be shown that the likelihood ratio test for testing H_0': $\rho_{k+1} = \cdots = \rho_p = 0$ is based on

$$W' = \prod_{i=k+1}^{p} (1 - \hat{r}_i^{\,2}).$$

Under H_0', $-n \, \log \, W' \to_d \chi^2_{(p-k)(q-k)}$.

6.9   Conditionally Uncorrelated Components

The partial correlation coefficient between three variables X and Y given Z is

$$\rho_{XY \cdot Z} = \frac{\rho_{XY} - \rho_{XZ} \rho_{YZ}}{\sqrt{1 - \rho_{XZ}^2} \sqrt{1 - \rho_{YZ}^2}}.$$

For a vector valued $\mathbf{Z}$, we write

$$\rho_{XY \cdot \mathbf{Z}} = cor(X - E(X|\mathbf{Z}), Y - E(Y|\mathbf{Z})).$$

Let $\mathbf{X} = (X_1, ..., X_p)$ be a vector valued random variable. Then the partial correlation between X_i and X_j given the rest of the variables is $-(P^{-1})_{ij}$, $i \neq j$. As

$$\sqrt{n}(\mathbf{R} - \mathbf{P}) \to_d \mathcal{R} = S - \frac{1}{2}[\text{diag}(S)\Sigma + \Sigma\text{diag}(S)],$$

we obtain

$$\sqrt{n}(-\mathbf{R}^{-1} + \mathbf{P}^{-1}) \to_d \mathbf{P}^{-1}\mathcal{R}\mathbf{P}^{-1}.$$

The partial correlation matrix can be estimated by the negative inverse of the correlation matrix. Partial correlations may be estimated also one-by-one using linear regression as guided by its definition. Note that partial correlation forms the main tools in *graphical models*.

6.10   Summary

This chapter provides an overview of multivariate correlation measures based on the covariance and correlation matrix of a *p*-variate random vector. The classical principal component analysis (PCA), canonical correlation analysis (CCA), and testing for independence between subvectors are described in more detail. Asymptotic behavior of the related tests and estimates are discussed as well. For further discussion and theory, see, for example, Anderson (2003), Bilodeau and Brenner (1999), Johnson and Wichern (1998), and Mardia *et al.* (1979).

References

Anderson TW 1999 Asymptotic theory for canonical correlation analysis. *J. Multivariate Analysis* **70**, 1–29.

Anderson TW 2003 *An Introduction to Multivariate Statistical Analysis*. Third Edition, Wiley, New York.

Bilodeau M and Brenner D 1999 *Theory of Multivariate Statistics*. Springer-Verlag, New York.

Hotelling H 1936 Relations between two sets of variables. *Biometrika* **28**, 321–377.

Hsu PL 1941 On the limiting distribution of the canonical correlations. *Biometrika* **32**, 38–45.

Hyvärinen A, Karhunen J, and Oja E 2001 *Independent Component Analysis*. Wiley, New York.

Johnson RA and Wichern DW 1998 *Multivariate Statistical Analysis*. Prentice Hall, Upper Saddle River, NJ.

Mardia KV 1970 Measures of multivariate skewness and kurtosis with applications. *Biometrika* **57**, 519–530.

Mardia KV, Kent JT, and Bibby JM 1979 *Multivariate Analysis*. Academic Press, Orlando, FL.

Muirhead RJ 1982 *Aspects of Multivariate Statistical Theory*. Wiley, New York.

Nelsen RB 1999 *An Introduction to Copulas*. Springer, New York.

Oja H 2010 *Multivariate Nonparametric Methods with R*. Springer, New York.

Pillai KCS 1955 Some new test criteria in multivariate analysis. *Annals of Mathematical Statistics* **26**, 117–121.

Tyler DE 1982 Radial estimates and the test for the sphericity. *Biometrika* **69**, 429–436.

Wilks SS 1935 On the independence of k sets of normally distributed statistical variates. *Econometrica* **3**, 309–326.

7

Robust Estimation of Scatter and Correlation Matrices

In this chapter, correlation measures and inference tools that are based on various robust covariance matrix functionals and estimates are discussed.

7.1 Preliminaries

In the previous chapters we considered pairwise correlation functionals and estimates, that is, separate functionals and estimates for all pairs of random variables X_i and X_j, $i \neq j = 1, \ldots, p$. The pairwise correlation values can then of course be collected as a correlation matrix but the resulting matrix is not necessarily positive definite. One possibility is to project the observed projection matrix to the space of positive definite matrices. The matrix may then lose the scale invariance property and its limiting distribution may be difficult to find.

In this chapter we adopt an alternative approach where we first find a robust estimate of the covariance (or scatter) matrix and then use this matrix to estimate the correlation matrix as well.

7.2 Multivariate Location and Scatter Functionals

The characteristics of univariate distributions most commonly considered are location, scale, skewness, and kurtosis. These concepts are often identified with the mean, standard deviation, and standardized third and fourth moments. Skewness and kurtosis are often seen only as secondary statistics indicating the stability of the primary

Robust Correlation: Theory and Applications, First Edition. Georgy L. Shevlyakov and Hannu Oja.
© 2016 John Wiley & Sons, Ltd. Published 2016 by John Wiley & Sons, Ltd.
Companion Website: www.wiley.com/go/Shevlyakov/Robust

statistics, location, and scale. Skewness and kurtosis are also used in (parametric) model selection. In parametric or semiparametric models one often has natural parameters for location and scale, that is, μ and Σ. In wide nonparametric models functionals for location and scale must be used instead. The functionals or measures are then supposed to satisfy certain natural equivariance or invariance properties.

We first define what we mean by a location vector and a scatter matrix defined as vector- and matrix-valued functionals in wide nonparametric families of multivariate distributions (often including empirical distributions as well). Let $\mathbf{X}$ be a p-variate random variable with cumulative distribution function (cdf) $F_{\mathbf{X}}$.

Definition 7.2.1.

   (i) A p-vector $\mathbf{T}(F)$ is a location vector *if it is affine equivariant; that is,*

$$\mathbf{T}(F_{\mathbf{AX+b}}) = \mathbf{AT}(F_{\mathbf{X}}) + \mathbf{b}$$

   for all random vectors $\mathbf{X}$, *all full-rank* $p \times p$-matrices $\mathbf{A}$, *and all* p-vectors $\mathbf{b}$.

   (ii) A symmetric $p \times p$ matrix $\mathbf{S}(F) \geq 0$ is a scatter matrix *if it is positive definite and affine equivariant in the sense that*

$$\mathbf{S}(F_{\mathbf{AX+b}}) = \mathbf{AS}(F_{\mathbf{X}})\mathbf{A}^T$$

   for all random vectors $\mathbf{X}$, *all full-rank* $p \times p$-matrices $\mathbf{A}$, *and all* p-vectors $\mathbf{b}$.

   (iii) A symmetric $p \times p$ matrix $\mathbf{V}(F) \geq 0$ is a shape matrix *if it is positive definite with* $det(\mathbf{V}) = 1$ *and affine equivariant in the sense that*

$$\mathbf{V}(F_{\mathbf{AX+b}}) \propto \mathbf{AV}(F_{\mathbf{X}})\mathbf{A}^T$$

   for all random vectors $\mathbf{X}$, *all full-rank* $p \times p$-matrices $\mathbf{A}$, *and all* p-vectors $\mathbf{b}$.

The mean vector and the covariance matrix

$$E(\mathbf{X}) \text{ and } cov(\mathbf{X}) = E[(\mathbf{X} - E(\mathbf{X}))(\mathbf{X} - E(\mathbf{X}))^T],$$

considered in detail in Chapter 6, are first examples of location vector and scatter matrix. If $\mathbf{S}(F)$ is a scatter functional,

$$\mathbf{V}(F_{\mathbf{X}}) = [det(\mathbf{S}(F_{\mathbf{X}}))]^{-1/p}\mathbf{S}(F_{\mathbf{X}})$$

is a related shape functional but the shape matrix can, however, be defined without any reference to a scatter matrix. The shape matrix can be seen as a "standardized" version of a scatter matrix. The condition $det(\mathbf{V}) = 1$ is sometimes replaced by the condition $tr(\mathbf{V}) = k$ (see Paindaveine 2008). Note that the covariance matrix provides a shape functional

$$\mathbf{V}(F_{\mathbf{X}}) = [det(cov(\mathbf{X}))]^{-1/p} cov(\mathbf{X})$$

The covariance matrix $cov(\mathbf{X})$ has the following important properties.

1. Affine equivariance: $cov(\mathbf{AX} + \mathbf{b}) = \mathbf{A}cov(\mathbf{X})\mathbf{A}^T$.

2. Additivity: for independent $\mathbf{X}_1$ and $\mathbf{X}_2$,
   $cov(\mathbf{X}_1 + \mathbf{X}_2) = cov(\mathbf{X}_1) + cov(\mathbf{X}_2)$.

3. Independence property: if $\mathbf{X}$ has independent components, then
   $cov(\mathbf{X})$ is diagonal.

The theory of location and scatter functionals has been developed mainly to find new tools for robust estimation of the regular mean vector and covariance matrix in a neighborhood of the multivariate normal model or in the wider model of elliptically symmetric distributions. The competitors of the regular covariance matrix are not additive and they do not usually have the independence property. Naturally the independence property is not important in the case of elliptical distributions as the multivariate normal distribution is the only elliptical distribution that can have independent margins. On the other hand, the independence property is crucial if one is working in the independent component model mentioned in Chapter 6 (the ICA problem).

Using the affine equivariance properties, one easily gets the following.

Theorem 7.2.1. *Assume that random p-vector* $\mathbf{Z}$ *has a spherical distribution, that is,* $\mathbf{UZ} \sim \mathbf{Z}$ *for all orthogonal matrices* $\mathbf{U}$. *Then, for all location vectors* $\mathbf{T}$, *for all scatter matrices* $\mathbf{S}$, *and for all shape matrices* $\mathbf{V}$,

$$\mathbf{T}(F_{\mathbf{Z}}) = \mathbf{0}, \quad \mathbf{S}(F_{\mathbf{Z}}) \propto \mathbf{I}_p, \ and \ \mathbf{V}(F_{\mathbf{Z}}) = \mathbf{I}_p.$$

The proof easily follows for the affine equivariance of location and scatter functionals. It also implies that if $\mathbf{X} \sim E_p(\mu, \mathbf{\Sigma}, \rho)$ is elliptic, then, again for all location vectors $\mathbf{T}$, for all scatter matrices $\mathbf{S}$, and for all shape matrices $\mathbf{V}$,

$$\mathbf{T}(F_{\mathbf{X}}) = \mu, \quad \mathbf{S}(F_{\mathbf{X}}) \propto \mathbf{\Sigma}, \ and \ \mathbf{V}(F_{\mathbf{X}}) = [det(\mathbf{\Sigma})]^{-1/p}\mathbf{\Sigma}.$$

Thus, in the elliptic case, all location vectors as well as all shape matrices estimate the same population quantities. The scatter matrices are not directly comparable, and a correction factor depending both on S and the spherical distribution of $\mathbf{Z}$ is needed for the Fisher-consistency to $\mathbf{\Sigma}$.

In several applications, like in principal component or canonical correlation analyses, it is sufficient to estimate the covariance matrix only up to a constant. The determinant det($\mathbf{S}$) or trace $tr(\mathbf{S})$ is often used as a global univariate measure of multivariate scatter. In fact, $[\det(\mathbf{S})]^{1/p}$ is the geometric mean and $tr(\mathbf{S})/p$ the arithmetic mean of the eigenvalues of $\mathbf{S}$. The functional $\det(cov(\mathbf{X}))$ is sometimes called the *generalized variance*; the functional $tr(cov(\mathbf{X}))$ may be seen as a multivariate extension of the variance as well because $tr(cov(\mathbf{X})) = E(\|\mathbf{X} - E(\mathbf{X})\|^2)$.

7.3   Influence Functions and Asymptotics

The influence function measures the robustness of a functional against a single outlier, that is, the effect of an infinitesimal contamination located at point $\mathbf{z}$. For a robust functional the influence function is hoped to be bounded and continuous. Consider now the contaminated distribution

$$F_\epsilon = (1 - \epsilon)F + \epsilon\Delta_{\mathbf{z}},$$

where $\Delta_{\mathbf{z}}$ denotes the CDF of a distribution placing all its mass at $\mathbf{z}$. The influence function (Hampel *et al.* 1986) of a functional $\mathbf{T}$ at F is then defined as

$$\mathrm{IF}(\mathbf{z}, \mathbf{T}, F) = \lim_{\epsilon \to 0} \frac{\mathbf{T}(F_\epsilon) - \mathbf{T}(F)}{\epsilon}.$$

Let now F be a spherical distribution and, write $\mathbf{z} = r\mathbf{u}$, where $r = \|\mathbf{z}\|$ is the length and $\mathbf{u} = \|\mathbf{z}\|^{-1}\mathbf{z}$ is the direction of the contamination vector $\mathbf{z}$. The influence functions of location, scatter, and shape functionals are derived in Hampel *et al.* (1986), Croux and Haesbrock (2000), and Ollila *et al.* (2003), for example. The influence function of location functional $\mathbf{T}$ is given by

$$\mathrm{IF}(\mathbf{z}, \mathbf{T}, F) = \gamma_T(r)\mathbf{u},$$

where $\gamma_T(r)$ is a real-valued function depending on $\mathbf{T}$ and F. Further, the influence functions of scatter and shape matrix functionals are

$$\mathrm{IF}(\mathbf{z}; \mathbf{S}, F) = \alpha_S(r)\mathbf{u}\mathbf{u}^T - \beta_S(r)\mathbf{I}_p$$

and

$$\mathrm{IF}(\mathbf{z}; \mathbf{V}, F) = \alpha_V(r)\left[\mathbf{u}\mathbf{u}^T - \frac{1}{p}\mathbf{I}_p\right]$$

for some real-valued functions $\alpha_S(r)$, $\beta_S(r)$, and $\alpha_V(r)$.

As an example of influence functions, consider the population mean vector and covariance matrix functionals

$$T(F) = E_F(\mathbf{X}) \quad \text{and} \quad S(F) = E_F[(\mathbf{X} - E_F(\mathbf{X}))(\mathbf{X} - E_F(\mathbf{X}))^T].$$

At spherical F, the corresponding influence functions are

$$\mathrm{IF}(\mathbf{z}, \mathbf{T}, F) = r\mathbf{u}, \quad \text{and} \quad \mathrm{IF}(\mathbf{z}; \mathbf{S}, F) = r^2\mathbf{u}\mathbf{u}^T - \frac{E_F(\|\mathbf{X}\|^2)}{p}\mathbf{I}_p.$$

Since the functions are linear and quadratic in r, the estimates are not robust against outliers.

Influence functions can also be used to compute the asymptotic variances of the estimates. If $\mathbf{X}_1, \ldots, \mathbf{X}_n$ is a random sample from a spherical distribution F with $\mathbf{T}(F) = \mathbf{0}$ and $\mathbf{S}(F) = \mathbf{I}_p$ and F_n is the corresponding empirical CDF, then natural location, scatter, and shape estimates are

$$\widehat{\mathbf{T}} = \mathbf{T}(F_n), \quad \widehat{\mathbf{S}} = \mathbf{S}(F_n), \text{ and } \widehat{\mathbf{V}} = \mathbf{V}(F_n).$$

Under general assumptions, the limiting distributions of $\sqrt{n}\widehat{\mathbf{T}}$, $\sqrt{n}\mathrm{vec}(\widehat{\mathbf{S}} - \mathbf{I}_p)$, and $\sqrt{n}\mathrm{vec}(\widehat{\mathbf{V}} - \mathbf{I}_p)$ are multivariate normal distributions with mean vectors zero and covariance matrices

$$\mathrm{ASV}(\widehat{\mathbf{T}}; F) = E_F[\mathrm{IF}(\mathbf{X}, \mathbf{T}, F)\mathrm{IF}(\mathbf{X}, \mathbf{T}, F)^T],$$

$$\mathrm{ASV}(\widehat{\mathbf{S}}; F) = E_F[\mathrm{vec}\{\mathrm{IF}(\mathbf{X}, \mathbf{S}, F)\}\mathrm{vec}\{\mathrm{IF}(\mathbf{X}, \mathbf{S}, F)\}^T],$$

and

$$\mathrm{ASV}(\widehat{\mathbf{V}}; F) = E_F[\mathrm{vec}\{\mathrm{IF}(\mathbf{X}, \mathbf{V}, F)\}\mathrm{vec}\{\mathrm{IF}(\mathbf{X}, \mathbf{V}, F)\}^T].$$

To simplify the expressions we can now use the following result. Note that, if $\sqrt{n}\widehat{\mathbf{T}} \to_d \mathcal{T}$ and $\sqrt{n}\mathrm{vec}(\widehat{\mathbf{S}} - \mathbf{I}_p) \to \mathcal{S}$ then, due to affine equivariance, also $\mathcal{T} \sim \mathbf{U}\mathcal{T}$ and $\mathcal{S} \sim \mathbf{U}\mathcal{S}\mathbf{U}^T$ for all orthogonal $p \times p$ matrices $\mathbf{U}$. For $\mathbf{I}_{p,p}$ $\mathbf{K}_{p,p}$, and $\mathbf{J}_{p,p}$, see Section 6.1. We then have the following result.

Theorem 7.3.1. *Assume that*

$$\mathbf{U}\mathcal{T} \sim \mathcal{T} \text{ and } \mathbf{U}\mathcal{S}\mathbf{U} \sim \mathcal{S} \quad \text{for all orthogonal } \mathbf{U}.$$

Then

$$E(\mathcal{T}) = \mathbf{0} \text{ and } E(\mathcal{S}) = E(\mathcal{S}_{11})\mathbf{I}_p$$

and

$$\mathrm{cov}(\mathcal{T}) = \mathrm{var}(\mathcal{T}_1)\mathbf{I}_p,$$

$$\mathrm{cov}(\mathrm{vec}(\mathcal{S})) = \mathrm{var}(\mathcal{S}_{12})(\mathbf{I}_{p,p} + \mathbf{K}_{p,p}) + \mathrm{cov}(\mathcal{S}_{11}, \mathcal{S}_{22})\mathbf{J}_{p,p}$$

and

$$\mathrm{cov}(\mathcal{T}, \mathrm{vec}(\mathcal{S})) = \mathbf{0}.$$

For a spherical distribution F with $\mathbf{T}(F) = \mathbf{0}$, $\mathbf{S}(F) = \mathbf{I}_p$, and $\mathbf{V}(F) = \mathbf{I}_p$ and $\sqrt{n}\widehat{\mathbf{T}} \to_d \mathcal{T}$, $\sqrt{n}\mathrm{vec}(\widehat{\mathbf{S}} - \mathbf{I}_p) \to_d \mathcal{S}$, and $\sqrt{n}\mathrm{vec}(\widehat{\mathbf{V}} - \mathbf{I}_p) \to_d \mathcal{V}$, the covariance matrices of the limiting distributions are then simply

$$cov(\mathcal{T}) = var(\mathcal{T}_1)\mathbf{I}_p,$$
$$cov(vec(S)) = var(S_{12})[\mathbf{I}_{p,p} + \mathbf{K}_{p,p} - 2\mathbf{J}_{p,p}] + var(S_{11})\mathbf{J}_{p,p}$$
$$= var(S_{12})[\mathbf{I}_{p,p} + \mathbf{K}_{p,p}] + cov(S_{11}, S_{22})\mathbf{J}_{p,p},$$

and

$$cov(vec(\mathcal{V})) = var(\mathcal{V}_{12}) \left[\mathbf{I}_{p,p} + \mathbf{K}_{p,p} - \frac{2}{p}\mathbf{J}_{p,p} \right]$$

where, using the influence functions with $R = \|\mathbf{X}\|$,

$$var(\mathcal{T}_1) = \frac{E_F[\gamma_T^2(R)]}{p}, \quad var(\mathcal{V}_{12}) = \frac{E_F[\alpha_V^2(R)]}{p(p+2)},$$
$$var(S_{12}) = \frac{E_F[\alpha_C^2(R)]}{p(p+2)} \tag{7.1}$$

and

$$var(S_{11}) = \frac{3E_F[\alpha_S^2(R)]}{p(p+2)} - \frac{2E_F[\alpha_S(R)\beta_C(R)]}{p} + E_F[\beta_S^2(R)]. \tag{7.2}$$

At general elliptical distributions with $T(F) = \mu$ and $S(F) = \Sigma$, the expressions for influence functions and limiting variances can be derived using the affine equivariance properties of functionals. Then, with $\mathcal{T}$ and S corresponding to the spherical case $\mu = \mathbf{0}$ and $\Sigma = \mathbf{I}_p$,

$$\sqrt{n}(\hat{\mathbf{T}} - \mu) \rightarrow_d N_p(\mathbf{0}, var(\mathcal{T}_1)\Sigma)$$

and the limiting distribution of $\sqrt{n}vec(\hat{\mathbf{S}} - \Sigma)$ is

$$N_{p^2}(\mathbf{0}, var(S_{12})[\mathbf{I}_{p,p} + \mathbf{K}_{p,p}]\Sigma \otimes \Sigma + cov(S_{11}, S_{22})vec(\Sigma)vec(\Sigma)^T).$$

This means that, in the general elliptic case, only two constants

$$var(S_{12}) \quad \text{and} \quad cov(S_{11}, S_{22})$$

are needed for the limiting efficiency comparisons between different scatter matrix estimates.

7.4   M-functionals for Location and Scatter

There are several alternative competing techniques to construct location and scatter functionals, for example, M-functionals, S-functionals, τ-functionals, projection-based functionals, CM- and MM-functionals, and so on. These functionals and related estimates are discussed throughout in numerous research and review papers, (see, for example, Maronna *et al.* 2006 and references therein). Next we consider M-functionals in more detail.

Definition 7.4.1. *Location and scatter M-functionals are functionals* $\mathbf{T} = \mathbf{T}(F_{\mathbf{X}})$ *and* $\mathbf{S} = \mathbf{S}(F_{\mathbf{X}})$, *which simultaneously satisfy two implicit equations*

$$\mathbf{T} = [E[w_1(R)]]^{-1} E[w_1(R)\mathbf{X}]$$

and

$$\mathbf{S} = [E[w_3(R)]]^{-1} E\left[w_2(R)(\mathbf{X} - \mathbf{T})(\mathbf{X} - \mathbf{T})^T\right]$$

for some suitably chosen weight functions $w_1(r)$, $w_2(r)$, *and* $w_3(r)$. *The random variable R is the (random) Mahalanobis distance between* $\mathbf{X}$ *and* $\mathbf{T}$; *that is,*

$$R = \sqrt{(\mathbf{X} - \mathbf{T})^T \mathbf{S}^{-1}(\mathbf{X} - \mathbf{T})}.$$

Consider an elliptic model with known $\rho(r)$ and its derivative function $\psi(r) = \rho'(r)$. If one then chooses $w_1(r) = w_2(r) = \psi(r)/r$ and $w_3(r) \equiv 1$, the M-functionals are called the *pseudo maximum likelihood (ML)* functionals corresponding to that specific distribution determined by ρ: $\mathbf{T} = \mathbf{T}(F)$ and $\mathbf{S} = \mathbf{S}(F)$ are then minimizers of negative pseudo log likehood

$$E\left[\rho((\mathbf{X} - \mathbf{T})^T \mathbf{\Sigma}^{-1}(\mathbf{X} - \mathbf{T}))\right] + \frac{1}{2}\log\det(\mathbf{\Sigma}).$$

In the multivariate normal case $w_1(r) \equiv w_2(r) \equiv 1$, and the corresponding functionals are the mean vector and the covariance matrix again. (see for example, Kent and Tyler 1996).

A classical M-estimate is *Huber's M-functional* with the choices

$$w_1(r) = \min\left(\frac{c}{r}, 1\right) \quad \text{and} \quad w_2(r) = d \cdot \min\left(\frac{c^2}{r^2}, 1\right)$$

(and $w_3(r) \equiv 1$) with some positive tuning constants c and d. The value of the functional does not depend strongly on the tails of the distribution; the tuning constant c controls this property. The constant d is just a scaling factor.

If $T_0(F)$ and $S_0(F)$ are any affine equivariant location and scatter functionals then so are the *one-step M-functionals*, starting from T_0 and S_0, and given by

$$T_1 = [E[w_1(R_0)]]^{-1} E[w_1(R_0)X]$$

and

$$S_1 = E\left[w_2(R_0)(X - T_0)(X - T_0)^T\right],$$

where now $R_0 = \sqrt{(X - T_0)^T S_0^{-1}(X - T_0)}$. It is easy to see that T_1 and S_1 are affine equivariant location and scatter functionals as well. Repeating this step until it converges yields the regular M-estimate (with $w_3(r) \equiv 1$).

The influence functions of the location and scatter functionals $T(F)$ and $S(F)$ at elliptical F were derived in Section 8.7 in Huber (1981). Recall that the influence function of location functional $T(F)$ is given by

$$IF(z, T, F) = \gamma_T(r)u,$$

with some $\gamma_T(r)$ and that the influence function of scatter functional $S(F)$ is

$$IF(z; S, F) = \alpha_S(r)uu^T - \beta_S(r)I_p$$

with some $\alpha_S(r)$ and some $\beta_S(r)$. We now have the following. In the spherical case, with $w_3(r) \equiv 1$,

$$\gamma_T(r) = c_1(F)w_1(r)r$$

and

$$\alpha_S(r) = c_2(F)w_2(r)r^2 \text{ and } \beta_S(r) = c_3(F) \left(w_2(r)r^2 - c_4(F)\right)$$

with constants $c_1(F), \ldots, c_4(F)$ depending on F and weight functions w_1 and w_2. For robust estimates one thus hopes that both $w_1(r)r$ and $w_2(r)r^2$ are bounded functions of r. Influence functions in general elliptic cases are again found using the affine equivariance properties of the functionals.

A companion location estimate is not necessary for finding scatter functionals: we say that $S(F)$ is a *scatter matrix functional with respect to the origin* if it is affine equivariant in the sense that

$$S(F_{AX}) = AS(F_X)A^T \text{ for all full-rank matrices } A.$$

Then, for any location functional $T(F)$, the functional

$$F_X \rightarrow S(F_{X-T(F_X)})$$

is a regular scatter functional. Moreover, as $X_1 - X_2$ is symmetric around zero for two independent copies of X, a *symmetrized version of* $S(F)$,

$$F_X \quad \rightarrow \quad S_{sym}(F_X) = S\left(F_{X_1-X_2}\right),$$

is a scatter matrix functional without any reference to a location functional. It is also easy to see that $S_{sym}(F)$ has the independence property that is important in the independent component analysis. (see, for example, Sirkiä *et al.* 2007).

Tyler's (Tyler 1987) famous shape functional $\mathbf{S}(F)$ with respect to the origin minimizes the criterion function

$$\mathbf{\Sigma} \rightarrow E \left[\log((\mathbf{x} - \mathbf{T})^T \mathbf{\Sigma}^{-1} (\mathbf{X} - \mathbf{T})) \right] + \frac{1}{2} \log \det(\mathbf{\Sigma}).$$

In fact, $\mathbf{S}(F_{\mathbf{X}})$ is the maximum likelihood estimate of $\mathbf{\Sigma}$ based on the directional data $\|\mathbf{X}_1\|^{-1}\mathbf{X}_1, \dots, \|\mathbf{X}_n\|^{-1}\mathbf{X}_n$ when $\mathbf{X}_1, \dots, \mathbf{X}_n$ is a random sample from an elliptical distribution with symmetry center at the origin and with the scatter matrix $\mathbf{\Sigma}$. Surprisingly, the exact (and asymptotic) distribution of $\widehat{S}$ is distribution-free in this elliptic model. Hettmansperger and Randles (2002) used for M-estimation the following weight functions:

$$w_1(r) = \frac{1}{r} \quad \text{and} \quad w_2(r) = \frac{p}{r^2};$$

then $\mathbf{T}(F)$ is an affine equivariant spatial median and $\mathbf{S}(F)$ is the Tyler's shape functional w.r.t. that center. Dümbgen (1998) introduced the symmetric version of Tyler's functional.

7.5   Breakdown Point

The breakdown point (BP) of a sample statistic $\widehat{\mathbf{T}} = \mathbf{T}(F_n)$ measures *global robustness*. We discuss the breakdown properties in the spirit of Donoho and Huber (1983). Hampel (1968) considered the breakdown point of a functional $\mathbf{T}(F)$ in an asymptotic sense.

Let $\mathbf{X} = (\mathbf{X}_1, \dots, \mathbf{X}_n)^T \in \mathbb{R}_n^p$ be an $n \times p$ data matrix. The breakdown point of a sample statistic $\mathbf{T}(\mathbf{X})$ is defined as the maximum proportion of contaminated "bad" observations in a dataset $\mathbf{X}$ that can make the observed value $\mathbf{T}(\mathbf{X})$ uninformative in the sense that it does not carry any information on the corresponding population quantity. To be more specific, define the m neighborhood $B_m(\mathbf{X}) \subset \mathbb{R}_n^p$ of $\mathbf{X}$ in sample space as

$$B_m(\mathbf{X}) = \{ \mathbf{X}^* \in \mathbb{R}_n^p : \mathbf{X} \text{ and } \mathbf{X}^* \text{ have at least } n - m \text{ joint obervations} \},$$

$m = 0, \dots, n$. A corrupted dataset $\mathbf{X}^* \in B_m(\mathbf{X})$ is obtained by replacing m observation vectors (rows) in $\mathbf{X}$ with arbitrary vectors. Clearly

$$\{\mathbf{X}\} = B_0(\mathbf{X}) \subset B_1(\mathbf{X}) \subset \cdots \subset B_n(\mathbf{X}) = \mathbb{R}_n^p.$$

To define the breakdown point, we also need a distance measure $\delta(\mathbf{T}, \mathbf{T}^*)$ between the observed value $\mathbf{T} = \mathbf{T}(\mathbf{X})$ and the corrupted value $\mathbf{T}^* = \mathbf{T}(\mathbf{x}^*)$ of the statistic. The maximum distance over all m-replacements is then

$$\delta_m(\mathbf{T};\mathbf{X}) = \sup\{\delta(\mathbf{T}(\mathbf{X}), \mathbf{T}(\mathbf{X}^*)) : \mathbf{X}^* \in B_m(\mathbf{X})\}, \quad m = 0, 1, \ldots, n,$$

$m = 0, \ldots, n.$

Definition 7.5.1. *The breakdown point of* $\mathbf{T}$ *at* $\mathbf{X}$ *is then*

$$BD(\mathbf{T};\mathbf{X}) = \min\left\{\frac{m}{n} : \delta_m(\mathbf{T};\mathbf{X}) = \infty\right\}.$$

For location statistics, a natural distance measure is $\delta(\mathbf{T}, \mathbf{T}^*) = \|\mathbf{T} - \mathbf{T}^*\|$. For scatter matrices, one can use $\delta(\mathbf{S}, \mathbf{S}^*) = \max(\|\mathbf{S}^{-1}\mathbf{S}^* - \mathbf{I}_p\|, \|(\mathbf{S}^*)^{-1}\mathbf{S} - \mathbf{I}_p\|)$, for example. The scatter matrix becomes at least partially "uninformative" if one of its eigenvalues goes to 0 or to ∞. Davies (1987) gave the following upper bound for the breakdown points of location vectors and scatter matrices.

Theorem 7.5.1. *For any location statistic* $\mathbf{T}$, *for any scatter statistic* $\mathbf{S}$, *and for any (genuinely p-variate) data matrix* $\mathbf{X}$,

$$BD(\mathbf{T};\mathbf{X}) \le \frac{n-p+1}{2n} \quad \text{and} \quad BD(\mathbf{S};\mathbf{X}) \le \frac{n-p+1}{2n}.$$

The sample mean vector and sample covariance matrix have the smallest possible breakdown point $1/n$. Maronna (1976) and Huber (1981) showed that M-statistics have relative low breakdown points, always below $1/(p+1)$. For high breakdown points, some alternative estimation techniques (e.g., S-estimates) should be used.

7.6 Use of Robust Scatter Matrices

7.6.1 Ellipticity assumption

In this chapter we assume that $\mathbf{X}_1, \ldots, \mathbf{X}_n$ is a random sample from an elliptical distribution F with location center $\mathbf{T}(F) = \mu$ and scatter matrix $\mathbf{S}(F) = \Sigma$ (with some choices $\mathbf{T}$ and $\mathbf{S}$). We also assume that, in the spherical case with $\mu = \mathbf{0}$ and $\Sigma = \mathbf{I}_p$, $\sqrt{n}vec(\hat{\mathbf{S}} - \mathbf{I}_p) \to_d S$. Then, under general assumptions, $vec(S)$ has a p^2-variate (singular) normal distribution with mean vector zero.

Robustness properties of the functionals as well as the procedures based on these functionals are considered using its influence function and the finite-sample breakdown point. Also the efficiency properties of correlation and principal component analysis tools based on the use of $\mathbf{S}$ inherit the efficiency properties of S, that is, the asymptotic efficiencies in the elliptic case depend only on

$$var(S_{12}) \quad \text{and} \quad cov(S_{11}, S_{22}).$$

We next consider these efficiencies.

7.6.2   Robust correlation matrices

Consider the limiting behavior of

$$\mathbf{R}_X = \text{diag}(\mathbf{S}_X)^{-1/2} \, \mathbf{S}_X \, \text{diag}(\mathbf{S}_X)^{-1/2}.$$

If $\mathbf{S}_X \rightarrow_p \mathbf{\Sigma}$ then naturally

$$\mathbf{R}_X \rightarrow_p \mathbf{P} = (\text{diag}(\mathbf{\Sigma}))^{-1/2} \mathbf{\Sigma} (\text{diag}(\mathbf{\Sigma}))^{-1/2}$$

where $\mathbf{P} = (\rho_{ij})$ is the population correlation matrix. It is not a restriction to assume that $\mathbf{P} = \mathbf{\Sigma}$, and then $\mathbf{\Sigma}_{ij} = \rho_{ij}$, the correlation between X_i and X_j, $i, j = 1, \ldots, p$. As before, if the fourth moments exist,

$$\sqrt{n} \mathbf{\Sigma}^{-1/2} (\mathbf{S}_X - \mathbf{P}) \mathbf{\Sigma}^{-1/2} \rightarrow_d S$$

where S is a multivariate normal matrix with zero expectation and with the same multivariate distribution as before. Using Slutsky's theorem, one then sees that

$$\sqrt{n}(\mathbf{R}_X - \mathbf{P}) \rightarrow_d \mathcal{R} \sim \mathbf{\Sigma}^{1/2} S \mathbf{\Sigma}^{1/2}$$
$$- \frac{1}{2}[\text{diag}(\mathbf{\Sigma}^{1/2} S \mathbf{\Sigma}^{1/2}) \mathbf{\Sigma} + \mathbf{\Sigma} \text{diag}(\mathbf{\Sigma}^{1/2} S \mathbf{\Sigma}^{1/2})].$$

Next note that

$$vec(\mathcal{R}) = \left[\mathbf{I}_{p,p} - \frac{1}{2}(\mathbf{\Sigma} \otimes \mathbf{I}_p + \mathbf{I}_p \otimes \mathbf{\Sigma}) \mathbf{D}_{p,p} \right] vec(\mathbf{\Sigma}^{1/2} S \mathbf{\Sigma}^{1/2})$$

and write $\mathbf{h}_{ij} = (\mathbf{e}_i \otimes \mathbf{e}_j - \frac{1}{2}\sigma_{ij}(\mathbf{e}_i \otimes \mathbf{e}_i + \mathbf{e}_j \otimes \mathbf{e}_j)^T$. Then

$$var(\mathcal{R}_{ij}) = \mathbf{h}_{ij}^T cov(vec(\mathbf{\Sigma}^{1/2} S \mathbf{\Sigma}^{1/2})) \mathbf{h}_{ij} = var(S_{12})(1 - \rho_{ij}^2)^2.$$

7.6.3   Principal component analysis

Next set

$$\mathbf{\Sigma} = \mathbf{U} \mathbf{\Lambda} \mathbf{U}^T \quad \text{and} \quad \hat{\mathbf{S}} = \hat{\mathbf{U}} \hat{\mathbf{\Lambda}} \hat{\mathbf{U}}^T$$

(eigenvector–eigenvalue decompositions) where $\mathbf{U} = (\mathbf{u}_1, \ldots, \mathbf{u}_p)^T$. Then we obtain the limiting variance of $\sqrt{n}(\hat{\lambda}_j - \lambda_j)$ and the limiting covariance matrix of $\sqrt{n}(\hat{\mathbf{u}}_j - \mathbf{u}_j)$

$$ASV(\hat{\lambda}_j) = var(S_{11}), \quad j = 1, \ldots, p,$$

and

$$ASV(\hat{\mathbf{u}}_j) = var(S_{12}) \sum_{i \neq j} \frac{\lambda_i \lambda_j}{\lambda_i - \lambda_j} \mathbf{u}_i \mathbf{u}_i^T, \quad j = 1, \ldots, p.$$

7.6.4   Canonical correlation analysis

Using any scatter matrix S, the limiting distributions of the canonical correlations and vectors with the assumption that the population correlations are distinct, that is, $\rho_1 > \cdots > \rho_p > 0$, were found by Taskinen *et al.* (2006). They showed, for example, that, for $p = q$, $\sqrt{n}(\hat{r}_i - \rho_i)$, $\sqrt{n}(\hat{\mathbf{a}}_i - \mathbf{a}_i)$, and $\sqrt{n}(\hat{\mathbf{b}}_i - \mathbf{b}_i)$ are asymptotically normal with zero mean and asymptotic variances

$$\mathrm{ASV}(\hat{r}_i) = (1 - \rho_i^2)^2 var(S_{12}),$$

$$\mathrm{ASV}(\hat{\mathbf{a}}_i) = \frac{1}{2} var(S_{11})\mathbf{a}_i\mathbf{a}_i^T$$

$$+ var(S_{12}) \sum_{k \neq i} \frac{(\rho_k^2 + \rho_i^2 - 2\rho_k^2\rho_i^2)(1 - \rho_i^2)}{(\rho_i^2 - \rho_k^2)^2} \mathbf{a}_k\mathbf{a}_k^T,$$

and

$$\mathrm{ASV}(\hat{\mathbf{b}}_i) = \frac{1}{2} var(S_{11})\mathbf{b}_i\mathbf{b}_i^T$$

$$+ var(S_{12}) \sum_{k \neq i} \frac{(\rho_k^2 + \rho_i^2 - 2\rho_k^2\rho_i^2)(1 - \rho_i^2)}{(\rho_i^2 - \rho_k^2)^2} \mathbf{b}_k\mathbf{b}_k^T.$$

7.7   Further Uses of Location and Scatter Functionals

The scatter matrices $\hat{S} = S(F_n)$ are often used to transform the dataset to a new coordinate system. If one writes (spectral or eigenvalue decomposition)

$$\hat{S} = \hat{U}\hat{\Lambda}\hat{U}^T,$$

where $\hat{U}$ is an orthogonal matrix and $\hat{\Lambda}$ is a diagonal matrix with positive diagonal elements in a decreasing order, then the components of the transformed data matrix

$$\mathbf{Z} = \mathbf{X}\hat{U}$$

are the estimated principal components provided by $\hat{S}$. The principal components are uncorrelated and ordered according to their dispersion. Principal components are often used to reduce the dimension of the data.

Scatter matrices are also often used to pre-standardize the data. The transformed standardized dataset

$$\mathbf{Z} = \mathbf{X}\hat{U}\hat{\Lambda}^{-1/2}$$

or

$$\mathbf{Z} = \mathbf{Y}\hat{S}^{-1/2}$$

then have standardized and uncorrelated components, and the observations $\mathbf{Z}_i$ tend to be spherically distributed in the elliptic case. The symmetric version of the square root matrix is usually chosen. Unfortunately, even in that case, the transformed dataset $\mathbf{Z}$ is not *affine invariant* as, for all $\mathbf{X}$, $\mathbf{A}$, and $\mathbf{S} = \mathbf{S}(\mathbf{X})$, there is an orthogonal matrix $\mathbf{V}$ such that

$$\mathbf{XA}^T[\mathbf{S}(\mathbf{XA}^T)]^{-1/2}\mathbf{V} = \mathbf{X}[\mathbf{S}(\mathbf{X})]^{-1/2}.$$

In the independent component analysis (ICA), most algorithms first standardize the data using the regular covariance matrix and then rotate the standardized data in such a way that the components of $\mathbf{Z} = \mathbf{X}\widehat{\mathbf{S}}^{-1/2}\mathbf{V}^T$ are "as independent as possible". In this procedure the regular covariance matrix may be replaced by any scatter matrix that has the independence property.

A location statistic $\mathbf{T} = \mathbf{T}(\mathbf{X})$ and a scatter matrix $\mathbf{S} = \mathbf{S}(\mathbf{X})$ may be used together to center and standardize the dataset. Then the transformed dataset is given by

$$\mathbf{Z} = (\mathbf{Y} - \mathbf{1}_n\mathbf{T}')\mathbf{S}^{-1/2}.$$

This is often called the *whitening* of the data. Then $\mathbf{T}(\mathbf{Z}) = \mathbf{0}$ and $\mathbf{S}(\mathbf{Z}) = \mathbf{I}_p$. Again, if you rotate the dataset using an orthogonal matrix $\mathbf{V}$, it is still true that $\mathbf{T}(\mathbf{ZV}) = \mathbf{0}$ and $\mathbf{S}(\mathbf{ZV}) = \mathbf{I}_p$. Recently, Tyler *et al.* (2009) developed an approach called the *invariant coordinate selection (ICS)*, which is based on the simultaneous use of two scatter matrices, $\mathbf{S}_1$ and $\mathbf{S}_2$. In this procedure the data are first standardized using $\mathbf{S}_1^{-1/2}$ and then the standardized data are rotated using the eigenvector matrix $\mathbf{V}$ that is based on $\mathbf{S}_2$ calculated for the transformed data $\mathbf{YS}_1^{-1/2}$. Then

$$\mathbf{S}_1(\mathbf{XS}_1^{-1/2}\mathbf{V}) = \mathbf{I}_p \quad \text{and} \quad \mathbf{S}_2(\mathbf{XS}_1^{-1/2}\mathbf{V}) = \mathbf{D},$$

where now $\mathbf{D}$ is the diagonal matrix of eigenvalues of $\mathbf{S}_1^{-1/2}\mathbf{S}_2$. If both $\mathbf{S}_1$ and $\mathbf{S}_2$ have the independence property, the procedure finds, under general assumptions, the independent components in the ICA problem.

Two location vectors and two scatter matrices may also be used simultaneously to describe the skewness and kurtosis properties of a multivariate distribution. Affine invariant multivariate *skewness statistics* may be defined as squared Mahalanobis distances between two location statistics

$$(\mathbf{T}_1 - \mathbf{T}_2)^T\mathbf{S}^{-1}(\mathbf{T}_1 - \mathbf{T}_2).$$

The eigenvalues of $\mathbf{S}_1^{-1}\mathbf{S}_2$, say $d_1 \geq \cdots \geq d_p$, may be used to describe the *multivariate kurtosis*. The measures of skewness and kurtosis can then be used in finding clusters in the data, in testing for symmetry, multivariate normality, or multivariate ellipticity as well as in the separation between different models (see Kankainen *et al.* 2007; Nordhausen *et al.* 2009; Ilmonen *et al.* 2012).

7.8   Summary

In this chapter robust multivariate location and scatter functionals and estimates are considered with a special focus on M-functionals. Robust versions of principal component analysis (PCA) and canonical correlation analysis (CCA) are described. Also the joint use of several scatter matrices in multivariate data analysis is briefly discussed.

References

Croux C and Haesbrock G 2000 Principal component analysis based on robust estimators of the covariance or correlation matrix: influence functions and efficiencies. *Biometrika* **87**, 603–618.

Davies PL 1987 Asymptotic behavior of S-estimates of multivariate location parameters and dispersion matrices. *Annals of Statistics* **15**, 1269–1292.

Donoho DL and Huber PJ 1983 The notion of breakdown point. In *A Festschrift for Erich L. Lehmann*, Bickel PJ, Doksum KA, and Hodges JL (eds), pp. 157–184, Belmont, Wadsworth.

Dümbgen L 1998 On Tyler's M-functional of scatter in high dimension. *Annals of the Institute of Statistal Mathematics* **50**, 471–491.

Hampel FR 1968 Contributions to the theory of robust estimation. Ph.D. Thesis, University of California, Berkeley.

Hampel FR, Rousseeuw PJ, Ronchetti EM, and Stahel WA 1986 *Robust Statistics: The Approach Based on Influence Functions*. Wiley, New York.

Hettmansperger TP and Randles RH 2002 A practical affine equivariant multivariate median. *Biometrica* **89**, 851–860.

Huber PJ 1981 *Robust Statistics*. Wiley, New York.

Ilmonen P, Oja H, and Serfling R 2012 On invariant coordinate system (ICS) functionals. *International Statistical Review* **80**, 93–110.

Kankainen A, Taskinen S and Oja H 2007 Tests of multinormality based on location vectors and scatter matrices. *Statistical Methods and Applications* **16**, 357–379.

Kent J and Tyler D 1996 Constrained M-estimation for multivariate location and scatter. *Annals of Statistics* **24**, 1346–1370.

Maronna RA 1976 Robust M-estimators of multivariate location and scatter. *Annals of Statistics* **4**, 51–67.

Maronna RA, Martin RD, and Yohai VY 2006 *Robust Statistics. Theory and Methods*. Wiley, New York.

Nordhausen K, Oja H, and Ollila E 2009 Multivariate models and the first four moments. In *Nonparametric Statistics and Mixture Models: A Festschrift in Honor of Thomas P. Hettmansperger*, Hunter, DR, Richards, DSR, and Rosenberger, JL (eds) pp. 267–287, World Scientific, Singapore.

Ollila E, Oja H, and Croux C 2003 The affine equivariant sign covariance matrix: asymptotic behavior and efficiency. *J. Mult. Analysis* **87**, 328–355.

Paindaveine D 2008 A canonical definition of shape. *Statistics and Probability Letters* **78**, 2240–2247.

Sirkiä S, Taskinen S, and Oja H 2007 Symmetrised M-estimators of scatter. *J. Mult. Analysis* **98**, 1611–1629.

Taskinen S, Croux C, Kankainen A, Ollila E, and Oja H 2006 Influence functions and efficiencies of the canonical correlation and vector estimates based on scatter and shape matrices. *J. Mult. Analysis* **97**, 359–384.

Tyler DE 1987 A distribution-free M-estimator of multivariate scatter. *Annals of Statistics* **15**, 234–251.

Tyler D, Critchley F, Dumbgen L, and Oja H 2009 Invariant co-ordinate selection. *J. Royal Statist. Soc.* B**71**, 549–592.

8

Nonparametric Measures of Multivariate Correlation

In this chapter, correlation measures and inference tools that are based on various concepts of univariate and multivariate signs and ranks are discussed.

8.1 Preliminaries

Classical multivariate statistical inference methods (Hotelling's T^2, multivariate analysis of variance, multivariate regression, tests for independence, canonical correlation analysis, principal component analysis, etc.) are based on the use of the regular sample mean vector and covariance matrix. See, for example, the monographs by Anderson (2003) and Mardia *et al.* (1979). These standard moment-based multivariate techniques are optimal under the assumption of multivariate normality but unfortunately poor in their efficiency for heavy-tailed distributions and are highly sensitive to outlying observations. In the previous chapter we discussed alternative location and scatter functionals as tools to robustify classical approaches.

In this chapter we consider the use of multivariate sign and rank methods in the analysis of multivariate dependence. The book by Puri and Sen (1971) gives a complete presentation of multivariate analysis methods based on marginal signs and ranks. In Oja (2010) nonparametric and robust competitors to standard multivariate inference methods based on spatial signs and ranks are discussed. See Oja (1999) for affine invariant/equivariant multivariate sign and rank methods.

Robust Correlation: Theory and Applications, First Edition. Georgy L. Shevlyakov and Hannu Oja.
© 2016 John Wiley & Sons, Ltd. Published 2016 by John Wiley & Sons, Ltd.
Companion Website: www.wiley.com/go/Shevlyakov/Robust

8.2   Univariate Signs and Ranks

We first recall the univariate concepts of *sign, rank, and signed-rank*. Let $x_1, \ldots, x_n$ be an univariate data set. The univariate sign function is

$$S(x) = \mathrm{sgn}(x) = \begin{cases} +1, & \text{for } x > 0; \\ 0, & \text{for } x = 0; \\ -1, & \text{for } x < 0. \end{cases}$$

The centered rank function is then

$$R_n(x) = \frac{1}{n} \sum_{i=1}^{n} S(x - x_i)$$

and finally the signed-rank function is

$$Q_n(x) = \frac{1}{2n} \sum_{i=1}^{n} [S(x - x_i) + S(x + x_i)] = \frac{1}{2}[R_n(x) - R_n(-x)].$$

Then the signed-rank function $Q_n(x)$ is thus the centered rank function $R_{2n}(x)$ calculated among the combined $2n$-dataset of the original observations $x_1, \ldots, x_n$ and their reflections $-x_1, \ldots, -x_n$. Note that the centered rank $R_n(x)$ and signed-rank $Q_n(x)$ provide both magnitude of x (robust distances from the median and from the origin, respectively) and direction of x (sign with respect to the median and sign with respect to the origin, respectively). The concepts of sign and rank and signed-rank are thus linked with the possibility to order the data (using the sign function $S(x)$). Unfortunately, there is no natural ordering in the multivariate case.

The numbers $s_i = S(x_i)$, $r_i = R(x_i)$, and $q_i = Q(x_i)$, $i = 1, \ldots, n$, are the observed signs, observed centered ranks, and observed signed-ranks. The possible values of observed centered ranks r_i and observed signed-ranks q_i then are

$$-\frac{n-1}{n}, \quad -\frac{n-3}{n}, \quad \ldots, \quad \frac{n-3}{n}, \quad \frac{n-1}{n}$$

and

$$-\frac{2n-1}{2n}, \quad -\frac{2n-3}{2n}, \quad \ldots, \quad \frac{2n-3}{2n}, \quad \frac{2n-1}{2n},$$

respectively. For the observed centered ranks, $\sum_{i=1}^{n} r_i = 0$. It is easy to find the connection between the regular rank (with values $1, 2, \ldots, n$) and the centered rank, namely,

$$\text{Centered rank} = \frac{2}{n} \left[\text{regular rank} - \frac{n+1}{2} \right].$$

For distinct values of x_i, there are $n!$ possible values of $(r_1, \ldots, r_n)$ and, for distinct values of $|x_i|$, $2^n n!$ possible values $(q_1, \ldots, q_n)$.

Notice that the ranks are invariant under any monotone transformations

$$x_i \to x_i^* = g(x_i) \text{ with strictly increasing } g,$$

while the signed-ranks are invariant only under

$$x_i \to x_i^* = s_i g(|x_i|) \text{ with strictly increasing } g : \mathbb{R}^+ \to \mathbb{R}.$$

The signs are naturally invariant under any transformations that preserve the sign.

The above definitions of univariate signs and ranks are based on the ordering of the data. However, in the multivariate case there is no natural ordering of the data points. The approach utilizing objective or criterion functions is then needed to extend the concepts to the multivariate case. The concepts of univariate sign and rank and signed-rank may be implicitly defined using the L_1 measures of distances

$$\frac{1}{n} \sum_{i=1}^{n} |x_i| = \frac{1}{n} \sum_{i=1}^{n} s_i x_i,$$

$$\frac{1}{2n^2} \sum_{i=1}^{n} \sum_{j=1}^{n} |x_i - x_j| = \frac{1}{n} \sum_{i=1}^{n} r_i x_i,$$

and

$$\frac{1}{4n^2} \sum_{i=1}^{n} \sum_{j=1}^{n} (|x_i - x_j| + |x_i + x_j|) = \frac{1}{n} \sum_{i=1}^{n} q_i x_i$$

(see Hettmansperger and Aubuchon 1988). The classical L_2 distance of the observations for the origin is similarly given by

$$\frac{1}{n} \sum_{i=1}^{n} |x_i|^2 = \frac{1}{n} \sum_{i=1}^{n} x_i x_i$$

and corresponds to the identity score function. The first L_1 measure is the *mean deviation* and provides the basis for the so-called least absolute deviation (LAD) methods; it yields different median-type estimates and sign tests if applied in the one-sample, two-sample, several-sample, and finally general linear model settings. The second measure is the *mean difference* between the observations. The second and third measures generate Hodges–Lehmann-type estimates and rank tests for different location problems. This formulation suggests, using various extensions of the L_1 measures, a natural way to generalize the concepts sign, rank, and signed-rank to the multivariate case (without defining a multivariate ordering).

Apply next the definitions to theoretical distributions, and assume that X, X_1, and X_2 are independent and identically distributed continuous random variables (copies) from a univariate distribution with CDF F. Then, similarly to empirical equations,

$$E|X| = E[S(X)X],$$

$$\frac{1}{2}E|X_1 - X_2| = E[(2F(X) - 1)X],$$

and

$$\frac{1}{4}E[|X_1 - X_2| + |X_1 + X_2|] = E[S(X)(F(|X|) - F(-|X|))X].$$

The theoretical (population) *centered rank function* and *signed-rank function* for F are then defined as

$$R_F(x) = 2F(x) - 1 \quad \text{and} \quad Q_F(x) = S(x)[F(|x|) - F(-|x|)].$$

If $X_1, \ldots, X_n$ is a random sample of size n from F then it is easy to see that

$$\sup_x |R_n(x) - R_F(x)| \to_P 0 \quad \text{and} \quad \sup_x |Q_n(x) - Q_F(x)| \to_P 0$$

as $n \to \infty$. The functions $Q_n(x)$ and $Q_F(x)$ are odd, that is, $Q_n(-x) = -Q_n(x)$ and $Q_F(-x) = -Q_F(x)$, and for distributions symmetric about the origin $Q_F(x) = R_F(x)$. Note that the inverse of the centered rank function (i.e., the inverse of the centered cumulative distribution function) is the univariate *quantile function*.

8.3 Marginal Signs and Ranks

Let $\mathbf{x}_1, \ldots, \mathbf{x}_n$ be a random sample from a p-variate distribution of $\mathbf{X}$. We start with the approach where the so-called *Manhattan norm*

$$\|\mathbf{x}\| = |x_1| + \cdots + |x_p|$$

is used in the multivariate extension of the sign and rank concepts. Using our strategy, the signs, centered ranks, and signed-ranks are implicitly defined through multivariate L_1-type objective functions

$$\frac{1}{n}\sum_{i=1}^{n} \|\mathbf{x}_i\| = \frac{1}{n}\sum_{i=1}^{n} \mathbf{s}_i^T \mathbf{x}_i,$$

$$\frac{1}{2n^2}\sum_{i=1}^{n}\sum_{j=1}^{n} \|\mathbf{x}_i - \mathbf{x}_j\| = \frac{1}{n}\sum_{i=1}^{n} \mathbf{r}_i^T \mathbf{x}_i,$$

and

$$\frac{1}{4n^2}\sum_{i=1}^{n}\sum_{j=1}^{n}[\|\mathbf{x}_i - \mathbf{x}_j\| + \|\mathbf{x}_i + \mathbf{x}_j\|] = \frac{1}{n}\sum_{i=1}^{n} \mathbf{q}_i^T \mathbf{x}_i.$$

It is then easy to see that $\mathbf{s}_i$, $\mathbf{r}_i$, and $\mathbf{q}_i$, $i = 1, \ldots, n$, are now simply the vectors of *marginal signs, ranks, and centered ranks*.

Note that the vectors of marginal signs, ranks, and centered ranks are not affine invariant or equivariant, that is, their behavior under the transformation $\mathbf{x}_i \rightarrow \mathbf{x}_i^* = \mathbf{A}\mathbf{x}_i + \mathbf{b}$ is unpredictable. Notice, however, that the vector of marginal ranks is invariant under component-wise monotone transformations

$$x_{ij} \rightarrow x_{ij}^* = g_j(x_{ij}) \text{ with strictly increasing } g_j, \ j = 1, \ldots, p,$$

and the vector of the signed-ranks is invariant under

$$x_{ij} \rightarrow x_{ij}^* = s_{ij}g_j(|x_i|) \text{ with strictly increasing } g_j : \mathbb{R}^+ \rightarrow \mathbb{R}, \ j = 1, \ldots, p.$$

Further, the *sign, rank, and signed-rank covariance matrices* are defined as

$$\widehat{scov} = \frac{1}{n}\sum_{i=1}^{n}\mathbf{s}_i\mathbf{s}_i^T, \quad \widehat{rcov} = \frac{1}{n}\sum_{i=1}^{n}\mathbf{r}_i\mathbf{r}_i^T, \quad \text{and} \quad \widehat{qcov} = \frac{1}{n}\sum_{i=1}^{n}\mathbf{q}_i\mathbf{q}_i^T.$$

Note that $\widehat{rcov}$ is invariant under location shifts $\mathbf{x}_i \rightarrow \mathbf{x}_i^* = \mathbf{x}_i + \mathbf{b}$. This is not true for the other two matrices and therefore for $\widehat{scov}$ and for $\widehat{qcov}$ the observation vectors should be first centered by the vector of marginal medians and the vector of marginal Hodges–Lehmann estimates, respectively. The matrix $\widehat{rcov}$ provides the pairwise *Spearman's rho* rank correlation values. Pairwise *Kendall's tau* correlation values are

$$\widehat{tcov} = \frac{2}{n(n-1)}\sum_{i=1}^{n-1}\sum_{j=i+1}^{n}\mathbf{S}(\mathbf{x}_i - \mathbf{x}_j)\mathbf{S}(\mathbf{x}_i - \mathbf{x}_j)^T$$

where $\mathbf{S}(\mathbf{x})$ is the vector of marginal signs of $\mathbf{x}$.

The corresponding population matrices for the distribution of $\mathbf{X}$ are, after proper centering,

$$scov = E[\mathbf{S}(\mathbf{X})\mathbf{S}(\mathbf{X})^T], \quad rcov = E[\mathbf{R}(\mathbf{X})\mathbf{R}(\mathbf{X})^T],$$

and

$$qcov = E[\mathbf{Q}(\mathbf{X})\mathbf{Q}(\mathbf{X})^T]$$

where

$$\mathbf{R}(\mathbf{x}) = E[\mathbf{S}(\mathbf{x} - \mathbf{X})] \quad \text{and} \quad \mathbf{Q}(\mathbf{x}) = \frac{1}{2}(E[\mathbf{S}(\mathbf{x} - \mathbf{X})] + E[\mathbf{S}(\mathbf{x} + \mathbf{X})]),$$

and

$$tcov = E[\mathbf{S}(\mathbf{X}_1 - \mathbf{X}_2)\mathbf{S}(\mathbf{X}_1 - \mathbf{X}_2)^T]$$

where $\mathbf{X}_1$ and $\mathbf{X}_2$ are two independent copies of $\mathbf{X}$. In the elliptic case, surprisingly,

$$scov = tcov = \frac{2}{\pi}\sin^{-1}(\rho_{ij})$$

and in the multivariate normal case

$$rcov = qcov = \frac{6}{\pi}\sin^{-1}\left(\frac{\rho_{ij}}{2}\right),$$

and therefore in these cases each of $\widehat{scov}$, $\widehat{rcov}$, $\widehat{qcov}$, and $\widehat{icov}$ can be used to build a consistent estimate of the correlation matrix with a known limiting behavior. Unfortunately, there is no guarantee that the correlation matrix, if estimated in this way, is positive definite.

8.4   Spatial Signs and Ranks

Let $\mathbf{x}_1, \ldots, \mathbf{x}_n$ be again an observed p-variate dataset. This approach is based on the use of *Euclidean norm*

$$\|\mathbf{x}\| = (x_1^2 + \cdots + x_p^2)^{-1/2}.$$

The multivariate concepts of *spatial sign*, *spatial rank*, and *spatial signed-rank* are then obtained as in the case of Manhattan distances. We obtain the following.

Definition 8.4.1.

(i) *The spatial sign function is*

$$\mathbf{S}(\mathbf{x}) = \begin{cases} \|\mathbf{x}\|^{-1}\mathbf{x}, & \mathbf{x} \neq \mathbf{0}, \\ \mathbf{0}, & \mathbf{x} = \mathbf{0}. \end{cases}$$

(ii) *The observed spatial rank function is*

$$\mathbf{R}_n(\mathbf{x}) = \frac{1}{n}\sum_{i=1}^{n} \mathbf{S}(\mathbf{x} - \mathbf{x}_i).$$

(iii) *The observed spatial signed-rank function is*

$$\mathbf{Q}_n(\mathbf{x}) = \frac{1}{2}[\mathbf{R}_n(\mathbf{x}) - \mathbf{R}_n(-\mathbf{x})].$$

Observe that in the univariate case regular sign, rank, and signed-rank functions are obtained. Clearly the multivariate signed-rank function $\mathbf{Q}_n(\mathbf{x})$ is also odd; that is, $\mathbf{Q}_n(-\mathbf{x}) = -\mathbf{Q}(\mathbf{x})$.

The observed *spatial signs* are $\mathbf{s}_i = \mathbf{S}(\mathbf{x}_i)$, $i = 1, \ldots, n$. As in the univariate case, the observed *spatial ranks* are certain averages of signs of pairwise differences

$$\mathbf{r}_i = \mathbf{R}_n(\mathbf{x}_i) = \frac{1}{n}\sum_{j=1}^{n} \mathbf{S}(\mathbf{x}_i - \mathbf{x}_j), \quad i = 1, \ldots, n.$$

Finally, the observed *spatial signed-ranks* are given as

$$\mathbf{q}_i = \mathbf{Q}_n(\mathbf{x}_i) = \frac{1}{2n} \sum_{j=1}^{n} [\mathbf{S}(\mathbf{x}_i - \mathbf{x}_j) + \mathbf{S}(\mathbf{x}_i + \mathbf{x}_j)], \quad i = 1, \dots, n.$$

The spatial sign $\mathbf{s}_i$ is just a direction vector of length one (lying on the unit p-sphere S_p) whenever $\mathbf{x}_i \neq \mathbf{0}$. The centered ranks $\mathbf{r}_i$ and signed-ranks $\mathbf{q}_i$ lie in the unit p-ball B_p. The direction of $\mathbf{r}_i$ ($\mathbf{q}_i$) roughly tells the direction of $\mathbf{x}_i$ from the center of the data cloud (the origin), and its length roughly tells how far away this point is from the center (the origin).

We next collect some equivariance properties of the spatial signs and ranks and signed-ranks. Let $\mathbf{U}$ be an orthogonal $p \times p$ matrix. For all $\mathbf{U}$, the transformations

$$\mathbf{x}_i \quad \rightarrow \quad \mathbf{U}\mathbf{x}_i, \quad i = 1, \dots, n,$$

induce transformations

$$\mathbf{s}_i \rightarrow \mathbf{U}\mathbf{s}_i, \quad \mathbf{r}_i \rightarrow \mathbf{U}\mathbf{r}_i \quad \text{and} \quad \mathbf{q}_i \rightarrow \mathbf{U}\mathbf{q}_i, \quad i = 1, \dots, n.$$

We then say that the spatial sign and rank and signed-rank are *orthogonal equivariant*. Note also that the spatial ranks are location invariant, that is, the spatial ranks $\mathbf{r}_1, \dots, \mathbf{r}_n$ of two sets of observations $\mathbf{x}_1, \dots, \mathbf{x}_n$ and $\mathbf{x}_1 + \mathbf{b}, \dots, \mathbf{x}_n + \mathbf{b}$ are the same for all location shifts $\mathbf{b}$.

To repeat, the sign, centered rank, and signed-rank are implicitly defined through multivariate L_1-type objective functions

$$\frac{1}{n} \sum_{i=1}^{n} \|\mathbf{x}_i\| = \frac{1}{n} \sum_{i=1}^{n} \mathbf{s}_i^T \mathbf{x}_i,$$

$$\frac{1}{2n^2} \sum_{i=1}^{n} \sum_{j=1}^{n} \|\mathbf{x}_i - \mathbf{x}_j\| = \frac{1}{n} \sum_{i=1}^{n} \mathbf{r}_i^T \mathbf{x}_i,$$

and

$$\frac{1}{4n^2} \sum_{i=1}^{n} \sum_{j=1}^{n} [\|\mathbf{x}_i - \mathbf{x}_j\| + \|\mathbf{x}_i + \mathbf{x}_j\|] = \frac{1}{n} \sum_{i=1}^{n} \mathbf{q}_i^T \mathbf{x}_i,$$

with the Euclidean norm $\|\mathbf{x}\| = (x_1^2 + \cdots + x_p^2)^{1/2}$.

Next let $\mathbf{X}$, $\mathbf{X}_1$, and $\mathbf{X}_2$ be three independent copies from F. The *theoretical spatial sign, spatial rank function, and spatial signed-rank functions*, $\mathbf{S}(\mathbf{x})$, $\mathbf{R}_F(\mathbf{x})$, and $\mathbf{Q}_F(\mathbf{x})$, then satisfy the following equations.

Definition 8.4.2.

$$E\|\mathbf{X}\| = E[\mathbf{S}(\mathbf{X})^T X],$$

$$\frac{1}{2}E\|\mathbf{X}_1 - \mathbf{X}_2\| = E[\mathbf{R}(\mathbf{X})^T X],$$

and

$$\frac{1}{4}E[\|\mathbf{X}_1 - \mathbf{X}_2\| + \|\mathbf{X}_1 + \mathbf{X}_2\|] = E[\mathbf{Q}(\mathbf{X})^T\mathbf{X}]$$

and then again

$$\mathbf{R}(\mathbf{x}) = E[\mathbf{S}(\mathbf{x} - \mathbf{X})]$$

and

$$\mathbf{Q}(\mathbf{x}) = \frac{1}{2}E[\mathbf{S}(\mathbf{x} - \mathbf{X}) - \mathbf{S}(\mathbf{x} + \mathbf{X})].$$

The rank function $\mathbf{R}(\mathbf{y})$ characterizes the distribution F up to a location shift (see Koltchinskii 1997). For F symmetric around the origin, $\mathbf{Q}(\mathbf{x}) = \mathbf{R}(\mathbf{x})$, for all $\mathbf{x} \in \mathbb{R}^p$. The empirical functions converge uniformly in probability to the theoretical ones under mild assumptions. (For the proof, see Möttönen *et al.* 1997. For an early contribution, see also Chaudhuri 1992.)

Chaudhuri (1996) considered the inverse of the spatial rank function and called it the *spatial quantile function* (see also Koltchinskii 1997). Serfling (2004) gives a review of the inference methods based on the concept of the spatial quantile, and introduces and studies some nonparametric measures of multivariate location, spread, skewness, and kurtosis in terms of these quantiles. The quantile at $\mathbf{u} = \mathbf{0}$ is the so-called spatial median, which is discussed in more detail later.

If the distribution of $\mathbf{X}$ is spherically symmetric around the origin, the rank and signed-rank function has a simple form, namely

$$\mathbf{R}(\mathbf{x}) = \mathbf{Q}(\mathbf{x}) = q(r)\mathbf{u},$$

where $r = \|\mathbf{x}\|$ and $\mathbf{u} = \|\mathbf{x}\|^{-1}\mathbf{x}$ and

$$q(r) = E\left[\frac{r - X_1}{((r - X_1)^2 + X_2^2 + \cdots + X_p^2)^{1/2}}\right] \leq 1.$$

See Oja (2010) for $q_F(r)$ in the multivariate normal and t-distribution cases.

If spatial sign and ranks are used to analyze the data, the *spatial sign and rank covariance matrices* also naturally play an important role. Let $\mathbf{X}_1$ and $\mathbf{X}_2$ be two independent copies of $\mathbf{X}$. Then the *spatial sign covariance matrix* and *the spatial Kendall's tau matrix* are

$$scov = E[\mathbf{S}(\mathbf{X})\mathbf{S}(\mathbf{X})^T] \quad \text{and} \quad tcov = E[\mathbf{S}(\mathbf{X}_1 - \mathbf{X}_2)\mathbf{S}(\mathbf{X}_1 - \mathbf{X}_2)^T],$$

respectively, with the corresponding estimates

$$\widehat{scov} = \frac{1}{n}\sum_{i=1}^{n}[\mathbf{S}(\mathbf{X}_i)\mathbf{S}(\mathbf{X}_i)^T]$$

and

$$\widehat{tcov} = \frac{2}{n(n-1)} \sum_{i=1}^{n-1} \sum_{j=i+1}^{n} [\mathbf{S}(\mathbf{X}_i - \mathbf{X}_j)\mathbf{S}(\mathbf{X}_i - \mathbf{X}_j)^T].$$

Moreover, the *spatial rank covariance matrix* and the *spatial signed-rank covariance matrix* are

$$rcov = E[\mathbf{R}(\mathbf{X})\mathbf{R}(\mathbf{X})^T] \quad \text{and} \quad qcov = E[\mathbf{Q}(\mathbf{X})\mathbf{Q}(\mathbf{X})^T],$$

respectively. Natural estimates then are

$$\widehat{rcov}(F_n) = \frac{1}{n} \sum_{i=1}^{n} \mathbf{r}_i \mathbf{r}_i^T \quad \text{and} \quad \widehat{qcov} = \frac{1}{n} \sum_{i=1}^{n} \mathbf{q}_i \mathbf{q}_i^T.$$

The functionals *scov*, *tcov*, *rcov*, and *qcov* are not scatter matrix functionals as they are not affine equivariant. They are equivariant under orthogonal transformations only. The *tcov* and *rcov* are shift invariant but, again, for *scov* and *qcov* a proper centering of observation vectors is necessary. The so-called *spatial median* and *spatial Hodges–Lehmann estimate* are used for the centering. These estimates can be defined using the corresponding estimating equations, namely, after the centering

$$\sum_{i=1}^{n} \mathbf{s}_i = \mathbf{0} \quad \text{and} \quad \sum_{i=1}^{n} \mathbf{q}_i = \mathbf{0},$$

respectively (see Oja 2010). Note also that the sign covariance matrix and Kendall's tau matrix are scaled in the sense that $tr(scov) = tr(tcov) = 1$.

Assume next that $\mathbf{X}$ has an elliptical distribution and its covariance matrix *cov* has distinct eigenvalues. Then the following is true. First, the matrices *scov*, *rcov*, *qcov*, *tcov*, and *cov* have the same eigenvectors. Second, the eigenvalues of *scov* and *tcov* are the same and there is a one-to-one correspondence (up to a constant) between their eigenvalues and the eigenvalues of *cov*. Third, the eigenvalues of *rcov* and *qcov* are the same. This all implies that the spatial sign and rank covariance matrices can be used to *robust estimation of the eigenvectors* of the covariance matrix. Then the eigenvalues can be estimated from the corresponding marginal distribution, and these results can be collected together for robust covariance and correlation matrix estimates (see Visuri *et al.* 2000).

Affine equivariant scatter estimates based on spatial signs and ranks can be constructed as follows. Find $\widehat{\mathbf{S}}$ so that, after the standardization $\mathbf{x}_i \rightarrow \widehat{\mathbf{S}}^{-1/2}\mathbf{x}_i$,

$$\widehat{pscov} = \mathbf{I}_p \quad \text{or} \quad \widehat{pscov} = \mathbf{I}_p.$$

The first set of estimating equations $\widehat{pscov} = \mathbf{I}_p$ give the celebrated *Tyler's (Tyler 1987) scatter matrix* estimate and the second $\widehat{ptcov} = \mathbf{I}_p$ *Dümbgen's (Dümbgen 1998) scatter matrix* estimate. Note that these estimates are identified only up to a constant but this is sufficient in most cases including the correlation matrix estimation. Tyler's and Dümbgen's estimates thus provide highly robust correlation matrix estimates. Tyler's estimate naturally needs pre-centering of the dataset. Hettmansperger and Randles (2002) find the standardization $\mathbf{x}_i \to \widehat{\mathbf{S}}^{-1/2}(\mathbf{x}_i - \widehat{\mu})$ such that

$$\frac{1}{n} \sum_{i=1}^{n} \mathbf{s}_i = \mathbf{0} \quad \text{and} \quad \frac{p}{n} \sum_{i=1}^{n} \mathbf{s}_i \mathbf{s}_i^T = \mathbf{I}_p.$$

Then $\widehat{\mu}$ is the affine equivariant spatial median and $\widehat{S}$ is Tyler's estimate with respect to the spatial median. See these three papers for the computation of the estimates as well as for their asymptotic properties.

8.5   Affine Equivariant Signs and Ranks

In the univariate case the sign function with respect to μ is the derivative of

$$V(x; \mu) = \text{abs} \left\{ \det \begin{pmatrix} 1 & 1 \\ \mu & x \end{pmatrix} \right\} = \text{abs}\{x - \mu\}$$

with respect to x, that is, $S(x; \mu) = S(x - \mu)$ with a univariate sign function S. Here $\text{abs}\{x\} = |x|$ is the absolute value of x. The signs with respect to the sample median $\widehat{\mu}$, that is, $s_i = S(x_i; \widehat{\mu})$, $i = 1, \ldots, n$, are then centered, which means that $\sum_{i=1}^{n} s_i = 0$.

Next, we extend this definition to multivariate setting. Write $I = (i_1, \ldots, i_{p-1})$, $1 \leq i_1 < \cdots < i_{p-1} \leq n$, for an ordered $(p-1)$ set of indices. The index I then refers to a $p-1$ subset of observations with indices listed in I. The multivariate centered sign function is now defined as the gradient of

$$V(\mathbf{x}; \mu) = \binom{n}{p-1}^{-1} \sum_I \frac{1}{p!} \text{abs} \left\{ \det \begin{pmatrix} 1 & 1 & \cdots & 1 & 1 \\ \mu & \mathbf{x}_{i_1} & \cdots & \mathbf{x}_{i_{p-1}} & \mathbf{x} \end{pmatrix} \right\}$$

$$= \binom{n}{p-1}^{-1} \sum_I \frac{1}{p!} \text{abs} \left\{ \det \left(\mathbf{x}_{i_1} - \mu \ \cdots \ \mathbf{x}_{i_{p-1}} - \mu \ \mathbf{x} - \mu \right) \right\}$$

$$= \text{ave}_I \left[\frac{1}{p!} \text{abs}\{ \mathbf{e}^T (I; \mu)(\mathbf{x} - \mu) \} \right]$$

with respect to $\mathbf{x}$. The vector $\mathbf{e}(I; \mu)$ is simply the vector of cofactors of $\mathbf{x} - \mu$ in

$$\det \left(\mathbf{x}_{i_1} - \mu \ \cdots \ \mathbf{x}_{i_{p-1}} - \mu \ \mathbf{x} - \mu \right)$$

and the sums and average go over all possible $p-1$ subsets I. Then the multivariate sign function with respect to μ is

$$S(\mathbf{x}; \mu) = \nabla V(\mathbf{x}; \mu) = \text{ave}_I \{ p!^{-1} S_I(\mathbf{x}; \mu) \mathbf{e}(I; \mu) \},$$

where

$$S_I(\mathbf{x}; \mu) = S(\mathbf{e}^T(I; \mu)(\mathbf{x} - \mu))$$

indicates whether $\mathbf{x}$ is above or below hyperplane defined by μ and the $p - 1$ obervations listed in I. The empirical centered multivariate signs are then defined, as in the univariate case, by

$$\mathbf{S}_i = \mathbf{S}(\mathbf{x}_i; \widehat{\mu}) \qquad i = 1, \ldots, n,$$

where $\widehat{\mu}$ is the *multivariate Oja median* (Oja 1983). The affine equivariant Oja median thus minimizes the criterion function

$$\sum_I V(\mathbf{x}_i, \ldots, \mathbf{x}_{i_p}, \mu)$$

where

$$V(\mathbf{x}_1, \ldots, \mathbf{x}_p, \mathbf{x}) = \frac{1}{k!}\text{abs}\left\{\det\begin{pmatrix} 1 & \cdots & 1 & 1 \\ \mathbf{x}_1 & \cdots & \mathbf{x}_p & \mathbf{x} \end{pmatrix}\right\}$$

is the volume of the p-variate simplex determined by the vertices $\mathbf{x}_1, \ldots, \mathbf{x}_p$ along with $\mathbf{x}$.

The multivariate signs are thus centered, that is, $\text{ave}_i\{\mathbf{S}_i\} = \mathbf{0}$ and the *sign covariance matrix* is now simply

$$\widehat{scov} = \text{ave}_i\{\mathbf{S}_i\mathbf{S}_i^T\}.$$

The signs $\mathbf{S}_i$ and $\widehat{scov}$ enjoy the following affine equivariance property: the transformation $\mathbf{x}_i \rightarrow \mathbf{x}_i^* = \mathbf{A}\mathbf{x}_i + \mathbf{b}$, $i = 1, \ldots, n$, induces the transformations

$$\mathbf{S}_i \quad \rightarrow \quad \mathbf{S}_i^* = \text{abs}\{\det(\mathbf{A})\}(\mathbf{A}^{-1})^T\mathbf{S}_i, \quad i = 1, \ldots, n,$$

and

$$\widehat{scov} \quad \rightarrow \quad \widehat{scov}^* = (\det(\mathbf{A}))^2(\mathbf{A}^{-1})^T\widehat{scov}\mathbf{A}^{-1}.$$

The theoretical (or population) sign covariance matrix for the distribution of $\mathbf{X}$ is

$$scov = E[\mathbf{S}_X(\mathbf{X}; \mu)\mathbf{S}_X(\mathbf{X}; \mu)^T\},$$

where μ is the population Oja median and

$$\mathbf{S}_X(\mathbf{x}; \mu) = E[S_I(\mathbf{x}; \mu)\mathbf{e}(I; \mu)]$$

is the population sign function. The expectation is taken over the independent copies of $\mathbf{X}$ with indices listed in I. (Of course, the expectation is the same for all I.) One can then show that in the spherical case $scov$ is proportional to the identity matrix $\mathbf{I}_p$ and, in the elliptic case,

$$scov \propto cov^{-1}.$$

This result then implies that, using $\widetilde{scov}$, one can find consistent estimates for the eigenvectors and proportional eigenvalues of cov. Finally, consistent estimation of the correlation matrix is also possible. For the influence functions and limiting distributions of these estimates, see Ollila *et al.* (2003). The sign covariance matrix is not robust as its influence function at a spherical distribution is linear in $\|\mathbf{x}\|$. Recall that the influence function of the covariance matrix is quadratic in $\|\mathbf{x}\|$.

Consider next the notions of the affine equivariant sign and rank covariance matrix. Let again $\mathbf{x}_1, \ldots, \mathbf{x}_n$ be the observed p-variate dataset. For the ranks, we need p subsets of observation with index sets $J = (i_1, \ldots, i_p)$, $1 \le i_1 < i_2 < \cdots < i_p \le n$. The *hyperplane* going through the observations listed in J is

$$\left\{ \mathbf{x} \in \mathbb{R}^p \ : \ \det \begin{pmatrix} 1 & \cdots & 1 & 1 \\ \mathbf{x}_{i_1} & \cdots & \mathbf{x}_{i_p} & \mathbf{x} \end{pmatrix} = 0 \right\}.$$

The index J thus may be thought to refer to the above hyperplane as well. Define next $d_0(J)$ and $\mathbf{d}(J)$ by the implicit equation

$$\det \begin{pmatrix} 1 & \cdots & 1 & 1 \\ \mathbf{x}_{i_1} & \cdots & \mathbf{x}_{i_p} & \mathbf{x} \end{pmatrix} = d_0(J) + \mathbf{d}^T(J)\mathbf{x},$$

where

$$\mathbf{d}(J) = (d_1(J), \ldots, d_p(J))^T$$

and $d_0(J), d_1(J), \ldots, d_p(J)$ are the cofactors of the elements in the last column of the above matrix.

The volume of the p-variate simplex determined by the p-set J along with $\mathbf{x}$ is

$$V_J(\mathbf{x}) = \frac{1}{p!} \mathrm{abs}\{d_0(J) + \mathbf{d}^T(J)\mathbf{x}\}.$$

Then recall that the multivariate Oja median minimizes the average of the volumes of simplices, or the empirical depth

$$D(\mathbf{x}) = p! \ \mathrm{ave}_J\{V_J(\mathbf{x})\}.$$

As in the univariate case, the centered rank function (related to this generalization of the multivariate median) is defined as the gradient

$$\mathbf{R}(\mathbf{x}) = \nabla D(\mathbf{x}) = \mathrm{ave}_J[S_J(\mathbf{x})\mathbf{d}(J)]$$

where the sign

$$S_J(\mathbf{x}) = S(d_0(I) + \mathbf{d}^T(I)\mathbf{x})$$

tells whether $\mathbf{x}$ is above or below hyperplane J. Then the p-variate ranks are

$$\mathbf{r}_i = \text{ave}_J[S_J(\mathbf{x}_i)\mathbf{d}(J)]$$

with the rank covariance matrix

$$\widehat{rcov} = \frac{1}{n}\sum_{i=1}^{n}\mathbf{r}_i\mathbf{r}_i^T.$$

Note that the ranks are centered so that $\sum_{i=1}^{n}\mathbf{r}_i = \mathbf{0}$.

The ranks and rank covariance matrix enjoy the following equivariance property. If

$$\mathbf{x} \to \mathbf{x}_i^* = \mathbf{A}\mathbf{x}_i + \mathbf{b}, \quad i = 1,\ldots,n,$$

then

$$\mathbf{R}_i \quad \to \quad \mathbf{R}_i^* = \text{abs}\{\det(\mathbf{A})\}(\mathbf{A}^{-1})^T\mathbf{S}_i, \quad i = 1,\ldots,n,$$

and

$$\widehat{rcov} \quad \to \quad \widehat{rcov}^* = (\det(\mathbf{A}))^2(\mathbf{A}^{-1})^T\widehat{rcov}\mathbf{A}^{-1}.$$

The theoretical (or population) rank covariance matrix for the distribution of $\mathbf{X}$ is

$$rcov = E[\mathbf{R}_X(\mathbf{X})\mathbf{R}_X(\mathbf{X})^T\}$$

where

$$\mathbf{R}_X(\mathbf{x}) = E[S_J(\mathbf{x})\mathbf{d}(J)]$$

is the population rank function and the expectation is taken over the p independent copies of $\mathbf{X}$ with indices listed in J. As in the case of the sign covariance matrix, one can show that in the spherical case $rcov$ is proportional to the identity matrix $\mathbf{I}_p$ and, in the elliptic case, also

$$rcov \propto cov^{-1}$$

and $\widehat{rcov}$ can be used to find consistent estimates for the eigenvectors and proportional eigenvalues of cov. Again, consistent estimation of the correlation matrix is also possible. The influence function is approximately linear in $\|\mathbf{x}\|$. Consult Ollila et al. (2004) for the properties of $\widehat{rcov}$.

8.6 Summary

In this chapter, three different multivariate extensions of the concepts of sign and rank (and signed-rank) and their use in multivariate correlation analysis are discussed. For further discussion and theory, see, for example, Puri and Sen (1971), Visuri et al. (2000), and Oja (2010).

References

Anderson TW 2003 *An Introduction to Multivariate Statistical Analysis*, Third Edition, Wiley, New York.

Chaudhuri P 1992 Multivariate location estimation using extension of R-estimates through U-statistics type approach. *Annals of Statistics* **20**, 897–916.

Chaudhuri P 1996 On a geometric notion of quantiles for multivariate data. *J. Amer. Statist. Soc.* **91**, 862–872.

Dümbgen L 1998 On Tyler's M-functional of scatter in high dimension. *Annals of the Institute of Statistal Mathematics* **50**, 471–491.

Hettmansperger TP and Aubuchon JC 1988 Comment on "Rank-based robust analysis of linear models. I. Exposition and review" by David Draper. *Statistical Science* **3**, 262–263.

Hettmansperger TP and Randles RH 2002 A practical affine equivariant multivariate median. *Biometrica* **89**, 851–860.

Koltchinskii VI 1997 M-estimation, convexity and quantiles. *Annals of Statistics* **25**, 435–477.

Mardia KV, Kent JT, and Bibby JM 1979 *Multivariate Analysis*, Academic Press, Orlando, FL.

Möttönen J, Oja H, and Tienari J 1997 On the efficiency of multivariate spatial sign and rank tests. *Annals of Statistics* **25**, 542–552.

Oja H 1983 Descriptive statistics for multivariate distributions. *Statistics and Probability Letters* **1**, 327–332.

Oja H 1999 Affine invariant multivariate sign and rank tests and corresponding estimates: a review. *Scandinavian Journal of Statistics* **26**, 319–343.

Oja H 2010 *Multivariate Nonparametric Methods with R*, Springer, New York.

Ollila E, Croux C, and Oja H 2004 Influence function and asymptotic efficiency of the affine equivariant rank covariance matrix. *Statistica Sinica* **14**, 297–316.

Ollila E, Oja H, and Croux C 2003 The affine equivariant sign covariance matrix: asymptotic behavior and efficiency. *J. Mult. Anal.* **87**, 328–355.

Puri ML and Sen PK 1971 *Nonparametric Methods in Multivariate Analysis*, Wiley, New York.

Serfling RJ 2004 Nonparametric multivariate descriptive measures based on spatial quantiles. *J. Statist. Planning and Inference* **123**, 259–278.

Tyler DE 1987 A distribution-free M-estimator of multivariate scatter. *Annals of Statistics* **15**, 234–251.

Visuri S, Koivunen V, and Oja H 2000 Sign and rank covariance matrices. *J. Statist. Planning and Inference* **91**, 557–575.

9

Applications to Exploratory Data Analysis: Detection of Outliers

9.1 Preliminaries

The need for fast, on-line algorithms to analyze high data-rate measurements is a vital element in real-life settings. Given the ever-increasing number of data sources coupled with increasing complexity of applications, and workload patterns, anomaly detection methods should be of a low complexity and must operate in real-time. In many modern applications, data arrive in a streaming fashion. Therefore, the underlying assumption of classical methods that the data is a sample from an underlying distribution is not valid, and Gaussian and nonparametric-based methods are inadequate. Streaming data can be regarded as an ever-changing superposition of distributions. Detection of such changes in real-time is one of the fundamental challenges. In what follows, we propose low-complexity robust modifications to the conventional univariate Tukey boxplot and bivariate bagplot based on fast highly efficient robust estimates of scale and correlation. Results using synthetic as well as real-life data show that our methods outperform the conventional boxplots and methods based on Gaussian limits.

Concurrently, in parallel with robust statistics, practical methods for analyzing data evolved known as *Exploratory Data Analysis (EDA)*, which is known popularly as

Robust Correlation: Theory and Applications, First Edition. Georgy L. Shevlyakov and Hannu Oja.
© 2016 John Wiley & Sons, Ltd. Published 2016 by John Wiley & Sons, Ltd.
Companion Website: www.wiley.com/go/Shevlyakov/Robust

datamining. A significant feature of *EDA* is that it does not assume an underlying probability distribution for the data, which is typical in classical statistical methods, and therefore it is flexible in practical settings.

Our work represents new results in robust data analysis technologies, providing alternatives to the boxplot technique. The univariate Tukey method summarizes the characteristics of a data distribution allowing for a quick visual inspection of streams of data over windows. Despite being a simple data analysis tool, it concisely summarizes information about location, scale, asymmetry, tails, and outliers in the data distribution. In our study, we concentrate on visualization of distribution tails and on detection of outliers in the data.

The remainder of this chapter is organized as follows. In Section 9.2, boxplot techniques of exploratory data analysis are reviewed. In Section 9.3, the goals of our study are formulated. In Section 9.4, a new *H*-measure of outlier detection performance is proposed and studied. In Section 9.5, two new robust versions of the Tukey boxplot are proposed and used in outlier detection. In Section 9.6, new bivariate boxplots are applied for detection of bivariate outliers. In Section 9.7, some conclusions are drawn.

9.2   State of the Art

In this section, we overview the works in the two principal areas of research: univariate and bivariate boxplot techniques of exploratory data analysis.

9.2.1   Univariate boxplots

A univariate boxplot represents a box that has a length usually equal to the interquartile range, a width equal to $\sqrt{n}$, where n is the size of a dataset, and whiskers that indicate the lowest and the highest value in the data range (Tukey 1977). The data location is denoted by a line that divides the boxplot in to two rectangles.

From a descriptive statistics point of view, a boxplot assembles the following five parameters of a dataset: the two extremes, the upper and lower quartiles, and the median. Different batches of data could be compared through their respective boxplots in a fast and convenient way. It is a common practice to identify outliers for those points of data that are located beyond the extremes (maximum and minimum) and draw them together with the corresponding boxplots (see Fig. 9.1).

There are many modifications of boxplots depending on the kind of analyses that should be performed on given data.

In McGill *et al.* (1978), there is a good explanation on introducing a variable-width plot-box for different datasets. In this case, the width of a boxplot is proportional to the square root of the size of its corresponding dataset. Further, the authors introduce also a notched boxplot, which explicitly displays confidence intervals around the location parameter (median). Notched boxplots are useful when it

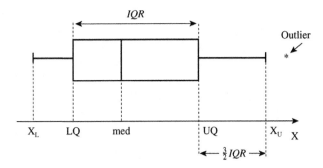

Figure 9.1   Tukey's univariate boxplot

comes to make a decision about considering the difference between the medians of respective datasets. A variable width notched boxplot combines the features of the aforementioned boxplots and this gives additional information to lessen the possibility of misinterpretations due to unwarranted assumptions. More details on calculating the box width and notch size of the variable width notched boxplot can be found in McGill *et al.* (1978).

More recent variations of boxplots are described in Potter *et al.* (2006). For most of the datasets it is necessary to have more than five descriptive parameters. When there is no prior information about the underlying data distribution, it becomes problematic to compare data with distributions, which differ in modalities. Adding the information on the underlying distributions to boxplots is a solution to this kind of problem. The histplot is a boxplot that is related to the variable width boxplots group. In the histplot, the data distribution density is estimated at the median and the two quartiles and this is reflected in the boxplot by changing it to a polygon. The histplot is still a simple approach used to display density information and it is easy to dismiss some important data features. A refined version of a histplot adds in estimated densities for every point between the upper and low quartiles. A hist-plot polygon resembles a vase and this new version of boxplot is called a vaseplot.

In Warren and Banfield (2003), the authors introduce the box-percentile plot, which is another method for adding the empirical cumulative distribution of the dataset into the boxplot. This type of boxplots could be considered as an extension of a vaseplot, since it displays densities along the whole range of values in a dataset. The median, 25th, and 75th percentiles are marked as in a vaseplot.

A violin plot is a combination of a boxplot with a box-percentile plot, where the boxplot is modified to a thick line (or solid black box) with the length proportional to the interquartile range and median denoted by a circle. Individual symbols for outlying data values are removed from the violin plot, since the outliers are contained in the density trace and individual points would clutter the diagram.

The last two modifications of boxplots bring valuable information from the descriptive statistics point of view. This summary of different variants of boxplots gives an idea about the efforts on visualizing information that makes it simple to analyze a batch of datasets.

For outlier identification purposes estimation methods play an important role in defining a boxplot. The classical estimates of the extremums of a boxplot are defined through

$$x_L = \text{"boxplot min"} = \max\left\{x_{(1)}, LQ - \frac{3}{2}\,\text{IQR}\right\} \tag{9.1}$$

and

$$x_R = \text{"boxplot max"} = \min\left\{x_{(n)}, UQ + \frac{3}{2}\,\text{IQR}\right\}, \tag{9.2}$$

where LQ and UQ are the lower and upper sample quartiles, respectively, and

$$\text{IQR} = UQ - LQ \tag{9.3}$$

is the interquartile range.

The values located beyond the minimum and maximum of a boxplot, calculated by these formulas, are considered as outliers. Since the sample interquartile range (IQR) is less resistant to outliers than, say, the highly robust median absolute deviation MAD $x = \text{med}\,|x - \text{med}\,x|$, a more robust detection rule based on the boxplot extremes

$$x_L = \max\{x_{(1)}, LQ - k_{MAD}\,\text{MAD}\,x\} \tag{9.4}$$

and

$$x_R = \min\{x_{(n)}, UQ + k_{MAD}\,\text{MAD}\,x\}, \tag{9.5}$$

can be proposed. The coefficient k_{MAD} is a threshold value chosen from additional considerations. The detection rule is defined as follows: an observation x is regarded as an outlier if

$$x < x_L \quad \text{or} \quad x > x_R.$$

Later, we will dwell on these issues in more detail. Here, we mainly focus on robust estimation of boxplot parameters.

9.2.2   Bivariate boxplots

Bivariate data offer new challenges when it comes to define a boxplot, which is used to be build on univariate datasets. According to Potter *et al.* (2006), the first attempt to extend a boxplot to two dimensions has been made by Becketti and Gould (1987) who proposed a rangefinder boxplot. The rangefinder boxplot is a simple extension of the univariate boxplot, where the five-number summary is calculated for each variable and a line is plotted for each corresponding variable perpendicular at the median point.

For bivariate data an important feature is correlation, which is not displayed in a rangefinder boxplot. Also this kind of boxplot fails to capture the bivariate shape of the data. According to Goldberg and Iglewicz (1992): "A bivariate data summary poses complications over univariate data: (a) The relationship between the variables is not captured by two univariate summaries, and (b) the number of directions for potential asymmetry and outliers is infinite rather than just "high" or "low"."

Many bivariate boxplot methods use the notion of *hinge* and *fence*. The hinge is defined as the line that encompasses 50% of the data and the fence is the line that separates the bulk of the data from the potential outliers.

For bivariate distributions of an elliptical shape, the *relplots* and *quelplots* are introduced in Goldberg and Iglewicz (1992). These versions of bivariate boxplot approaches hinge and fence by fitting ellipses.

Robust elliptical boxplots are used for small samples of data or assume that they follow a symmetric distribution. The *robust quarter elliptic plot* is defined as four quarters of ellipsis matched on their major and minor axes so that the graph is continuous and smooth. The quelplot reveals the asymmetry of data more accurately, thus making it a better nonparametric method to work with. Different estimates are used to compute the parameters of relplots and quelplots, such as the least squares, bivariate biweight M-estimate, one-step biweight, minimum volume ellipsoid, least median of squares, Theil-Sen, and three-group resistant line (Hoaglin *et al.* 1983). This choice includes traditional, robust, and nonparametric field of statistics groups of estimates. Bivariate biweight M-estimates prove to have a good performance compared to others.

From the authors' point of view (Goldberg and Iglewicz 1992), a univariate boxplot can be made more precise for symmetric data by using more efficient location and scale estimates, such as the F-mean and biweight. In this sense, the symmetric relplot is analogous to this version of univariate boxplot.

The method proposed in Goldberg and Iglewicz (1992) is focused on giving a degree of asymmetry based on bivariate biweight estimates but computationally is not effective. While retaining robust properties of relplots and quelplots, an attempt was made by Zani *et al.* (1998) to effectively calculate the hinge of bivariate data by constructing a spline superimposed on the 50% hull that does not require the estimation of parameters. The convex hull, which contains 50% of data, is smoothed then by using a B-spline procedure.

Similar to the idea of convex hull peeling but extended and based on the notion of the halfspace location depth, a bagplot is introduced by Rousseeuw *et al.* (1999). The halfspace location depth $ldepth(\theta, Z)$ of any point $\theta \in \mathbb{R}^2$ relative to a bivariate data cloud $Z = \{z_1, z_2, ..., z_n\}$ is the smallest number of z_i contained in any closed halfplane with the boundary line going through θ.

The depth region D_k is the set of all θ with $ldepth(\theta, Z) \geq k$. The depth regions are convex polygons and $D_{k+1} \subset D_k$. It should be noted that median in Rousseeuw *et al.* (1999) is defined as θ with the highest $ldepth(\theta, Z)$ if there is only one such θ; otherwise the median is defined as the center of gravity of the deepest region.

The bagplot is the hinge because it is calculated to encompass half of the points in the dataset. A fence is obtained by inflating the bagplot by the factor of 3. Sometimes around the depth median a *blotch* is drawn, which is a 95% confidence region for this parameter. It is important to note in Rousseeuw *et al.* (1999) the published performance of the algorithm, which computes the location depth of an arbitrary point in $O(nlogn)$ time while the construction of the vertices of a depth contour is $O(n^2logn)$ and $O(n^2(logn)^2)$ time is required for computing the depth median. The halfspace depth notion is of great advantage, as it allows the definition of a bagplot for multivariate data. The bagplot in three dimensions is a convex polyhedron, which still makes it difficult to reveal data features. A *bagplot matrix* that contains the bagplot of each pair of variables could be drawn for data in any dimension.

The two-dimensional boxplot introduced in Tongkumchum (2005) is similar to the rangefinder but extended to display linear relationships between variables by fitting a straight line uniquely to a bivariate data set using Tukey's line (Tukey 1977). The algorithm behind the two-dimensional boxplot begins by dividing points in the scatterplot in three nonoverlapping regions equal or nearly equal in size according to their x values. The medians of x and y values are calculated for outer regions and used then to find the parameters of the linear equation for Tukey's line. The hinge and fences are built based on interquartile ranges calculated for the x values and interquartile ranges for points along Tukey's line. Two-dimensional boxplots are simple to construct and due to their simplicity they are computationally effective in performance.

In Atkinson and Riani (1997), a fast very robust "forward" search was applied to a large body of multivariate problems. A scatterplot matrix is built for multivariate data, elements of which are bivariate boxplots. The "forward" search method consists of fitting m observations to $m + 1$ by choosing the $m + 1$ observations with the smallest least squares residuals. Choosing the initial outlier free group of observations is discussed in Atkinson and Riani (1997) in more detail.

A robust bivariate boxplot based on the median correlation coefficient is proposed in Shevlyakov and Vilchevski (2002). Assuming bivariate normal distribution of data and its elliptical structure it would be possible to inscribe a rectangle inside the elliptical contour. The inscribed rectangle with edges parallel to principal axes embodies all the necessary properties of a boxplot. The transition to the principal axes of the ellipse of equal probability for the bivariate normal distribution could be calculated based on the following relationships:

$$x' = (x - \mu_1) \cos \phi + (y - \mu_2) \sin \phi,$$

$$y' = -(x - \mu_1) \sin \phi + (y - \mu_2) \cos \phi,$$

where

$$\tan 2\phi = \frac{2\rho\sigma_1\sigma_2}{\sigma_1^2 - \sigma_1^2}.$$

Robust estimates are used for calculation of the component-wise location parameters $\widehat{\mu}_1 = $ med x and $\widehat{\mu}_2 = $ med y, the component-wise scale parameters $\widehat{\sigma}_1 = $ MAD x and $\widehat{\sigma}_2 = $ MAD x, and the median correlation coefficient r_{med} as a highly robust estimate of the correlation coefficient ρ (Shevlyakov and Vilchevski 2002). Then the angle ϕ is calculated as follows:

$$\tan 2\phi = \frac{2r_{med} \text{ MAD } x \text{ MAD } y}{(\text{MAD } x)^2 - (\text{MAD } y)^2}.$$

The hinge and the fence are represented by two rectangles, one into another, with the sides parallel to principal axes. The sides of the inner rectangle are equal to the sample interquartile ranges IQR x' and IQR y' evaluated from the samples $\{x'_i\}$ and $\{y'_i\}$. The boundaries of the fence are defined by the median absolute deviations

$$x'_L = \max\{x'_{(1)}, \text{med } x' - 5 \text{ MAD } x'\},$$

$$x'_R = \min\{x'_{(n)}, \text{med } x' + 5 \text{ MAD } x'\},$$

$$y'_L = \max\{y'_{(1)}, \text{med } y' - 5 \text{ MAD } y'\},$$

$$y'_R = \min\{y'_{(n)}, \text{med } y' + 5 \text{ MAD } y'\}.$$

The relative location of the inner rectangle to the outer one may indicate the departures from symmetry.

9.3   Problem Setting

The goals of our study can be formulated as follows:

- Modification of the conventional Tukey univariate boxplot based on fast highly efficient and robust estimates of scale.

- Optimization of the proposed methods of detection of outliers by a new measure of outlier detection performance, namely, the harmonic mean $H(P_D, 1 - P_F)$ between the power of detection P_D and the complementary value to the false alarm rate $1 - P_F$.

- Developing numerical algorithms for the proposed methods of data visualization with indications of data location, scale, correlation, tail areas, and outliers.

- Performance evaluation of the proposed robust boxplots and their classical analogs in various data distribution models and on real-life data.

- Developing new methods for construction of bivariate boxplots based on fast highly efficient and robust estimates of scale and correlation for elliptical data distributions.

Since in the application of boxplot techniques we mainly focus on detection of univariate outliers, next we specify the related methods and algorithms.

In statistics, an outlier is an observation that is numerically distant from the rest of the data (Barnett and Lewis 1978). Grubbs (1969) defines an outlier as follows: " An outlying observation, or outlier, is one that appears to deviate markedly from other members of the sample in which it occurs."

Outliers can occur by chance in any distribution, but they, as a rule, are the indications either of measurement errors or that the population has a heavy-tailed distribution. In the former case it is natural to discard them or use statistics that are robust to outliers, while in the latter case their presence indicates that the distribution has a high kurtosis and that one should be very cautious in using tools or intuitions that assume a normal distribution. A frequent cause of outliers is given by a mixture of two distributions, which may be two distinct sub populations, or may indicate "good data" versus "bad data"–that is, exploiting the main *paradigma* of robust statistics; this case is described by the Tukey gross error model (3.1).

Given a data set of n observations of a random variable X, let $\bar{x}$ be the mean and let s be the standard deviation of the data distribution. A common point of view is as follows: an observation x is declared as an outlier if it lies outside of the interval

$$(\bar{x} - k\,s, \bar{x} + k\,s), \tag{9.6}$$

where the value of k is usually taken between 2 and 3; in the latter case Equation (9.6) is called the 3-sigma rule. The justification for this value relies on the fact that assuming normal distribution one expects to have a certain, usually between 95% and 99% of the data, in the interval centered in the mean with a semi length equal to k standard deviations.

From Equation (9.6) it follows that the observation x is considered an outlier if

$$\frac{|x - \bar{x}|}{s} > k_\alpha, \tag{9.7}$$

where the threshold k_α is determined from the given false alarm rate (the probability of error of the first kind) at the normal distribution

$$P\left\{ \frac{|x - \bar{x}|}{s} > k_\alpha \right\} = \alpha, \tag{9.8}$$

with α usually taken equal to $0.05, 0.1$, or 0.2.

The problem with the above approach is that it assumes a normal distribution of the data, something that frequently does not occur. Moreover, the mean and standard deviation are highly sensitive to outliers.

However, the rule (9.7) for detection of outliers is recommended for use in statistical practice, since it is just the classical Grubbs test (Grubbs 1950, 1969).

The natural way to enhance the power of detection by the Grubbs test in case of the presence of outliers in the data is to use robust estimates of location and scale in its

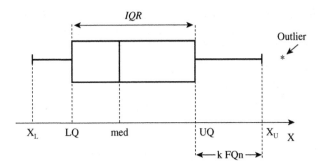

Figure 9.2   FQ-boxplot

structure, for instance, the sample median med x and the sample interquartile range IQR, as was done in the M.Sc. work of one of the authors (Shevlyakov 1973)–the obtained results, namely the percentiles and power of detection in the Tukey gross error model, were represented in Shevlyakov and Vilchevski (2002, Chapter 7).

Thus, a natural extension of the classical Grubbs test is to use its robust versions with the sample median med x as a robust estimate of location and robust estimates of scale, say, the highly robust median of absolute deviations MAD x or the fast highly robust and efficient FQ_n-estimate of scale (see Chapter 4) with the detection rule in the form of the MAD and FQ tests given by

$$\frac{|x - \text{med } x|}{\text{MAD } x} > k_\alpha \quad \text{or} \quad \frac{|x - \text{med } x|}{FQ_n \, x} > k_\alpha. \tag{9.9}$$

However, we prefer to use the robust *MAD*-BP and *FQ*-BP outlier detection rules based on the boxplot (BP) extremes (9.4) and (9.5). In the case of the *FQ*-boxplot (see Fig. 9.2), the corresponding thresholds for detection of outliers are of the form:

$$x_L = \max\{x_{(1)}, LQ - k_{FQ} FQ_n \, x\} \tag{9.10}$$

and

$$x_R = \min\{x_{(n)}, UQ + k_{FQ} FQ_n \, x\}. \tag{9.11}$$

Moreover, we use the fast robust highly efficient estimate FQ_n of scale and its derivatives for constructing new FQ versions of bivariate boxplots.

9.4   A New Measure of Outlier Detection Performance

Traditionally, the performance of statistical tests for outlier detection is evaluated by their power and false alarm rate. It requires ensuring the upper bound for false alarm rate while measuring the detection power, which proves to be a difficult task. In this section we introduce a new measure of outlier detection performance H_m as the harmonic mean of the power and unit minus false alarm rate. The H_m maximizes the detection power by minimizing the false alarm rate and enables an easier way for evaluation and parameter tuning of an outlier detection algorithm.

9.4.1 Introductory remarks

Let a random variable X be distributed according to the Tukey gross error model with the shift contamination:

$$f(x) = (1 - \epsilon)N(x; 0, 1) + \epsilon N(x; \mu, 1),$$

where $0 < \epsilon < 0.5$, $\mu > 0$. Then the problem of detection of a possible outlier X can be posed as the following hypotheses testing problem:

H_0: an observation X is an inlier when $X \sim N(0, 1)$,

H_1: an observation X is an outlier when $X \sim N(\mu, 1)$.

In this case, it can be easily shown that the optimal likelihood ratio test is of the form: the decision is made in favor of H_0 if $X < t$, and vice versa, in favor of H_1 if $X \geq t$, where t is a threshold, as to which value depends on the chosen approach: the Neyman–Pearson, Bayesian, minimax, etc (Lehmann and Romano 2008) (see Fig. 9.3).

The conventional characteristics of detection performance are the power of detection $P_D = P(X \geq t | H_1)$ and the false alarm rate $P_F = P(X \geq t | H_0)$.

In what follows, we mainly deal with the Neyman–Pearson approach, which presumes maximization of the detection power under the bounded false alarm rate

$$P_D \to \max_t, \qquad P_F \leq \alpha.$$

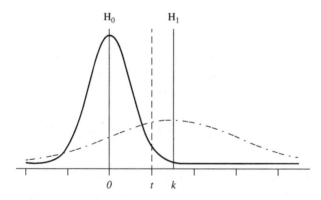

Figure 9.3   Hypotheses testing under shift contamination

In this case, the optimal threshold value t^* is obtained by equating the false alarm probability to α: $P_F(t^*) = \alpha$.

Generally, shifting the threshold parameter t, the power and false alarm rate values change accordingly. The graphical representation of the power and false alarm values defines the *ROC* curve (receiver operating characteristic) (Hastie *et al.* 2009). The area under the curve (*AUC*) is a common metric to compare different classification (detection) methods with each other. The closer to 1.0 the value of *AUC*, the better is the performance of the classification method under study.

9.4.2   *H*-mean: motivation, definition and properties

Within the Neyman–Pearson approach, the *ROC* curve gives the full information about the performance of a classification (detection) method or algorithm.

Motivation

However, in many cases it is almost impossible to ensure an upper bound on the false alarm rate; on the other side, it is important to find an optimal balance between these metrics, that is, the power and false alarm rate, for the method under study.

Moreover, in real-life applications, when one deals with small samples and Monte Carlo techniques, it is difficult to provide stability of small false alarm probability rates and thus stability of statistical inference.

Next, if a researcher is interested in optimization of the parameters of a classification (outlier detection) procedure, then it is more convenient to use a single scalar measure of performance, that is, rather the unconditional optimization tools than the conditional or multicriteria ones.

Definition of the *H*-mean

In information retrieval classification studies, the *F*-measure of classification performance (the harmonic mean between precision P and recall R—$H(P, R)$) is widely and successfully used (Hastie *et al.* 2009), so we propose to apply it to outlier detection as a measure of comparison and optimization (the harmonic mean between the detection power P_D and unit minus the false alarm rate $1 - P_F$:

$$H_{mean} = H(P_D, 1 - P_F) = 2\,\frac{P_D(1 - P_F)}{P_D + 1 - P_F}.$$

The properties of the *H*-mean

Recall the properties of the harmonic mean $H = H(x_1, x_2)$ with $0 < x_1 \leq 1, 0 < x_2 \leq 1$, the reciprocal of the average of reciprocals

$$H = \frac{1}{1/2(1/x_1 + 1/x_2)} = \frac{2x_1 x_2}{x_1 + x_2} \quad :$$

- H is one of the three Pythagorean means, the average $A = (x_1 + x_2)/2$, the geometric mean $G = \sqrt{x_1 x_2}$, and H with the following relation between them:

$$H \leq G \leq A.$$

- The harmonic mean mitigates the impact of a larger number and aggravates the impact of a smaller one.

- The harmonic mean aims at the processing of rates and ratios (numerous examples in physics, genetics, finance).

The relationship between the *ROC* curve and H-mean is shown in Fig. 9.4.

The following result states that sufficiently high values of the H-mean provide reasonable boundary values upon the detection power from below and the false alarm rate from above.

Theorem 9.4.1. *Given an H-mean value H, the following inequalities hold:*

$$P_D > P_D^{\min} = \frac{H}{2 - H}, \quad P_F < P_F^{\max} = 1 - P_D^{\min},$$

where P_D is the detection power and P_F is the false alarm rate.

Some numerical values of these boundaries are represented in Table 9.1.

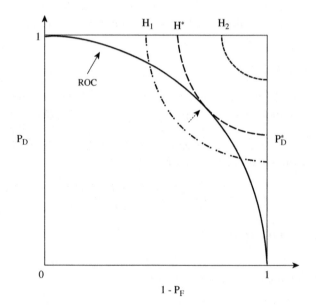

Figure 9.4   The relationship between the ROC curve and H-mean

Table 9.1   Boundaries of the power and false alarm rate

H	0.80	0.90	0.91	0.92	0.93	0.94	0.95	0.99	0.999
P_D^{min}	0.67	0.82	0.83	0.85	0.86	0.87	0.90	0.98	0.998
P_F^{max}	0.33	0.18	0.17	0.15	0.14	0.13	0.10	0.02	0.002

9.5   Robust Versions of the Tukey Boxplot with Their Application to Detection of Outliers

In this section, we mainly consider the boxplot (BP) detection tests of the form: an observation x is regarded as an outlier if $x < x_L$ or $x > x_U$, where x_L and x_U are the lower and upper extremes, respectively. In this setting, these thresholds also depend on a free parameter k, which is chosen from the false alarm rate $\alpha = 0.1$.

9.5.1   Data generation and performance measure

The Monte Carlo experiments are conducted by generating 300 samples of observations from the mixture of normal distributions (Tukey's gross error model) (Tukey 1960)

$$f(x) = (1 - \epsilon)N(x; 0, 1) + \epsilon N(x; \mu, \sigma), \qquad (9.12)$$

where $0 \le \epsilon < 0.5$ is the probability of outliers (the fraction of contamination) in the data and $\sigma > 1$ is their scale.

For evaluating the performance of different tests for detection of outliers, the H-mean is used, since, in this case, tests with the different values of the false alarm probability can be effectively compared. In our study, we just have this case: the false alarm rates for the Tukey and modified boxplots are $\alpha = 0.06$ and $\alpha = 0.1$, respectively.

9.5.2   Scale and shift contamination

The results of the Monte Carlo experiment are given in Tables 9.2 and 9.3 with the best performing statistics in boldface.

From Tables 9.2 and 9.3 it follows that both under scale and shift contamination, the performances of boxplot tests, generally, are close to each other, and all of them outperform the classical Grubbs test, which is catastrophically bad. This effect can be explained by nonrobustness of the Grubbs test forming statistics, the sample mean and standard deviation, under contamination.

Further, the robust *MAD* and *FQ BP* versions are slightly but systematically better than the Tukey boxplot test. Similar results are obtained for the gross error models with $\epsilon = 0.2$.

Table 9.2   *H*-means for detection tests under scale contamination: $\mu = 0$, $\sigma = 3$

$\epsilon = 0.1$	20	50	100	1,000	10,000
Tukey BP	0.64	**0.72**	0.72	0.72	0.72
MAD-BP	**0.67**	**0.72**	**0.73**	**0.73**	**0.73**
FQ-BP	0.66	**0.72**	0.72	0.72	**0.73**
Grubbs test	0.17	0.29	0.30	0.30	0.30

Table 9.3   *H*-means for detection tests under shift contamination: $\mu = 3$, $\sigma = 1$

$\epsilon = 0.1$	20	50	100	1,000	10,000
Tukey BP	**0.75**	0.79	0.80	0.80	0.80
MAD-BP	0.73	**0.80**	0.80	0.80	0.80
FQ-BP	0.73	0.79	**0.81**	**0.81**	**0.81**
Grubbs test	0.32	0.39	0.40	0.39	0.39

Table 9.4   *H*-means for detection tests under shift contamination with the different values of ϵ: $\mu = 3, \sigma = 1, n = 100$

ϵ	0.05	0.10	0.20	0.30	0.40	0.50
Tukey BP	0.63	0.62	0.59	0.55	0.51	0.43
MAD-BP	0.65	0.65	0.60	**0.56**	**0.52**	**0.44**
FQ-BP	**0.67**	**0.67**	**0.61**	**0.56**	0.50	0.40
Grubbs test	0.65	0.56	0.41	0.31	0.25	0.21

In Table 9.4, the dependence of detection performance w.r.t. the contamination parameter ϵ is studied. It is observed that with small and moderate levels of shift contamination, the *FQ*-boxplot is marginally better than its competitors. For larger fractions of contamination ($\epsilon \geq 0.3$), the *MAD*-boxplot outperforms its competitors. This can be explained by the fact that the *MAD* is a minimax bias estimate of scale under the Tukey gross error model (Hampel *et al.* 1986).

9.5.3   Real-life data results

We tested our algorithms on a real-world dataset obtained from an experimental set-up with representative cloud applications. It is a data intensive application implemented on Hadoop based on a distributed set of auctioning services. The analyzed data consist of 10 hours worth of service requests collected at 30 second intervals, into

Table 9.5   *H*-means for boxplot tests applied to server data

	Tukey BP	*MAD*-BP	*FQ*-BP
% idle	0.51	0.55	**0.58**
tpc	0.47	**0.57**	**0.57**
bread/s+bwrtn/s	0.47	**0.56**	**0.56**
rxpck/s+txpck/s	0.20	**0.33**	0.31

which 50 anomalies are injected over the duration of the experimental time-period. The anomalies are major failures or performance issues. We consider the metrics such as the server idle time (*% idle*), the traffic per second (*tpc*), the speed of reading and writing of data blocks (*bread/s+bwrtn/s*), and the speed of receiving and transmitting data blocks (*rxpck/s+txpck/s*).

From Table 9.5, it follows that the *MAD*-, and *FQ*-boxplots considerably outperform the Tukey boxplot in terms of *H*-mean. Similar results are also observed if the false alarm rate α and the power of detection P_D are used. For instance, in the case of the traffic per second (*tpc*), we have $\alpha = 0.06$, $P_D = 0.31$ for the Tukey boxplot and $\alpha = 0.1$, $P_D = 0.42$ for the *MAD*- and *FQ*-boxplots. Generally, all those results exhibit a rather low level of detection power; to raise it, we should increase the false alarm rate.

9.5.4   Concluding remarks

The two robust versions of the Tukey boxplot are proposed. Both versions aim at the symmetric distribution as their classical counterpart, the first *MAD*-BP being preferable under heavy contamination, while the second is *FQ*-BP – under moderate contamination. The thresholds k can be adjusted to the adopted level of the false alarm probability α when detecting outliers. We recommend the numerically obtained optimal values $k_{MAD} = 1.44$ and $k_{FQ} = 0.97$ corresponding to the rate $\alpha = 0.1$ under normality. All the boxplot tests considerably outperform the classical Grubbs test, which is catastrophically bad under contamination.

9.6   Robust Bivariate Boxplots and Their Performance Evaluation

9.6.1   Bivariate *FQ*-boxplot

In this section, a new bivariate boxplot is described. An elliptical model-based approach is considered given that data are distributed according to the bivariate normal distribution. The construction of the *FQ*-boxplot represented in this section is based on the fast robust highly efficient FQ_n estimates of scale and correlation (see Chapters 4 and 5).

Suppose a set of bivariate data $\{(x_1, y_1), \ldots, (x_n, y_n)\}$ is given.

Step 1. Compute the FQ estimate r_{FQ} of the correlation coefficient ρ:

$$r_{FQ} = \frac{FQ_n^2(\tilde{x} + \tilde{y}) - FQ_n^2(\tilde{x} - \tilde{y})}{FQ_n^2(\tilde{x} + \tilde{y}) + FQ_n^2(\tilde{x} - \tilde{y})}, \qquad (9.13)$$

where FQ_n is given by (4.37) and $\tilde{x}$ and $\tilde{y}$ are standardized variables

$$\tilde{x} = \frac{x - \text{med } x}{\text{MAD } x}, \quad \tilde{y} = \frac{y - \text{med } y}{\text{MAD } y}.$$

Step 2. Compute the angle α

$$\alpha = \arctan\left(\frac{2r_{FQ}\hat{\sigma}_1\hat{\sigma}_2}{\hat{\sigma}_1^2 - \hat{\sigma}_2^2}\right),$$

where $\hat{\sigma}_1 = FQ_n(x)$ and $\hat{\sigma}_2 = FQ_n(y)$.

Step 3. Compute the location of the data (x_C, y_C) using the affine equivariant spatial median (use the R-package) or the component-wise medians $x_C = \text{med } x$ and $y_C = \text{med } y$.

Step 4. Shift the initial axes (x, y) by (x_C, y_C) and rotate them by the angle α, transforming them to the principal component axes (x', y') (Cramér 1946)

$$x' = (x - x_C)\cos\alpha + (y - y_C)\sin\alpha,$$
$$y' = -(x - x_C)\sin\alpha + (y - y_C)\cos\alpha.$$

Step 5. Compute the robust version of the Mahalanobis distance $d_M(Z_i)$ of each point $Z_i = (x_i, y_i)$ from the estimated location center $\mu_C = (x_C, y_C)$

$$d_M(Z_i) = \sqrt{(Z_i - \mu_C)^T S^{-1}(Z_i - \mu_C)},$$

where the robust covariance matrix S has the following components:

$$S_{11} = \hat{\sigma}_1^2, \quad S_{22} = \hat{\sigma}_2^2,$$
$$S_{12} = S_{21} = r_{FQ}\hat{\sigma}_1\hat{\sigma}_2.$$

Step 6. Choose 50% of the closest points to the estimated center and build the inner region (hinge) as the convex hull of those points.

Step 7. For building the fence (the outer region), do the following substeps:

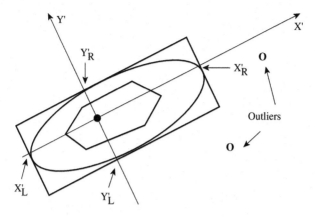

Figure 9.5   FQ-bivariate boxplot

7.1: compute the boundaries

$$x'_L = \max\{x'_{(1)}, x'_C - k\ FQ_n(x')\},$$

$$y'_L = \max\{y'_{(1)}, y'_C - k\ FQ_n(y')\},$$

$$x'_R = \min\{x'_{(n)}, x'_C + k\ FQ_n(x')\},$$

$$y'_R = \max\{y'_{(n)}, y'_C + k\ FQ_n(y')\},$$

where k is a free parameter lying in the interval $(2, 3)$ to be specified from the additional considerations;

7.2: build the rectangular with these boundaries (its sides are parallel to the axes (x', y')) (see Fig. 9.5);

7.3: inscribe the ellipse into that of a rectangle;

7.4: build the convex hull of the points in that ellipse and take it as the fence.

9.6.2   Bivariate *FQ*-boxplot performance

Here, we compare the performance of the proposed FQ-boxplot with the Tukey bagplot (Rousseeuw *et al.* 1999). First, we focus on how well the boxplot reveals suspicious points in the data cloud (see Fig. 9.6).

The bivariate normal distribution with density $N(x, y; \mu_1, \mu_2, \sigma_1, \sigma_2, \rho)$, where the parameters μ_1 and μ_2 are the means and σ_1 and σ_2 are the standard deviations of the r.v.'s X and Y, respectively, is used to model regular data.

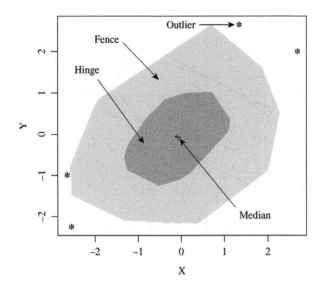

Figure 9.6   FQ-bivariate boxplot realization: $N(x, y; 0, 0, 1, 1, 0.5)$, $n = 200$, 4 suspicious observations

To model the contaminated data, we use the Tukey gross error model

$$f(x, y) = (1 - \epsilon)N(x, y; 0, 0, 1, 1, \rho) + \epsilon N(x, y; \mu_1, \mu_2, k, k, -\rho), \quad k > 1.$$

The experiment is designed as in Filzmoser (2005), where data points are modeled injecting the given percent of contaminated data from 5 to 50%. For both types of bivariate boxplots, the percent of correct outliers and the percent of data points identified erroneously as outliers are computed. The legend in Fig. 9.7 explains the graphics.

The first experiment is performed on the shift contaminated data: the results are exhibited in Figs 9.8 to 9.10.

The second experiment models the scale contamination of data: the results are exhibited in Figs 9.11 to 9.13.

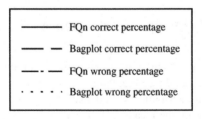

Figure 9.7   Legend

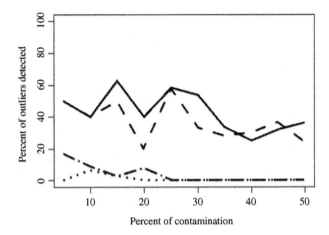

Figure 9.8   Shift contamination: $\rho = 0, \mu_1 = 3, \mu_2 = 3, k = 1, \rho' = 0; n = 50$

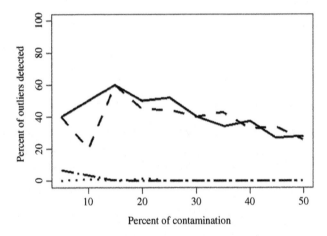

Figure 9.9   Shift contamination: $\rho = 0, \mu_1 = 3, \mu_2 = 3, k = 1, \rho' = 0; n = 100$

9.6.3   Measuring the elliptical deviation from the convex hull

We introduced the elliptical model-based boxplot to identify outliers that do not fit this model. It is of interest to compare the FQ-boxplot with the other boxplots, which assume bivariate normal distribution of data. Here, we propose to estimate the deviation from an elliptical shape by measuring the relative error of its deviation from the convex hull of the outer region of the FQ-boxplot (see Fig. 9.14). The mean square

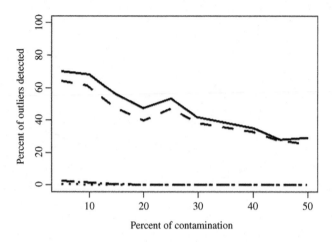

Figure 9.10   Shift contamination: $\rho = 0, \mu_1 = 3, \mu_2 = 3, k = 1, \rho' = 0; n = 1,000$

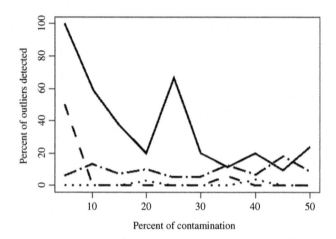

Figure 9.11   Scale contamination: $\rho = 0.9, \mu_1 = 0, \mu_2 = 0, k = 3, \rho' = 0; n = 50$

relative error is computed by the following formula:

$$I = \frac{1}{2\pi} \int_0^{2\pi} \left(\frac{R(\phi) - R_e(\phi)}{R_e(\phi)} \right)^2 d\phi \tag{9.14}$$

The experiment is repeated 100 times with different parameter settings for the contaminated Gaussian model. The cloud of points consists of 500 points. In our case, on the whole, a cloud with a larger number of points resembles the elliptical structure.

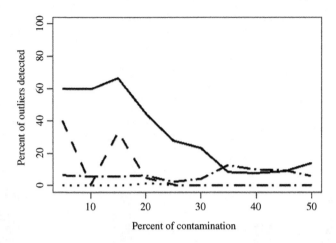

Figure 9.12   Scale contamination: $\rho = 0.9, \mu_1 = 0, \mu_2 = 0, k = 3, \rho' = 0; n = 100$

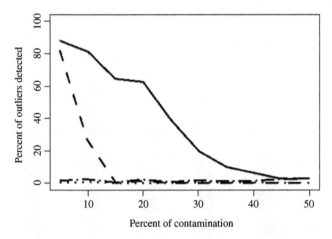

Figure 9.13   Scale contamination: $\rho = 0.9, \mu_1 = 0, \mu_2 = 0, k = 3, \rho' = 0; n = 1,000$

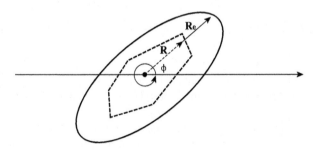

Figure 9.14   Ellipse deviation estimate

Table 9.6   Types of contaminated Gaussian distribution densities

Model number	Contaminated Gaussian distribution densities
1	$0.9N(x, y; 0, 0, 1, 0.5, 0) + 0.1N(x, y; 0, 0, 3, 3, 0)$
2	$0.9N(x, y; 0, 0, 1, 0.5, 0) + 0.1N(x, y; 0, 0, 10, 10, 0)$
3	$0.5N(x, y; 0, 0, 1, 0.5, 0.5) + 0.5N(x, y; 0, 0, 3, 3, -0.5)$
4	$0.5N(x, y; 0, 0, 1, 0.5, 0.5) + 0.5N(x, y; 0, 0, 3, 3, 0)$
5	$0.5N(x, y; 0, 0, 1, 0.5, 0.9) + 0.1N(x, y; 0, 0, 10, 10, -0.9)$

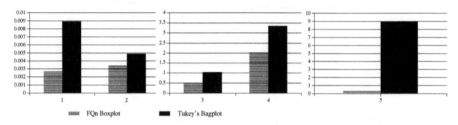

Figure 9.15   Ellipse shape deviations of the Tukey bagplot and FQ-boxplot

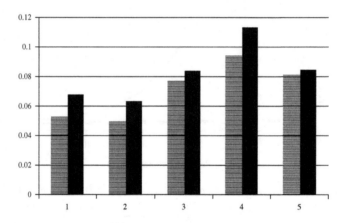

Figure 9.16   The variances of location estimates: grey—FQ-boxplot, black—Tukey's bagplot

Parameter settings for contaminated data are displayed in Table 9.6. These models are used as input data for comparing both types of bivariate boxplots. In Fig. 9.15, the comparison is performed with the Tukey bagplot.

The results in Fig. 9.15 exhibit only the quality of reproducing the elliptical shape, not taking into account the error of estimation of the center of the data cloud. The comparison of the variances of the estimates of location is given in Fig. 9.16.

9.7 Summary

In this chapter, we apply the highly robust and efficient estimates of scale and correlation proposed and studied in Chapter 4 and Chapter 5 to the tools of exploratory data analysis. These applications comprise new versions of the boxplot techniques aimed at the visualization of both univariate and bivariate data and new methods and algorithms of detection of outliers in the data, also univariate and bivariate. For multivariate data, the topic of detection of outliers is most challengeable, and at present it is the objective of intensive statistical research. There are many researchers actively working in this area (see, for example, Atkinson *et al.* 2004, Riemann *et al.* 2008). Amongst them, we would like to note the researches on processing multivariate and functional data by the advantageous statistical and numerical techniques based on the concepts of data depths and distances (Hubert *et al.* 2015, Rousseeuw and Hubert 2011).

References

Atkinson AC and Riani M 1997 Bivariate boxplots, multiple outliers, multivariate transformations and discriminant analysis. *Environmetrics* **8**, 583–602.

Atkinson A, Riani M, and Cerioli A 2004 *Exploring Multivariate Data with the Forward Search*, Springer.

Barnett V and Lewis T 1978 *Outliers in Statistical Data*, Wiley.

Becketti S and Gould W 1987 Rangefinder box plots: a note. *The American Statistician* **41**, 149.

Cramér H 1946 *Mathematical Methods of Statistics*, Princeton University Press, Princeton.

Filzmoser P 2005 Identification of multivariate outliers: a performance study. *Austrian J. of Statistics* **34**, 127–138.

Goldberg KM and Iglewicz B 1992 Bivariate extensions of the boxplot. *Technometrics* **34**, 307–320.

Grubbs FE 1950 Sample criteria for testing outlying observations. *Ann. Math. Stat.* **21**, 27–58.

Grubbs FE 1969 Procedures for detecting outlying observations in samples. *Techometrics* **11**, 1–21.

Hampel FR, Ronchetti E, Rousseeuw PJ, and Stahel WA 1986 *Robust Statistics. The Approach Based on Influence Functions*, Wiley.

Hastie T, Tibshirani R, and Friedman J 2009 *The Elements of Statistical Learning*, Springer.

Hoaglin DC, Mosteller F and Tukey JW 1983 *Understanding Robust and Exploratory Data Analysis*, Wiley.

Hubert M, Rousseeuw PJ, and Segaert P 2015 Multivariate and functional classification using depth and distance. *arXiv.1504.01128v1[stat.ME]*.

Lehmann EL and Romano JP 2008 *Understanding Robust and Exploratory Data Analysis*, Wiley.

McGill R, Tukey JW, and Larsen WA 1978 Variations of box plots. *The American Statistician* **32**, 12–16.

Potter K, Hagen H, Kerren A, and Dannenman P 2006 Methods for presenting statistical information: the box plot. *Visualization of Large and Unstructured Data Sets* **S-4**, 97–106.

Riemann C, Filzmoser P, Garrett R, and Dutter R 2008 *Statistical Data Analysis Explained*, Wiley.

Rousseeuw PJ and Hubert M 2011 Robust statistics for outlier detection. *Data Mining Knowledge Discovery*, 73–79.

Rousseeuw PJ, Ruts I, and Tukey JW 1999 The bagplot, a bivariate boxplot. *The American Statistician* **53**, 382–387.

Shevlyakov GL and Vilchevski NO 2002 *Robustness in Data Analysis: criteria and methods*, VSP, Utrecht.

Tongkumchum P 2005 Two-dimensional box plot. *Songklanakarin J. Sci. Technol.* **27**, 859–866.

Tukey JW 1960 A survey of sampling from contaminated distributions. In *Contributions to Probability and Statistics*. (ed. Olkin I), pp. 448–485, Stanford Univ. Press.

Tukey JW 1977 *Exploratory Data Analysis*, Addison–Wesley.

Warren WE and Banfield J 2003 The box-percentile plot. *J. Statist. Software* **8**, 1–14.

Zani S, Riani M, and Corbellini A 1998 Robust bivariate boxplots and multiple outlier detection. *Comput. Statistics and Data Analysis* **28**, 257–270.

10

Applications to Time Series Analysis: Robust Spectrum Estimation

Conventional methods of power spectrum estimation are very sensitive to the presence of outliers in the data (Kleiner *et al.* 1979); thus generally the issues of robustness are of vital importance within this area of statistics of random processes. In this chapter various robust versions of the conventional methods of power spectra estimation are considered. Their performance evaluation is studied in autoregressive models, not only contaminated by large amplitude outliers but also in a new model of violation of an inner structure of a random process. It is found that the best robust estimates of power spectrum are based on robust highly efficient estimates of autocovariances. Several open problems for future research are formulated.

10.1 Preliminaries

Much less attention is devoted in the literature to robust estimation of data spectra as compared to robust estimation of location, scale, regression, and covariance (Hampel *et al.* 1986; Huber 1981; Maronna *et al.* 2006). However, it is necessary to study these problems due to both their theoretical and practical importance (estimation of time series power spectra in various applications, such as communication, geophysics, medicine, etc.), and also because of the instability of classical methods of power spectrum estimation in the presence of outliers in the data (Kleiner *et al.* 1979).

Robust Correlation: Theory and Applications, First Edition. Georgy L. Shevlyakov and Hannu Oja.
© 2016 John Wiley & Sons, Ltd. Published 2016 by John Wiley & Sons, Ltd.
Companion Website: www.wiley.com/go/Shevlyakov/Robust

There are several classical approaches to estimation of the power spectrum of time series, for example, via the nonparametric periodogram and the Blackman–Tukey formula methods, as well as via the parametric Yule–Walker and filter-based methods (Blackman and Tukey 1958; Bloomfield 1976; Brockwell and Davis 1991). Thereafter, we will consider their various robust versions: to the best of our knowledge, their first systematic study is made in the dissertation thesis of Bernhard Spangl (2008).

In what follows, we partially use the aforementioned study as a baseline, but most follow the classification of robust methods of power spectrum estimation given in Spangl and Dutter (2005) and Spangl (2008), specify them, and propose several new approaches with their comparative performance evaluation. Basically, to obtain good robust estimates of the power spectrum, we use highly efficient robust estimates of scale and correlation (Shevlyakov and Smirnov 2011; Shevlyakov et al. 2012; Smirnov and Shevlyakov 2014).

Our main goals are both to outline the existing approaches to robust estimation of power spectrum and to indicate open problems, so our work is partially a review and partially a program for future research.

The remainder of this chapter is as follows. In Section 10.2, classical methods of power spectrum estimation are briefly enlisted. In Section 10.3, robust modifications of classical approaches are formulated. In Section 10.4, a few preliminary results on the comparative study of the performance evaluation of various robust methods are represented. In Section 10.5, some conclusions and open problems for future research are drawn.

10.2   Classical Estimation of a Power Spectrum

10.2.1   Introductory remarks

Consider the second-order representation for a stationary time series $\{x_t\}$ by its mean $\mu = E(x_t)$ and its autocovariance sequence

$$c_{xx}(\tau) = cov(x_t, x_{t+\tau}), \quad \tau = 0, 1, 2, \dots.$$

The second-order representation completely characterizes the Gaussian case, but this is not so for an arbitrary stationary time series. However, since the specification of higher-order properties of a time series is a rather difficult task, we confine ourselves to estimating second-order properties such as the mean and the covariances.

The nonparametric power spectrum definition

Any second-order stationary process $\{x_t\}$ has the following spectral representation:

$$x_t = \int_{-1/2}^{1/2} \exp(i2\pi tf)dZ(f), \tag{10.1}$$

where $Z(f)$, $-1/2 \leq f \leq 1/2$ is a process with orthogonal increments. The process $Z(f)$ defines a monotone increasing function $F(f)$ called the *spectral distribution function* (Doob 1953)

$$F(f) = E(|Z(f)|^2), \quad dF(f) = E(|dZ(f)|^2),$$

$$F\left(-\frac{1}{2}\right) = 0, \quad F\left(\frac{1}{2}\right) = \sigma^2 = c_{xx}(0). \tag{10.2}$$

When $F(f)$ exists, its derivative $S(f) = F'(f)$ is called the *spectral density function* of $\{x_t\}$ of either the *power spectral density* or the *power spectrum*.

From (10.1) and the orthogonal increments property of $Z(f)$ it follows that

$$c_{xx}(\tau) = \int_{-1/2}^{1/2} \exp(i2\pi\tau f)S(f)df, \quad \tau = 0, \pm 1, \dots, \tag{10.3}$$

when the spectrum $S(f)$ exists. Therefore, the covariances $c_{xx}(\tau), \tau = 0, \pm 1, \dots$, are the Fourier coefficients of $S(f)$, namely

$$S(f) = \sum_{\tau=-\infty}^{\infty} c_{xx}(\tau) \exp(-i2\pi f\tau). \tag{10.4}$$

The Fourier series transform (10.4) means that any stationary time series can be regarded as the limit of finite sums of sinusoids of the form $A_i \cos(2\pi f_i t + \Phi_i)$ over all frequencies $f = f_i$. The amplitudes $A_i = A(f_i)$ and phases $\Phi_i = \Phi(f_i)$ are uncorrelated random variables with $S(f) \approx E(A^2(f))$ (Kleiner *et al.* 1979).

The autoregressive power spectrum

Among the main parametric models for stationary time series, namely the autoregressive (AR) model, the moving average (MA) model, and the combined autoregressive moving average ($ARMA$) model, we are especially interested in the p-order $AR(p)$ process

$$x_t = \sum_{\tau=1}^{p} \phi_\tau x_{t-\tau} + \epsilon_t, \tag{10.5}$$

where the white noise innovation process $\{\epsilon_t\}$ is formed by Gaussian uncorrelated random variables with mean 0 and finite variance σ_ϵ^2.

First, finite-order AR processes are extremely important in the representation of physically significant linear random processes. For instance, a continuous $AR(p)$ process may be written as a p-order differential equation of the form

$$k_p \frac{d^p x(t)}{dt^p} + k_{p-1} \frac{d^{p-1}x(t)}{dt^{p-1}} + \dots + x(t) = \epsilon(t),$$

In other words, the $AR(p)$ processes described by (10.5) naturally generalize the model of ordinary differential equations for the case of random noise perturbations. In the particular case $p = 2$, this equation describes either the behavior of a damped linear second-order mechanical system or an electrical circuit that is excited by random impulses.

Second, the $AR(p)$ process (10.5) is a maximum entropy approximation to an arbitrary stationary random process (Van den Bos 1971). Thus in this sense, the finite-order AR process can be regarded as the least favorable approximation to a stationary random process.

The power spectrum of the $AR(p)$ process (10.5) is given by (Bloomfield 1979)

$$S(f) = \frac{\sigma_\epsilon^2}{\left| 1 - \sum_{\tau=1}^{p} \phi_\tau \exp\{-i2\pi f\tau\} \right|^2}. \tag{10.6}$$

10.2.2   Classical nonparametric estimation of a power spectrum

Power spectrum periodograms

The nonparametric approach to estimation of a power spectrum is based on smoothing periodograms (Bloomfield 1979).

Let $x_t, t = 1,\ldots,n$, be a second-order stationary time-series with zero mean. Assume that the time intervals between two consecutive observations are equally spaced with duration Δt. Then the periodogram is defined as follows:

$$\widehat{S}_P(f) = \frac{\Delta t}{n} \left| \sum_{t=1}^{n} x_t \exp\{-i2\pi ft\Delta t\} \right|^2 \tag{10.7}$$

over the interval $(-f_N, f_N)$, where f_N is the Nyquist frequency: $f_N = 1/(2\Delta t)$.

It can be seen that the periodogram $\widehat{S}_P(f)$ computed at the frequency $f = f_k = k/(n\Delta t)$, where k is an integer such that $k \leq \lfloor n/2 \rfloor$ is equal to the squared absolute value of the discrete Fourier transform $X(f_k)$ of the sequence $x_1,\ldots,x_n$ given by the following formula:

$$X(f_k) = \Delta t \sum_{t=1}^{n} x_t \exp\{-i2\pi f_k t\Delta t\}. \tag{10.8}$$

To provide consistency of the periodogram estimate $\widehat{S}_P(f)$ and to reduce its bias and variance, the conventional techniques based on tapering and averaging of periodograms are used (Welch 1967; Bloomfield 1979).

The Blackman–Tukey formula

The Blackman–Tukey formula gives the representation of formula (10.7) via the sample autocovariances $\widehat{c}_{xx}(\tau)$ of the time series x_t (Blackman and Tukey 1958)

$$\widehat{S}_P(f) = \widehat{S}_{BT}(f) = \Delta t \sum_{\tau=-(n-1)}^{n-1} \widehat{c}_{xx}(\tau) \exp\{-i2\pi f \tau \Delta t\} \tag{10.9}$$

where

$$\widehat{c}_{xx}(\tau) = \frac{1}{n} \sum_{t=1}^{n-|\tau|} (x_{t+\tau} - m)(x_t - m) \tag{10.10}$$

with

$$m = \frac{1}{n} \sum_{t=1}^{n} x_t.$$

Certainly, there are other methods of estimating the autocovariances, but the estimate (10.10) has a generally smaller mean squared error that that of other estimates (Jenkins and Watts 1968).

10.2.3   Parametric estimation of a power spectrum

The widely used form of a parametric power spectrum estimation procedure exploits the autoregressive $AR(p)$ model (10.5) for the underlying power spectrum $S(f)$. The power spectrum estimate $\widehat{S}_{AR}(f)$ has the form (Bloomfield 1979)

$$\widehat{S}_{AR}(f) = \frac{\Delta t \widehat{\sigma}_\epsilon^2}{\left| 1 - \sum_{j=1}^{p} \widehat{\phi}_j \exp\{-i2\pi f j \Delta t\} \right|^2}, \quad |f| \leq f_N, \tag{10.11}$$

where $\widehat{\phi}_1, \ldots, \widehat{\phi}_p$ and $\widehat{\sigma}_\epsilon^2$ are the maximum likelihood estimates of the model parameters.

10.3   Robust Estimation of a Power Spectrum

10.3.1   Introductory remarks

A natural way to provide robustness of the classical estimates of a power spectrum is based on using highly robust and efficient estimates of location, scale, and correlation in the classical estimates. Here we enlist several highly robust and efficient estimates of location, scale, and correlation, which are used in further constructions.

Robust estimate of location

To estimate means, the highly robust estimate of location, namely the sample median, is used.

Robust estimates of scale

To estimate standard deviations, we use two estimates: the median absolute deviation and the fast low-complexity FQ_n-estimate. The median absolute deviation $\mathrm{MAD}_n\, x = \mathrm{med}\, |x - \mathrm{med}\, x|$ is a highly robust estimate of scale with the maximal value of the breakdown point 0.5, but its efficiency is only 0.37 at the normal distribution. In Chapter 4, a highly robust and efficient M-estimate of scale denoted by FQ_n is proposed and studied

$$FQ_n(x) = 1.483\ \mathrm{MAD}_n\, x\ \left(1 - \frac{Z_0 - n/\sqrt{2}}{Z_2}\right),\qquad (10.12)$$

$$Z_k = \sum_{i=1}^{n} u_i^k e^{-u_i^2/2},\quad u_i = \frac{x_i - \mathrm{med}\, x}{1.483\ \mathrm{MAD}_n\, x},\quad k = 0, 2;\quad i = 1,\ldots,n.$$

Recall that the efficiency and breakdown point of the FQ_n-estimate are equal to 0.81 and to 0.5, respectively.

Robust estimates of correlation

To estimate covariances, we use the MAD- and FQ-correlation coefficients.

A remarkable robust minimax bias and variance MAD-correlation coefficient with the breakdown point 0.5 and efficiency 0.37 is given by (5.11)

$$r_{MAD}(x, y) = \frac{\mathrm{MAD}^2 u - \mathrm{MAD}^2 v}{\mathrm{MAD}^2 u + \mathrm{MAD}^2 v},\qquad (10.13)$$

where u and v are the robust principal variables (see Chapter 3)

$$u = \frac{x - \mathrm{med}\, x}{\sqrt{2}\ \mathrm{MAD}\, x} + \frac{y - \mathrm{med}\, y}{\sqrt{2}\ \mathrm{MAD}\, y},\quad v = \frac{x - \mathrm{med}\, x}{\sqrt{2}\ \mathrm{MAD}\, x} - \frac{y - \mathrm{med}\, y}{\sqrt{2}\ \mathrm{MAD}\, y}.$$

A much higher efficiency of 0.81 with the same breakdown point 0.5 can be provided by using the FQ-correlation coefficient

$$r_{FQ}(x, y) = \frac{FQ^2(u) - FQ^2(v)}{FQ^2(u) + FQ^2(v)}.\qquad (10.14)$$

10.3.2   Robust analogs of the discrete Fourier transform

Robust L_p-norm analogs of the *DFT*

Since the computation of the discrete Fourier transform (*DFT*) (10.8) is the first step in periodogram estimation of a power spectrum, consider the following robust L_p-norm analogs of the *DFT*.

As the classical *DFT* $X(f)$ can be obtained through the L_2-norm approximation

$$X(f) \propto \arg\min_{Z} \sum_{t=1}^{n} |y_t(f) - Z|^2$$

to the Fourier transformed data

$$y_t(f) = x_t \exp\{-i2\pi f\, t\Delta t, \quad t = 1, \dots, n,$$

the L_p-norm analog of $X(f)$ (up to the scale factor) is defined as follows:

$$X_{L_p}(f) \propto \arg\min_{Z} \left\{ \sum_{t=1}^{n} |y_t(f) - Z|^p \right\}^{1/p}, \quad 1 \le p < \infty. \tag{10.15}$$

The case of $1 \le p < 2$, and especially the L_1-norm or the median Fourier transform, are of our particular interest (Pashkevich and Shevlyakov 1995; Spangl and Dutter 2005; Spangl 2008):

$$X_{L_1}(f) \propto \arg\min_{Z} \left\{ \sum_{t=1}^{n} |y_t(f) - Z| \right\}. \tag{10.16}$$

The other possibilities such as the component-wise, spatial medians, and trimmed mean analogs of the DFT are also considered in Pashkevich and Shevlyakov (1995) and Spangl (2008).

Robust cross-product DFT analogs

Here we exploit the well-known relation connecting the cross-product, covariance and means:

$$\sum x_t z_t = n\, cov(x, z) + n\, \bar{x}\, \bar{z}. \tag{10.17}$$

Since the *DFT* is decomposed into the real (cosine) and imaginary (sine) parts as $X(f) = X^c(f) + iX^s(f)$, we apply Equation (10.17) to them using robust estimates of covariances and means. Denoting the results of application of (10.17) to the cosine and sine parts as $X_{CP}^c(f)$ and $X_{CP}^s(f)$, respectively, we define the robust cross-product analog of the classical *DFT* as follows:

$$X_{CP}(f) = X_{CP}^c(f) + iX_{CP}^s(f). \tag{10.18}$$

In the case of the conventional estimate of the covariance in (10.17), namely, the sample covariance $n^{-1}\sum(x_t - \bar{x})(z_t - \bar{z})$, we get the classical definition of the *DFT*.

10.3.3   Robust nonparametric estimation

Now we apply the aforementioned robust analogs of the*DFT* as well as the highly robust and efficient estimates of scale and correlation to the classical nonparametric estimation of power spectrum.

Robust estimation by robust periodograms

Here we apply the robust L_p-norm analogs of the DFT to the classical periodogram $\widehat{S}_P(f)$ (10.7):

$$\widehat{S}_{L_p}(f) \propto \left|X_{L_p}(f)\right|^2. \tag{10.19}$$

In what follows, the L_1- or the median periodogram is our particular interest.

Robust estimation by the robust version of the Blackman–Tukey formula

In order to construct robust modifications of the Blackman–Tukey formula, we have to consider robust estimates of autocovariances $\widehat{c}_{xx}(\tau)$ instead of the conventional ones used in (10.9). These robust estimates are based on the highly robust *MAD* and *FQ* estimates of scale and correlation (10.13) and (10.14)

$$\widehat{c}_{MAD}(t, t-\tau) = r_{MAD}(x_t, x_{t-\tau})\text{MAD}^2 x_t,$$
$$\widehat{c}_{FQ}(t, t-\tau) = r_{FQ}(x_t, x_{t-\tau})FQ^2(x_t). \tag{10.20}$$

To provide the required Toeplitz property (symmetry, equal elements on subdiagonals, semipositive definiteness) of the autocovariance matrix $\widehat{C}_{xx}$ built of the pair-wise robust autocovariances (10.20), a new effective iterative transformation algorithm applied to the robust autocorrelation matrix $\widehat{P}_{xx}$ (Letac 2011) is used:

Step 1: average the pair-wise autocorrelations $r_{MAD}(x_t, x_{t-\tau})$ and $r_{FQ}(x_t, x_{t-\tau})$ on the subdiagonals of the robust autocorrelation matrix $\widehat{P}_{xx}$ and denote the results as $r_{MAD}(\tau)$ and $r_{FQ}(\tau)$, respectively;

Step 2: check the positive definiteness of the obtained correlation matrix $\widehat{P}$: if it is positive definite then set the elements of the robust autocovariance matrix as follows:

$$\widehat{c}_{MAD}(\tau) = r_{MAD}(\tau)\text{MAD}^2 x_t,$$
$$\widehat{c}_{FQ}(\tau) = r_{FQ}(\tau)FQ^2(x_t); \tag{10.21}$$

otherwise transform all the elements $\hat{\rho}_{ij}$, $i < j$ of the autocorrelation matrix $\hat{P}_{xx}$ by the compressing factor

$$\hat{\rho}_{ij}^{(1)} = \sin\left(\frac{\pi}{2}\,\hat{\rho}_{ij}\right), \quad i < j. \tag{10.22}$$

As a rule, 3–4 iterations are sufficient to get positive definiteness of the correlation matrix. Thus, the Toeplitz transformed estimates are substituted into Equation (10.9) and the corresponding robust estimates of power spectrum are denoted as $\hat{S}_{MAD}(f)$ and $\hat{S}_{FQ}(f)$, respectively.

10.3.4 Robust estimation of power spectrum through the Yule–Walker equations

A classical approach to estimation of autoregressive parameters $\phi_1, \ldots, \phi_p$ in (10.5) is based on the solution of the linear system of the Yule–Walker equations (Bloomfield 1976):

$$
\begin{cases}
\hat{c}(1) = & \hat{c}(0)\hat{\phi}_1 + \hat{c}(1)\hat{\phi}_2 & + \cdots + & \hat{c}(p-1)\hat{\phi}_p \\
\hat{c}(2) = & \hat{c}(1)\hat{\phi}_1 + \hat{c}(2)\hat{\phi}_2 & + \cdots + & \hat{c}(p-2)\hat{\phi}_p \\
\quad \cdot \quad \cdot \quad \cdot \quad \cdot \quad \cdot \quad \cdot \quad \cdot \\
\hat{c}(p) = \hat{c}(p-1)\hat{\phi}_1 + \hat{c}(p-2)\hat{\phi}_2 & + \cdots + & \hat{c}(0)\hat{\phi}_p.
\end{cases} \tag{10.23}
$$

The estimate of the innovation noise variance is defined by

$$\hat{c}(0) = \hat{c}(1)\hat{\phi}_1 + \hat{c}(2)\hat{\phi}_2 + \cdots + \hat{c}(p)\hat{\phi}_p + \hat{\sigma}_\epsilon^2. \tag{10.24}$$

Substituting robust estimates of autocovariances (10.21) into (10.23) and (10.24), we get the robust analogs of the Yule–Walker equations. Solving these equations, we arrive at the robust estimate of power spectrum in the form of (10.11).

10.3.5 Robust estimation through robust filtering

A wide collection of robust methods of power spectra estimation is given by various robust filters (Kalman, Masreliez, ACM-type, robust least squares, filter-cleaners, etc.) providing preliminary cleaning the data with the subsequent power spectra estimation. An extended comparative experimental study of robust filters is made in Spangl and Dutter (2005) and Spangl (2008); below we compare some of those results with ours (Shevlyakov et al. 2014).

10.4   Performance Evaluation

10.4.1   Robustness of the median Fourier transform power spectra

The median Fourier transform power spectrum estimate $\widehat{S}_{L_1}(f) \propto |X_{L_1}(f)|^2$ inherits the maximum value of the sample median breakdown point $\epsilon^* = 1/2$.

Theorem 10.4.1. *The breakdown point of* $\widehat{S}_{L_1}(f)$ *is equal to* $1/2$. *Here, the breakdown point* ϵ^* *is understood as the maximal ratio of the number of unbounded observations in the data sample under which the estimate still remains bounded (Hampel et al. 1986).*

Figures 10.1 to 10.3 illustrate this phenomenon: the observed realization is the mixture of $\sin(\pi t/4)$ and $\sin(\pi t/8)$ observed on the 40% and 60% intervals of duration, respectively. In this case, the classical periodogram indicates the presence of both peaks whereas the median periodogram indicates only one spectrum peak, which corresponds to the domination of the observation time signal $\sin(\pi t/8)$.

10.4.2   Additive outlier contamination model

In the Monte Carlo experiment, as has already been mentioned, an autoregressive model is used because of the following two reasons: first, it is a direct stochastic counterpart of an ordinary differential equation; second, an autoregressive model is the maximum entropy parametric approximation to an arbitrary strictly stationary random process (Cover and Thomas 1991).

In this experiment, we use the autoregressive model of the second-order AR(2)

$$x_t = x_{t-1} - 0.9\, x_{t-2} + \epsilon_t$$

and of the fourth order AR(4)

$$x_t = x_{t-1} - 0.9\, x_{t-2} + 0.5\, x_{t-3} - 0.1\, x_{t-4} + \epsilon_t,$$

together with Gaussian additive outliers (AO) generated by the normal law with density $N(x; 0, 10)$. The comparative study is performed on different sample sizes n and numbers of trials M (see Figs 10.4 to 10.7).

10.4.3   Disorder contamination model

In this chapter we propose a contamination model dubbed as a disorder contamination describing the violations of the thin structure of a random process, when an AR-process is shortly changed for another and then it returns to the previous state.

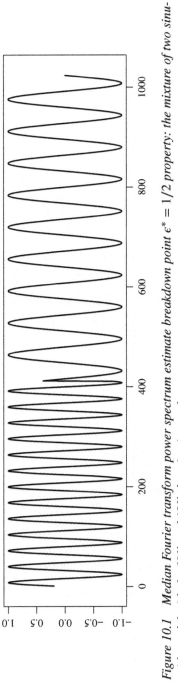

Figure 10.1   Median Fourier transform power spectrum estimate breakdown point $\epsilon^* = 1/2$ property: the mixture of two sinusoids model with the 60% and 40% duration intervals.

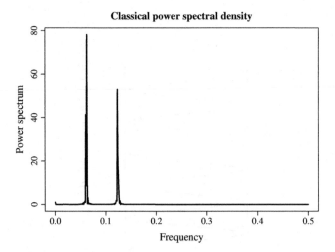

Figure 10.2   Median Fourier transform power spectrum estimate breakdown point
$\epsilon^* = 1/2$ *property: the classical raw periodogram power spectrum estimate.*

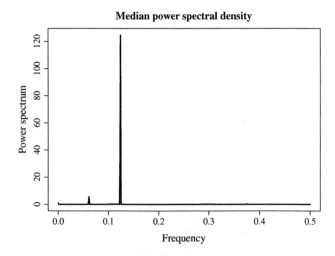

Figure 10.3   Median Fourier transform power spectrum estimate breakdown point
$\epsilon^* = 1/2$ *property: the median periodogram power spectrum estimate.*

Below, the following disorder model is used:

$$x_t = -0.6\, x_{t-1} - 0.6\, x_{t-2} + \epsilon_t$$

as the main process observed at $t = 0, 1, \ldots, 400$ and at $t = 512, \ldots, 1024$;

$$x_t = x_{t-1} - 0.9\, x_{t-2} + \epsilon_t$$

as the disorder process at $t = 401, \ldots, 511$.

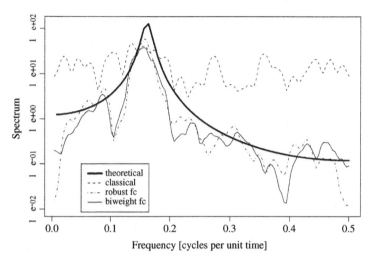

Figure 10.4 Power spectrum estimation with robust filter-cleaners: AR(2) model with 10% AO *contamination; n* = 100, *M* = 400 *(Spangl 2008).*

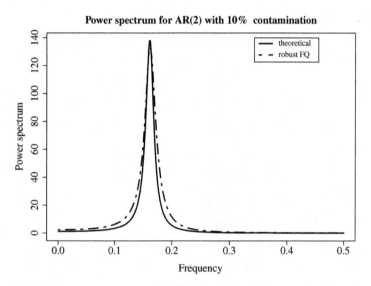

Figure 10.5 Power spectra estimation by the Yule–Walker method: AR(2) model with 10% AO *contamination; n* = 128, *M* = 2000.

The results of signal processing are exhibited in Figs 10.8 and 10.9: the classical periodogram indicates two spectrum peaks of the main and contamination processes, whereas the median periodogram indicates only one peak of the main process.

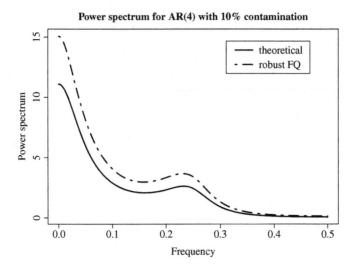

Figure 10.6   Power spectra estimation by the Yule–Walker method: AR(4) model with 10% AO *contamination;* $n = 128$, $M = 2000$.

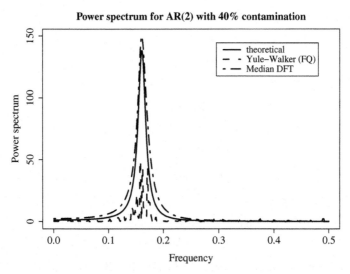

Figure 10.7   Power spectra estimation: AR(2) model with 40% AO *contamination;* $n = 1024$, $M = 50$.

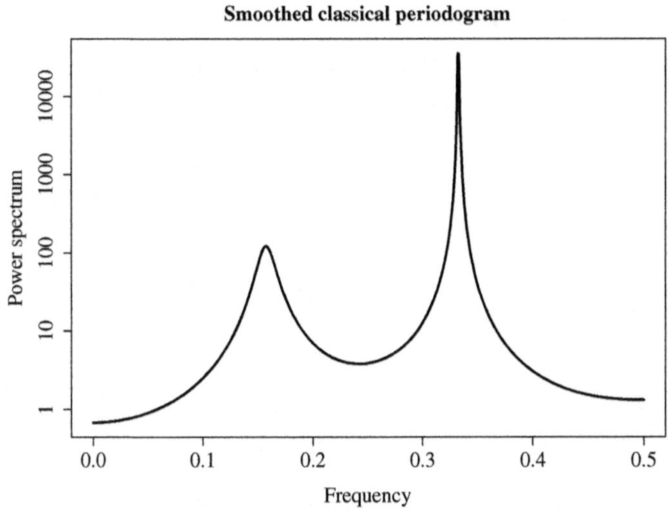

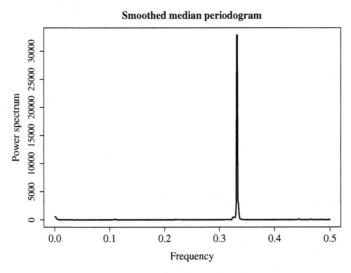

Figure 10.8   Smoothed classical power spectrum estimation in the disorder model with 10% *contamination;* $n = 1024$, $M = 50$.

Figure 10.9   Smoothed robust power spectrum estimation in the disorder model with 10% *contamination;* $n = 1024$, $M = 50$.

10.4.4   Concluding remarks

1. From Figs 10.4 to 10.6 it follows that the classical periodogram is catastrophically bad under contamination and that the robust FQ Yule–Walker estimate considerably outperforms robust filter methods.

2. From Fig. 10.7 it follows that the bias of estimation by the FQ Yule–Walker method increases with growing dimension and contamination. It can also be shown that under heavy contamination, the median periodogram and the robust Blackman–Tukey method outperform the FQ Yule–Walker method in estimating the peak location, although they have a considerable bias in amplitude.

3. The median periodogram exhibits high robustness both with respect to amplitude outliers and to disorder contamination.

10.5   Summary

All the presented results are of a preliminary character indicating many open problems: first of all, the statistical analysis of the asymptotic properties of the proposed estimates, the reduction of their bias and variance on finite samples, and the study of the properties of the direct and inverse L_p-norm analogs of the Fourier transform.

References

Blackman RB and Tukey JW 1958 *The Measurement of Power Spectra*, Dover, New York.

Bloomfield P 1976 *Fourier Analysis of Time Series: An Introduction*, Wiley.

Brockwell PJ and Davis RA 1991 *Time Series: Theory and Methods*, Springer, New York.

Cover TM and Thomas JA 1991 *Elements of Information*, Wiley, New York.

Doob JL 1953 *Stochastic Processes*, Wiley, New York.

Hampel FR, Ronchetti E, Rousseeuw PJ, and Stahel WA 1986 *Robust Statistics. The Approach Based on Influence Functions*, Wiley.

Huber PJ 1981 *Robust Statistics*, Wiley.

Jenkins GM and Watts DG 1968 *Spectral Analysis and Its Applications*, Holden-Day, Michigan University.

Kleiner B, Martin RD, and Thomson DJ Robust estimation of power spectra. *J. Royal Statist. Soc.* **B41**, 313–351.

Letac G 2011 Does there exist a copula in n dimensions with a given correlation matrix? *Int. Conf. on Analytical Methods in Statistics (AMISTAT 2011)*. October 27–30, Prague, the Czech Republic.

Maronna R, Martin D, and Yohai V 2006 *Robust Statistics. Theory and Methods*, Wiley.

Pashkevich ME and Shevlyakov GL 1995 The median analog of the Fourier transform. In *Proc. CDAM 1995*. Minsk, Belarus (in Russian).

Shevlyakov GL and Smirnov PO 2011 Robust estimation of the correlation coefficient: an attempt of survey. *Austrian J. of Statistics* **40**, 147–157.

Shevlyakov GL, Smirnov PO, Shin VI, and Kim K 2012 Asymptotically minimax bias estimation of the correlation coefficient for bivariate independent component distributions. *J. Mult. Anal.* **111**, 59–65.

Smirnov PO and Shevlyakov GL 2014 Fast highly efficient and robust one-step M-estimators of scale based on FQ_n. *Computational Statistics and Data Analysis* **78**, 153–158.

Spangl B 2008 On robust estimation of power spectra. *On Robust Spectral Density Estimation*. Dissertation, Technical University of Vienna.

Spangl B and Dutter R 2005 On robust estimation of power spectra. *Austrian J. of Statistics* **34**, 199–210.

Van den Bos A 1971 Alternative interpretation of maximum entropy spectral analysis. *IEEE Trans. Inform. Theory* **17**, 493–494.

Welch PD 1967 The use of Fast Fourier Transform for the estimation of power spectra: a method based on time averaging over short, modified periodograms. *IEEE Trans. Audio and Electroacoustics* **AU-15**, 70–73.

11

Applications to Signal Processing: Robust Detection

In this chapter, first, we give a brief review of robust hypothesis testing with the focus on Huber's minimax approach; second, we apply the results of robust minimax estimation of location to the problem of robust detection of a known signal; and, third, we propose to use redescending M-estimates of location including stable ones for robust detection of weak signals. In the description of typical problem settings in robust detection of signals, we partially follow Shevlyakov and Vilchevski (2002, 2011).

11.1 Preliminaries

11.1.1 Classical approach to detection

Consider the problem of detection of a known signal

$$H_0 : \ x_i = e_i \quad \text{versus} \quad H_1 : \ x_i = s_i + e_i, \quad i = 1, 2, \ldots, n, \qquad (11.1)$$

where $\{s_i\}_1^n$ is a known signal with independent noises e_i from a common distribution F with a density f.

Given prior information on noise distributions, the classical theory of hypothesis testing suggests various optimal (in the Bayesian, minimax, Neyman–Pearson senses) methods to solve problem (11.1) (Kendall and Stuart 1963; Lehmann 1959). In this case, all optimal methods are based on the likelihood ratio statistic (*LHR*): for solution, one must compute the value of this statistic and compare it with a certain

Robust Correlation: Theory and Applications, First Edition. Georgy L. Shevlyakov and Hannu Oja.
© 2016 John Wiley & Sons, Ltd. Published 2016 by John Wiley & Sons, Ltd.
Companion Website: www.wiley.com/go/Shevlyakov/Robust

threshold. The differences between the aforementioned approaches result only in the values of that threshold.

Let us look back at problem (11.1); the *LHR*-statistic has the form

$$LHR(\mathbf{x}) = \prod_{i=1}^{n} \frac{f(x_i - s_i)}{f(x_i)}, \quad \mathbf{x} = (x_1, x_2, \ldots, x_n).$$

Note that in this case it is necessary to know the true density f.

Commonly, in real-life problems of signal processing, noise distributions are only partially known. For instance, it may be known that the underlying distribution is close to normal, or/and there is some information on its behavior in the central zone, and nothing is known about the distribution tails, etc. Similarly, observed signals may be distorted by impulsive noises, which is typical for many applications: acoustics, sound-location, radio-location, and communication (Shevlyakov and Vilchevski 2002, 2011).

The classical methods based on the *LHR*-statistic behave poorly in the above situations. There are many works (e.g. Krishnaiah and Sen 1984) on nonparametric and rank procedures of detection that provide protection from impulsive noise and gross errors in the data, but with the considerable lack of efficiency as compared to *LHR*-statistic detection rules in the case of absence of outliers. Thus, design of highly efficient and low-complexity nonparametric tests for detection of signals still remains an important problem.

11.1.2 Robust minimax approach to hypothesis testing

There are other alternatives to the classical approach based on the *LHR*-statistic. Huber (1965) notices that the *LHR*-statistic is not robust, since observations x_i, which make the ratio $f(x_i - s_i)/f(x_i)$ close to zero or very large, may destroy the *LHR*-statistic. For example, this happens under heavy-tailed noise distributions.

Consider the problem of detection of a known constant signal in the additive noise

$$H_0: x_i = e_i \quad \text{versus} \quad H_1: x_i = \theta + e_i, \quad i = 1, 2, \ldots, n, \tag{11.2}$$

where $\theta > 0$ is a known signal.

Huber (1965) proposes the trimmed version of the *LHR*-statistic

$$TLHR(\mathbf{x}) = \prod_{i=1}^{n} \pi(x_i),$$

$$\pi(x) = \begin{cases} c', & x < c', \\ \dfrac{f_1(x)}{f_0(x)}, & c' \le x \le c'', \\ c'', & x > c'', \end{cases}$$

where $f_0(x)$ and $f_1(x)$ are the distribution densities of observations under the hypotheses H_0 and H_1, respectively; $0 \le c' < c'' < \infty$ are some parameters. Here the uncertainty of information about noise distributions defines the use of the composite hypotheses H_j, $j = 0, 1$; thus, the problems of detection become nonparametric. Note that the parameters c' and c'' are uniquely defined only for a sufficiently large parameter θ.

Another implementation of the minimax approach is used in the following problem of the composite hypothesis testing (Kuznetsov 1976):

$$H_j = \{f_j(x): \underline{f}_j(x) \le f_j(x) \le \overline{f}_j(x)\}, \quad j = 0, 1,$$

where $\underline{f}_j(x)$ and $\overline{f}_j(x)$ are given functions, which determine the lower and upper bounds for the true density $f_j(x)$, respectively. Note that these bounds are not densities: they can be obtained with the use of the confidence limits for the kernel density estimates. In this setting, the minimax decision rule that minimizes the Bayes risk in the least favorable case is obtained so that it can be applied to detection of any approximately known signal.

11.1.3   Asymptotically optimal robust detection of a weak signal

The case of detection of a weak signal is both of theoretical and practical importance, as the robust Huber's test cannot be applied to weak signals, for which the classes of distribution densities for the hypotheses overlap.

Consider the following setting of the hypothesis testing problem

$$H_0 : x_i = e_i \quad \text{versus} \quad H_1 : x_i = \theta s_i + e_i, \quad i = 1, 2, \ldots, n, \qquad (11.3)$$

for an arbitrary $\theta > 0$. Let f be a noise distribution density from a certain class $\mathcal{F}$, and d be a decision rule from the class $\mathcal{D}$ of randomized decision rules.

In this case, the power of detection P_D for $f \in \mathcal{F}$ is defined as

$$P_D = \beta_d(\theta \mid f) = E_\theta(d(\mathbf{x}) \mid f), \quad f \in \mathcal{F},$$

and the probability of false alarm is given by $P_F = \alpha = \beta_d(0 \mid f)$.

Next we specify the problem (11.3) of detection and set it as a maximin problem: search for a decision rule that provides the asymptotic level of the power of detection independently of the chosen f from the class $\mathcal{F}$ (El-Sawy and Vandelinde 1977)

$$\max_{d \in \mathcal{D}} \min_{f \in \mathcal{F}} \beta_d(\theta \mid f) \qquad (11.4)$$

under the side condition

$$\beta_d(0 \mid f) \le \alpha \quad \text{for all} \quad f \in \mathcal{F}.$$

Furthermore, $n^{1/2}\theta_n$ is used as the test statistic in the decision rule d_ρ, where θ_n is an M-estimate of location

$$\theta_n = \arg \min_\theta \sum_{i=1}^{n} \rho(x_i - \theta s_i). \tag{11.5}$$

The function of contrast ρ should satisfy the following regularity conditions:

(C1) the convex and symmetric function of contrast $\rho(u)$ strictly increases with positive u;

(C2) the score function $\psi(u) = \rho'(u)$ is continuous for all u;

(C3) $E_F(\psi^2) < \infty$ for all $f \in \mathcal{F}$;

(C4) $\frac{\partial E_F(\psi(x-\theta))}{\partial \theta}$ exists and is nonzero in some neighborhood of θ.

The decision rule has the form

$$d(\mathbf{x}) = \begin{cases} 1, & n^{1/2}\theta_n > \gamma, \\ \dfrac{1}{2}, & n^{1/2}\theta_n = \gamma, \\ 0, & n^{1/2}\theta_n < \gamma, \end{cases}$$

where γ is a threshold defined by the false alarm probability α.

El-Sawy and Vandelinde (1977) show that under conditions (C1) to (C4) the following relations hold for each v greater than the threshold γ:

$$\beta_d(0 \mid f) \leq \beta_{d_{\rho*}}(0 \mid f^*) \quad \text{for} \quad \text{all} \quad f \in \mathcal{F},$$

$$\inf_{f \in \mathcal{F}} \beta_{d_{\rho*}}(v \mid f) = \beta_{d_{\rho*}}(v \mid f^*) = \sup_{d \in D} \beta_d(v \mid f^*).$$

These relations mean that the pair $(d_{\rho*}, f^*)$ is the saddle point of the function $\beta_d(\theta \mid f)$, and thus it is the solution of problem (11.4).

11.2 Robust Minimax Detection Based on a Distance Rule

11.2.1 Introductory remarks

Consider coherent binary detection of a known signal in additive noise

$$H_0: x_i = e_i \quad \text{versus} \quad H_1: x_i = s_i + e_i, \quad i = 1, 2, \ldots, n, \tag{11.6}$$

where $\mathbf{s} = (s_1, \ldots, s_n)$ is a known signal.

We propose the following decision rule for problem (11.6):

$$d(\mathbf{x}) = \begin{cases} 1, & \sum_{i=1}^{n} \rho(x_i) > \sum_{i=1}^{n} \rho(x_i - s_i), \\[2mm] \dfrac{1}{2}, & \sum_{i=1}^{n} \rho(x_i) = \sum_{i=1}^{n} \rho(x_i - s_i), \\[2mm] 0, & \sum_{i=1}^{n} \rho(x_i) < \sum_{i=1}^{n} \rho(x_i - s_i), \end{cases} \tag{11.7}$$

where $\rho(u)$ is a distance criterion or a function of contrast (Chelpanov and Shevlyakov 1983; Shevlyakov 1976).

The randomized decision rule (11.7) means that the choice of the hypotheses H_0 or H_1 depends on the distance of the signals $\mathbf{0}$ and $\mathbf{s}$ from the observed data $\mathbf{x}$ measured by the value of this distance.

In the general case of multi alternative detection of the known signals $s_1, \ldots, s_k$

$$H_1: \ x_i = s_1 + e_i,$$

$$\vdots \tag{11.8}$$

$$H_k: \ x_i = s_k + e_i, \quad i = 1, \ldots, n,$$

the decision is made in favor of the signal s_j (the hypothesis H_j) that minimizes the distance from the observed data

$$s_j = \arg \min_{\theta = s_1, \ldots, s_k} \sum_{i=1}^{n} \rho(x_i - \theta). \tag{11.9}$$

The proposed non traditional detection rule (11.9) has a natural connection with the problem of estimation of a location parameter if to consider the transition from the problem of multi alternative detection (11.8) to the problem of estimation of the location parameter θ with the continuum set of hypotheses

$$\theta_n = \arg \min_{\theta} \sum_{i=1}^{n} \rho(x_i - \theta).$$

There is another motivation for the introduced detection rule. In statistics, there are two approaches to parameter estimation: point and interval estimation. Evidently, the above procedure of hypothesis testing may be referred to as "point hypothesis testing". Henceforth we call this procedure a ρ-distance rule.

11.2.2 Asymptotic robust minimax detection of a known constant signal with the ρ-distance rule

Consider the particular case of detection problem (11.6) when $s_1 = \cdots = s_n = \theta$, that is, detection of a known constant signal $\theta > 0$ (11.2). To solve it, we use the minimum distance rule (11.7), where $\rho(u)$ is a function of contrast or a loss function characterizing the assumed form of a distance (Shevlyakov and Kim 2006). This choice of a detection rule is mostly determined by the fact that it allows for the direct use of Huber's minimax theory of M-estimates of location (Huber 1964, 1981). It can be easily seen that the choice $\rho(u) = -\log f(u)$ makes the optimal *LHR* test statistic minimizing the Bayesian risk with equal costs and prior probabilities of hypotheses. Note that, in this case, first, it is necessary to know exactly the shape of a noise distribution density f to figure out the ρ-distance rule and, second, the *LHR*-statistics usually behave poorly under departures from the assumed model of a noise density f.

The following cases of detection rule (11.7) are of a particular interest:

1. $\rho(u) = u^2$ defines the *LS* or L_2-*distance rule*, which is optimal under normal noise with the sample mean as the *LHR* statistic

$$\sum_{i=1}^{n} x_i^2 < \sum_{i=1}^{n} (x_i - \theta)^2 \iff \frac{1}{n} \sum_{i=1}^{n} x_i < \frac{\theta}{2};$$

2. $\rho(u) = |u|$ yields the *LAV* or L_1-*distance rule*, which is optimal under the double-exponential or Laplace noise distribution with the Huber-type test statistic

$$\sum_{i=1}^{n} |x_i| < \sum_{i=1}^{n} |x_i - \theta| \iff \frac{1}{n} \sum_{i=1}^{n} \psi_H(x_i; 0, \theta) < \frac{\theta}{2},$$

where $\psi(x; a, b) = \max(a, \min(x, b))$;

3. the Chebyshev distance $\rho(u) = \max|u|$ defines the L_∞-*distance rule*, which is optimal under the uniform noise distribution.

In what follows, we consider the asymptotic weak signal approach when the signal θ decreases with sample size as $\theta = \theta_n = A/\sqrt{n}$ given a constant $A > 0$. For reasonable decision rules, the error probability then converges as $n \to \infty$ to a nonzero limit (Hossjer and Mettiji 1993). Moreover, within this approach, the error probability is closely related to the Pitman efficacy of the detector test statistic and, therefore, Huber's minimax theory can be used to analyze the detector performance (Kassam and Poor 1985; Kassam 1988). Finally, since weak signals are on the border of not being distinguishable, it is therefore especially important to know the error probabilities.

Under the regularity conditions $(\mathcal{F}1)$, $(\mathcal{F}2)$, $(\Psi1)$ to $(\Psi4)$ of Section 3.2, the asymptotic normality of the ρ-distance detection rule statistic is established in Shevlyakov and Vilchevski (2002, for details, see p. 272).

Moreover, the following result holds

$$P_D = \Phi\left(\frac{A}{2} \frac{E_F(\psi')}{\sqrt{E_F(\psi^2)}} \right). \tag{11.10}$$

From (11.10) it follows that the maximin problem

$$\max_{\psi \in \Psi} \min_{f \in F} P_D(\psi,f)$$

is equivalent to the minimax problem

$$\min_{f \in F} \max_{\psi \in \Psi} V(\psi,f),$$

where $V(\psi,f) = E_F(\psi^2)/(E_F(\psi'))^2$ is the asymptotic variance for M-estimates of a location parameter.

Thus, all results on minimax estimation of location are also true in this case, i.e., they provide the guaranteed level of the power function P_D

$$P_D(\psi^*,f) \geq P_D(\psi^*,f^*) \quad \text{for} \quad \text{all} \quad f \in F.$$

11.2.3   Detection performance in asymptotics and on finite samples

Error probability of detection

A natural performance evaluation characteristic of the ρ-distance rule (11.7) is given by the error probability

$$P_E = Q\left(\frac{A}{2\sqrt{V(\psi,f)}} \right), \tag{11.11}$$

where $Q(x) = (2\pi)^{-1/2} \int_x^\infty e^{-t^2/2} dt$. Since the signal energy E is equal to $\theta^2 n$, we have $A = \sqrt{E}$. Moreover, in the particular case of the unit noise variance, it can be written as $A = \sqrt{SNR}$, where SNR is the signal-to-noise ratio. In the case of a symmetric detection rule, the probabilities of the first and second kind are equal, or in other words, the false alarm probability P_F and the probability of missing P_M are equal, and they both are equal to the error probability P_E: $P_F = P_M = P_E$. Therefore, the error probability P_E is an exhaustive characteristic of detection performance in this case.

From (11.11) it follows that the minimax problem with respect to the error probability

$$\min_{\psi \in \Psi} \max_{f \in F} P_E(\psi,f)$$

is equivalent to Huber's minimax problem

$$\min_{\psi \in \Psi} \max_{f \in F} V(\psi,f).$$

Thus, all the results on the minimax estimation of location are also applicable in this case: the optimal function of contrast ρ^* in the minimum distance rule (11.7) is defined by the maximum likelihood choice for the least informative noise density f^* minimizing Fisher information: $\rho^*(u) = -\log f^*(u)$. Moreover, similar to the guaranteed value of detection power P_D, the error probability is upper-bounded in class $\mathcal{F}$:

$$P_E(\psi^*, f) \leq P_E(\psi^*, f^*) \quad \text{for all} \quad f \in \mathcal{F}.$$

Noise distribution densities and detection rules

We compute error probabilities for the Gaussian noise with the density $f(x) = N(x; 0, 1)$, the Cauchy noise with the density $C(x; 0, 1) = 1/[\pi(1 + x^2)]$, the Cauchy ε-contaminated Gaussian noise with the density

$$f_{CN}(x) = (1 - \varepsilon) N(x; 0, 1) + \varepsilon C(x; 0, 1)$$

where ε is the contamination parameter ($0 \leq \varepsilon < 1$), and for the generalized Gaussian noises with the density

$$f_{GG}(x; \beta, q) = \frac{q}{2\beta\Gamma(1/q)} \exp\left(-\frac{|x|^q}{\beta^q}\right)$$

where β and q are the parameters of scale and shape, respectively.

Also we set the unit distribution variance with the following exceptions–for the Cauchy and ε-contaminated Gaussian densities.

Moreover, we consider the class $\mathcal{F}_{12}$ (3.44) of noise distribution densities with a bounded variance

$$\sigma^2(f) = \int_{-\infty}^{\infty} x^2 f(x)\, dx \leq \overline{\sigma}^2$$

and the density value at the center of symmetry (for details, see subsection 3.4.2)

$$f(0) \geq \frac{1}{2a} > 0 \, .$$

Recall that the least informative density f_{12}^* given by (3.45) in this class has three branches, namely, the normal with relatively small variances when $\overline{\sigma}^2(f) < 2a^2/\pi$, the Laplace with relatively large variances when $\overline{\sigma}^2 > 2a^2$, and the Weber–Hermite density with relatively moderate variances when $2/\pi \leq \overline{\sigma}^2/a^2 \leq 2$.

The corresponding minimax detection ρ-distance rule (11.7) can be described as follows: (i) with relatively small variances for $\overline{\sigma}^2(f) < 2a^2/\pi$ (relatively short tails), it is the L_2-distance rule with $\rho(u) = u^2$; (ii) with relatively large variances $\overline{\sigma}^2 > 2a^2$ (relatively heavy tails), it is the L_1-distance rule with $\rho(u) = |u|$; (iii) with relatively moderate variances, it is a compromise between the L_1-distance and L_2-distance rules. In the latter case, the exact function of contrast is $\rho(u) = -\log f_{12}^*(u)$, which can be rather effectively (with at most 2.5 % relative error) approximated by the low-complexity L_{p^*}-distance function of contrast with the power $p^* \in (1, 2)$ given by (for details, see Shevlyakov and Vilchevski 2002, pp. 76–79)

$$
p^* = \begin{cases} 5.33 - 7.61x + 3.73x^2, & 2/\pi < x \le 1.35, \\ 2.66 - 1.65x + 0.41x^2, & 1.35 < x < 2, \end{cases} \tag{11.12}
$$

where $x = \overline{\sigma}^2 / a^2$.

Further, we compare the performance of the low-complexity minimax L_{p^*}-distance rule with $\rho(u) = |u|^{p^*}$, in which the power p^* is chosen from (11.12) with the L_1-, L_2-distance rules and Huber's detectors. In the latter case, we consider the minimum distance detection rule (11.7) with the conventional Huber loss function $\rho_H^*(u) = -\log f_H^*(u)$ where the least informative distribution is described by the Gaussian central part and exponential tails (for details, see Huber 1981) and the contamination parameter ε is set equal to 0.1. The chosen value of ε seems to be a reasonable upper bound upon the contamination parameter in applications (Hampel et $al.$ 1986).

Performance evaluation

To clarify the procedure of computing the error probability, let us consider the Gaussian case in detail. To simplify the analysis, we rewrite Equation (11.11) as follows:

$$
P_E = Q\left(\frac{A\, I_1}{2\, \sqrt{I_2}} \right), \tag{11.13}
$$

where $I_1 = \int_{-\infty}^{\infty} \psi'(x) f(x)\, dx$ and $I_2 = \int_{-\infty}^{\infty} \psi^2(x) f(x)\, dx$.

Example 11.2.1. *Consider the error probability for the minimax L_{p^*}-distance and L_2-distance rules in the Gaussian noise. Then the choice of the optimal structure is defined by the ratio σ^2/a^2. Subsequently, we have $f(0) = 1/\sqrt{2\,\pi}$, $\sigma^2(f) = 1$, $a = \sqrt{\pi/2}$, $\sigma^2/a^2 = 2/\pi$. Hence, the minimax detection rule is based on the L_2-distance with $p^* = 2$, the score function is linear $\psi^*(x) = x$, the integrals are $I_1 = 1$ and $I_2 = 1$, and thus its error probability is given by $P_E = Q(0.5\,A)$. For the L_2-distance rule, apparently the choice is the same: $\psi(x) = x$ and $P_E = Q(0.5\,A)$.*

Example 11.2.2. *Consider the error probability for Huber's and L_1-distance rules in the Gaussian noise. For $\varepsilon = 0.1$, the score function is $\psi(x) = \psi_H^*(x) = \max\,[-1.14, \min\,(x, 1.14)]$, the required integrals cannot be evaluated in a closed form, hence they were computed numerically, and $P_E = Q(0.461\,A)$.*

For the L_1-distance with $p^* = 1$, the score function is the sign function $\psi(x) = \mathrm{sgn}(x)$, the integrals are $I_1 = 2f(0) = \sqrt{2/\pi}$ and $I_2 = 1$, and thus we get $P_E = Q(A/\sqrt{2\pi}) = Q(0.399\,A)$.

The results of computing for the Gaussian, extremely heavy-tailed Cauchy, heavy-tailed ε-contaminated Gaussian with $\varepsilon = 0.1$, and the close to the uniform exponential-power ($q = 100$) density are exhibited in Table 11.1 and Figs 11.1 to 11.3.

The structure of the minimax L_{p*}-distance rule is determined by the ratio $\overline{\sigma}^2/a^2$, and contrary to Huber's rule, the parameters $\overline{\sigma}^2$ and a^2 of class $\mathcal{F}_{12}$ can be directly estimated from the sample. In general, for estimating $\overline{\sigma}^2$ we can use, for example, the upper confidence limit for the estimate of variance and for $1/(2a)$ the lower confidence limit for the nonparametric estimate of a distribution density at the center of symmetry: $\hat{f}(0) \leq \hat{f}(0)$. Taking into account the L_{p*}-distance rule with the parameter $1 \leq p^* \leq 2$ when both restrictions of class $\mathcal{F}_{12}$ hold as equalities, we choose the estimates of variance and density as the characteristics of this class. For a variance, this is the customary sample variance $\widehat{\overline{\sigma}}^2 = n^{-1} \sum_{i=1}^{n} (x_i - \overline{x})^2$.

To avoid the difficulties of nonparametric estimation of a distribution density, we estimate the value of the underlying density at its center of symmetry and the related parameter a using the following simple formula based on the central order statistics $x_{(k)}$ and $x_{(k+1)}$ ($n = 2k$ or $n = 2k + 1$) (Shevlyakov and Vilchevski 2002)

$$\hat{a} = 1/[2\hat{f}(0)] = [(n + 1)(x_{(k+1)} - x_{(k)})]/2.$$

Table 11.1   The factor $I_1/(2\sqrt{I_2})$ in (11.13) for various noises.

	L_1	L_2	Huber's: $\varepsilon = 0.1$	L_{p*}
	Ex. 2	Ex. 1	Ex. 2	Ex. 1
Gaussian: $N(x; 0, 1)$	0.399	0.5	0.461	0.5
				$p^* = 2$
Cauchy: $C(x; 0, 1)$	0.318	0	0.274	0.318
				$p^* = 1$
Contaminated Gaussian: $0.9\,N(x; 0, 1) + 0.1\,C(x; 0, 1)$	0.393	0	0.461	0.393 $p^* = 1$
Generalized Gaussian: $q = 1.5$	0.476	0.5	0.554	0.649 $p^* = 1.5$
Generalized Gaussian: $q = 4$	0.320	0.5	0.428	0.5 $p^* = 2$
Generalized Gaussian: $q = 100$	0.289	0.5	0.386	0.5 $p^* = 2$

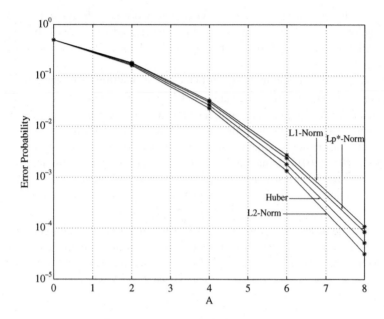

Figure 11.1 Error probability in the Gaussian noise: $n = 20$, $A = \sqrt{SNR}$. Source: *Shevlyakov (2006), Reproduced with permission from IEEE.*

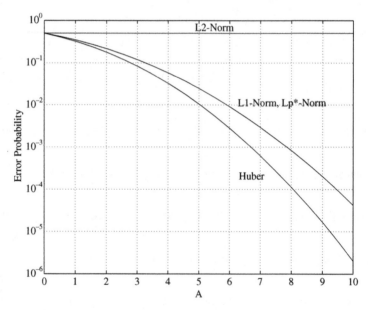

Figure 11.2 Error probability in the contaminated Gaussian noise: asymptotics, $\varepsilon = 0.1$, $A = \sqrt{E}$. Source: *Shevlyakov (2006), Reproduced with permission from IEEE.*

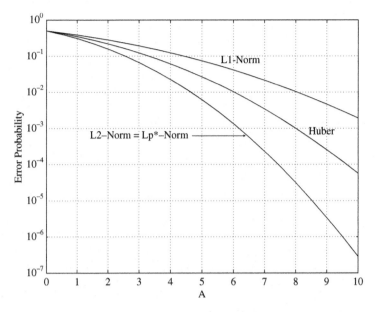

Figure 11.3   Error probability in the generalized Gaussian noise close to uniform: asymptotics, $q = 100$, $A = \sqrt{SNR}$. Source: Shevlyakov (2006), Reproduced with permission from IEEE.

On finite samples ($n = 20,100$), the performance of the L_{p*}-distance, Huber's, L_1- and L_2-distance rules under the Gaussian, Cauchy, ε-contaminated Gaussian, and uniform noises was studied by the Monte Carlo technique. The detection model was chosen consistently with the initial assumption of detection of a weak signal: H_0 : $x_i = e_i$ versus $H_1 : x_i = \theta + e_i$, $i = 1, \ldots, n$, where the useful signal $\theta = \theta_N = A/\sqrt{n}$.

On small samples with $n = 20$, the results of modeling in the Gaussian noise are displayed in Fig. 11.1; the other results are discussed below. On large samples with $n = 100$, the results of modeling are close to the asymptotic results given by (11.13).

11.2.4   Concluding remarks

Gaussian noise

On large samples and in asymptotics, the minimax L_{p*}-distance rule coincides with the optimal L_2-distance rule, both being better than Huber's and the L_1-distance rule.

On the contrary, on small samples, Fig. 11.1 shows that the minimax L_{p*}-distance rule is close in performance to the robust L_1-distance rule, being slightly inferior to Huber's on small samples.

Sample size effects

There are some peculiarities of the dependence of the minimax detection rule on the sample size n; this deserves separate attention. Since the estimates $\widehat{\sigma}^2$ and $\widehat{a}$ of the parameters of class $\tilde{F}$ are consistent, the sample minimax detection performance tends in probability to the exact minimax detection performance as $n \to \infty$. This is confirmed by Monte Carlo modeling on samples of size $n = 100$.

The results exhibited in Fig. 11.1 can be explained by the bias of the sample distribution of the threshold statistic $\widehat{\widehat{\sigma}}^2 / \widehat{a}^2$ with small n: its values determine the choice of the appropriate branch of the algorithm. For Gaussian samples of size $n = 20$, the choice of the L_2-distance rule occurs approximately for 10% of cases ($P[\widehat{\widehat{\sigma}}^2 / \widehat{a}^2 < 2/\pi] \approx 0.1$), the choice of the L_1-distance rule–for the 20% of cases ($P[\widehat{\widehat{\sigma}}^2 / \widehat{a}^2 > 2] \approx 0.2$), and the L_{p^*}-distance rules with $1 < p^* < 2$ are chosen for the remaining 70% of cases with the average value of the power parameter $\overline{p^*} \approx 1.25$. For samples of size $n = 100$, we have the opposite situation: approximately for 45% of cases, the L_2-distance branch of the minimax test is realized, whereas the L_1-distance branch occurs only for the 5% with the average value of the power as $\overline{p^*} \approx 1.85$.

Cauchy and ε-contaminated Gaussian noise

Although these models describe extremely heavy-tailed noises, nevertheless they deserve attention as, say, the Cauchy distribution may arise through the distribution of the ratio of Gaussian random variables. From Fig. 11.2 and Table 11.1 it can be seen that the L_2-distance rule with the linear score statistic naturally has extremely poor performance both in asymptotics and on finite samples; the minimax, L_1-distance, and Huber's detectors exhibit their good robust properties with the latter being evidently better than the former in the mixture models, and vice versa for the Cauchy noise. Note that we also examined the detection performance in other heavy-tailed distribution models, namely, the Laplace, the mixture models with the Gaussian contamination not so heavy as the Cauchy, and the generalized Gaussian densities with $1 < q < 2$, but the obtained results were qualitatively the same.

Short-tailed noise

The detection performance again reverts not once in short-tailed noise models described by the generalized Gaussian distribution density with $q = 100$ close to the uniform (see Fig. 11.3). In asymptotics, the L_2-distance and minimax detectors proved their superiority over Huber's and the L_1-distance rules, but on small samples, the aforementioned small size sample effect reveals itself: the minimax detector is close in performance to the L_1-distance rule, and thus it is slightly inferior to Huber's. The qualitatively similar effects were also observed in some other examined short-tailed and finite noise density models, for example, for the generalized Gaussian density with $q = 4$.

Final remark

Our main aim was to show some new possibilities of Huber's minimax approach to robust detection associated with the usage of a new class of densities with an upper-bounded variance. The proposed low-complexity minimax power detector exhibits both high robustness in heavy-tailed noise and good efficiency in short-tailed noise on small and large samples. A similar approach can be applied to much more complicated models of signal detection than the simple binary detection model analyzed in this paper.

11.3   Robust Detection of a Weak Signal with Redescending M-Estimates

11.3.1   Introductory remarks

To design highly efficient and robust detectors of a weak signal, an asymptotic approach to robust estimation exploiting redescending score functions is used. Two new indicators of robustness of detection, the detection error sensitivity and detection stability, are introduced. The optimal Neyman–Pearson rules maximizing detection efficiency under the guaranteed level of detection stability are written out. Under heavy-tailed noise distributions, the proposed asymptotically robust detection rules based on redescending score functions, namely, the minimum error sensitivity and the radical ones, outperform conventional linear bounded Huber's and redescending Hampel's rules both on small and large samples.

Consider the following binary hypothesis testing problem:

$$H_0: \ x_i = e_i \quad \text{versus} \quad H_1: \ x_i = \theta \, s_i + e_i, \quad i = 1, \dots, n, \tag{11.14}$$

where $\{x_i\}$ are observations; $\{s_i\}$ is a known signal of finite power

$$P = \lim_{n \to \infty} \frac{1}{n} \sum_{i=1}^{n} s_i^2 < \infty \ ;$$

$\theta = A/\sqrt{n}$ is a signal amplitude tending to zero as the sample size $n \to \infty$ with a finite constant $A > 0$ thus defining an asymptotic weak signal approach (Hossjer and Mettiji 1993); $\{e_i\}$ is i.i.d. random noise with a distribution density f.

To solve problem (11.14), we apply the Neyman–Pearson detection rule

$$\sqrt{n} \ \theta_n(x_1, \dots, x_n) > \lambda_{1-\alpha} \tag{11.15}$$

with Huber's M-estimate θ_n of the signal amplitude θ as a test statistic and $\lambda_{1-\alpha}$ as a critical value.

Recall that an M-estimate θ_n minimizes the empirical risk:

$$\theta_n = \arg \min_\theta \sum_i \rho(x_i - \theta \, s_i)$$

with a function of contrast (a loss function) $\rho(x)$ and thus satisfies the estimating equation

$$\sum_{i=1}^{n} s_i \, \psi(x_i - \theta_n \, s_i) = 0 \, , \tag{11.16}$$

where $\psi(x) = \rho'(x)$ is a score function. Note that formula (11.16) represents a generalized version of the M-estimate of location defined by Huber (1964). To provide the false alarm probability $P_F = \alpha$, the threshold $\lambda_{1-\alpha}$ should be taken as in El-Sawy and Vandelinde (1977)

$$\lambda_{1-\alpha} = \Phi^{-1}(1 - \alpha) \, \sqrt{V(\psi, f)/P} \tag{11.17}$$

(here $\Phi(z)$ is the standard Gaussian cumulative) with the asymptotic variance of the test statistic θ_n from (11.16) defined by the well-known Huber formula (Huber 1964)

$$V(\psi, f) = \int_{-\infty}^{\infty} \psi^2(x) f(x) \, dx \Big/ \left(\int_{-\infty}^{\infty} \psi'(x) f(x) \, dx \right)^2 . \tag{11.18}$$

In this case, the probability of detection can be written as follows:

$$P_D = 1 - \Phi(\Phi^{-1}(1 - \alpha) - \sqrt{E/V(\psi, f)} \,) \, , \tag{11.19}$$

where

$$E = \lim_{n \to \infty} \sum_{i=1}^{n} (\theta s_i)^2 = A^2 P$$

is the signal energy (El-Sawy and Vandelinde 1977).

In Shevlyakov *et al.* (2010), it is shown that the Neyman–Pearson detection rule (11.15) based on conventional redescending heuristic M-estimates with the three-part score function ψ_{25A}, Huber's skipped ψ_{sk}, Tukey's biweight ψ_{bi}, and Smith's ψ_{Sm} (for details, see subsection 3.3.6) outperform Huber's minimax detectors under heavy-tailed noise on small and large samples.

In Section 3.5 we introduced Shurygin's variational optimization approach to robust stable estimation of location with redescending M-estimates, which generally outperform the conventional minimax Huber's and the aforementioned redescending M-estimates of location.

In what follows, we adapt the results obtained for stable estimation of a location parameter to robust Neyman–Pearson detection of weak signals with Huber's M-estimates as test statistics in the detection rule (11.15).

Thus, our main goal is to ground the choice of a redescending score function recommended for designing robust detectors.

11.3.2 Detection error sensitivity and stability

A brief recollection from stable estimation

For the sake of convenience and consistency of narration, we briefly recall a few necessary results from the theory of stable estimation (for details, see Section 3.5).

In Shurygin (1994a), a new global measure of an estimate's sensitivity, namely, *the variance sensitivity*, is introduced as the Lagrange functional derivative of the asymptotic variance $V(\psi,f)$ (11.18) formally given by

$$VS(\psi,f) = \frac{\partial V(\psi,f)}{\partial f} = \frac{\int \psi^2(x)\,dx}{\left(\int \psi'(x)f(x)\,dx\right)^2}. \tag{11.20}$$

It can be shown that the boundness of the Lagrange derivative (11.20) holds under the condition of square integrability of score functions $\psi(x)$. Such score functions are necessarily redescending, as the integral $\int \psi^2(x)\,dx$ converges only if $\psi(x) \to 0$ for $|x| \to \infty$.

Next, the notion of an estimate's stability is defined. Since it is natural to search for the minimum variance sensitive score function ψ under a given distribution density f, the optimal score function that minimizes the variance sensitivity (11.20) is given by

$$\psi_{MVS}(x) = -f'(x) . \tag{11.21}$$

This and further similar results are obtained by the calculus of variations techniques through the Euler–Lagrange equations for the appropriate functionals.

The estimate with the score function (11.21) is called the estimate of *the minimum variance sensitivity*. The minimum sensitivity functional has the form

$$VS_{min}(f) = \left(\int \psi_{MVS}^2 dx\right)^{-1} = \left(\int (f'(x))^2 dx\right)^{-1}. \tag{11.22}$$

By comparing an estimate's variance sensitivity with the above minimum, the *stability* of any M-estimate is defined as follows:

$$0 \le stb(\psi,f) = \frac{VS_{min}(f)}{VS(\psi,f)} \le 1 . \tag{11.23}$$

Thus we get the two well-balanced complementary indicators of the global accuracy and robustness of an estimate, namely, its efficiency and stability, both lying in the segment $[0,1]$. Setting different weights for efficiency and stability, various optimality criteria can be used (Shevlyakov *et al.* 2008; Shurygin 1994a, 1994b). A reasonable choice is made by equating efficiency and stability: $eff(\psi,f) = stb(\psi,f)$. The corresponding asymptotically stable estimate is called *radical*, and its score function is given by

$$\psi_{rad}(x) = \psi_{ML}(x)\,\sqrt{f(x)}. \tag{11.24}$$

As already mentioned, the proposed approach to designing stable robust M-estimates of location necessarily leads to redescending score functions. Here a natural question arises: can optimal robust M-estimates designed in the framework of the conventional approaches to robustness, Huber's or Hampel's, be also redescending? In the Princeton experiment (Andrews *et al.* 1972) by Monte Carlo modeling of a variety of M-estimates in a variety of situations, it was found that the redescending M-estimate, dubbed 25A, dominated over the others; further, the corresponding theory based on the influence functions tools was deeply and thoroughly developed in Hampel *et al.* (1986). Moreover, in Shevlyakov *et al.* (2010) it is shown that conventional redescending score functions described and studied in Hampel *et al.* (1986) can be effectively used in robust detection, outperforming Huber's linear bounded detectors.

Definition of detection error sensitivity and stability

For detection rules (11.15), instead of the asymptotic variance as a measure of the accuracy of a decision rule statistic, we suggest to use the probability of error of the second kind (the probability of missing)

$$P_E(\psi,f) = 1 - P_D(\psi,f) = \Phi(\xi_{1-\alpha} - \sqrt{E/V(\psi,f)}), \qquad (11.25)$$

where $\xi_{1-\alpha} = \Phi^{-1}(1 - \alpha)$ (11.19). Then its Lagrange functional derivative takes the form:

$$\partial P_E(\psi,f)/\partial f = g(V)\, VS(\psi,f),$$

where $g(V) = \sqrt{E}\varphi(\xi_{1-\alpha} - \sqrt{E/V})/(2\, V^{3/2})$ and $\varphi(x) = \Phi'(x)$.

Given the signal energy E, consider the dependence of the factor $g(V)$ on the variance V in the segment $V_L \leq V \leq V_U$ of its admissible values. The lower constraint V_L is just the bound $V_L = 1/I(f)$ taken from the Cramér–Rao inequality (here $I(f) = \int [f'(x)/f(x)]^2 f(x)\, dx$ is the Fisher information for location) whereas the upper bound V_U is defined by the consistency condition for the detection rule (11.15)

$$\xi_{1-\alpha} - \sqrt{E/V(\psi,f)} < 0,$$

taking the form $V \leq E/\xi_{1-\alpha}^2$. Thus, any value V_U such that $V_U \leq E/\xi_{1-\alpha}^2$ and $V_U \geq V_L$ can be taken as the upper bound. It can be easily shown that $g'(V) > 0$ in the segment $V_L \leq V \leq V_U$; hence the maximum of $g(V)$ is attained at the upper bound $V = V_U$: $g_{max} = g(V_U)$.

Thus we naturally arrive at the following definition for the detection error sensitivity $ES(\psi,f)$ as an indicator proportional to the variance sensitivity $VS(\psi,f)$ of the corresponding test statistic:

Definition 11.3.1. *The error sensitivity of detection rule (11.15) is defined as*

$$ES(\psi,f) = g(V_U)\, VS(\psi,f)\,.$$

Further, similarly to the tools developed for stable estimation, we search for the optimal *minimum detection error sensitivity* score minimizing $ES(\psi,f)$, which evidently coincides with the minimum variance sensitivity score (3.65)

$$\psi_{MES}(x) = -f'(x) . \tag{11.26}$$

The corresponding *minimum detection error sensitivity* functional is of the form:

$$ES_{\min}(f) = g(V_U) \left(\int (f'(x))^2 dx \right)^{-1} .$$

Next we introduce the stability of detection as follows:

Definition 11.3.2. *The stability of detection rule (11.15) is defined as*

$$\text{stb}_{det}(\psi,f) = \frac{ES_{\min}(f)}{ES(\psi,f)} ,$$

where $ES_{\min}(f) = ES(\psi_{MES},f)$ *is the minimum error sensitivity.*

From Definition 11.3.2 it directly follows that the stability of detection also coincides with the stability of the corresponding test statistic:

$$\text{stb}_{det}(\psi,f) = \text{stb}(\psi,f) .$$

Thus, due to the simple structure of detection rule (11.15), to the natural choice of the probability of error of the second kind (11.25) as a measure of detection performance, and to the proposed definition of error sensitivity, we can directly apply all the tools developed for stable estimation to the analysis of detection performance.

11.3.3   Performance evaluation: a comparative study

Detection rules

Consider the detection rules given by (11.15) with the M-estimates θ_n as test statistics defined by the following score functions:

- the maximum likelihood score $\psi_{ML}(x) = -f'(x)/f(x)$,

- the linear score $\psi_{Mean}(x) = x$ with the weighted mean as a test statistic,

- the sign score $\psi_{Med}(x) = \text{sign}(x)$ with the median as a test statistic,

- Huber's linear bounded score (Huber 1964)

$$\psi_H(x) = \max\left[-1.14, \min(x, 1.14)\right]$$

  optimal for the model of contaminated standard Gaussian distributions with the contamination parameter $\varepsilon = 0.1$,

- Hampel's redescending three-part score $\psi_{25A}(x)$ given by

$$
\psi_{25A}(x) = \begin{cases}
x, & \text{for } 0 \leq |x| \leq a, \\
a \, \text{sign}(x), & \text{for } a \leq |x| \leq b, \\
a \, \dfrac{r - |x|}{r - b} \, \text{sign}(x), & \text{for } b \leq |x| \leq r, \\
0, & \text{for } r \leq |x|,
\end{cases}
$$

with the following values of the parameters: $a = 1.31$, $b = 2.039$, and $r = 4$ (see Hampel *et al.* 1986, p. 167),

- the minimum detection error sensitivity score given by (11.26) $\psi_{MES}(x) = -f'(x)$,

- the radical score function given by (11.24) $\psi_{Rad}(x) = -f'(x)/\sqrt{f(x)}$.

Hampel's three-part M-estimate revealed its advantageous properties in the Princeton experiment (Andrews *et al.* 1972). Later, it was shown that the optimal V-robust redescending hyperbolic tangent M-estimate is very close to it in performance, being only about 1% more efficient (see Hampel *et al.* 1986, Table 3, p. 167). In this study, we prefer Hampel's three-part score due to its much simpler implementation as compared to the optimal tanh-score.

Noise distributions

In the comparative analysis of detection rules, the following noise distribution densities $f(x)$ are used:

- the standard Gaussian $\phi(x) = (2\pi)^{-1/2} \exp(-x^2/2)$;

- the Laplace $f(x) = 2^{-1/2} \exp(-|x|\sqrt{2})$;

- the Cauchy $f(x) = C(x) = \pi^{-1}(1 + x^2)^{-1}$;

- the Cauchy contaminated standard Gaussian $f(x) = 0.9 \, \phi(x) + 0.1 \, C(x)$.

Now we briefly justify our choice. The presence of a Gaussian in this list is quite conventional (Kim and Shevlyakov 2008)–see also Section 3.6. The choice of the Cauchy and Cauchy contaminated Gaussian noise distributions is caused by their extremely heavy tails, the detection performance resistance to which is desired. Although the Laplace distribution has moderately heavy exponential tails, it is the least favorable (the least informative) distribution minimizing Fisher information in the wide class of the nondegenerate distributions (see Section 3.4), and thus it deserves consideration.

Evaluation criteria

To compare the performances of detection in noise models, we use the following indicators of the quality of detection:

- Pitman's asymptotic efficiency, which in case of weak signals coincides with the asymptotic efficiency of a test statistic (Noether 1967)

$$\text{eff}(\psi,f) = \frac{V(\psi_{ML},f)}{V(\psi,f)} = \frac{1}{I(f)\,V(\psi,f)} \,,$$

since $V(\psi_{ML},f) = 1/I(f)$, where $I(f) = \int (f'(x)/f(x))^2 f(x)\,dx$ is Fisher information;

- the detection stability

$$\text{stb}(\psi,f) = \frac{\left(\int \psi'(x)f(x)\,dx\right)^2}{\int \psi^2(x)\,dx \int (f'(x))^2\,dx} \,;$$

- the probability of missing $P_M = 1 - P_D$, where the probability of detection P_D is given by (11.19).

From (11.19) it follows that the probability of missing depends on the following two parameters: the signal energy E and the asymptotic variance $V(\psi,f)$ (11.18) as follows:

$$P_M = \Phi\left(\Phi^{-1}(1-\alpha) - \kappa(\psi,f)\,\sqrt{E}\right) \tag{11.27}$$

with the factor $\kappa(\psi,f) = 1/\sqrt{V(\psi,f)}$. Given signal energy, the comparative quality of detection is characterized by that factor: the larger κ, the better quality of detection.

Asymptotic performance

Tables 11.2 and 11.3 exhibit detection efficiency, detection stability, and the factor $\kappa(\psi,f)$: the best detector performances for each noise distribution are boldfaced except the maximum likelihood and the minimum error sensitivity cases, which indicate the points of reference for comparison.

11.3.4   Concluding remarks

Stability

Since the corresponding score functions are not square integrable on the real line, the maximum likelihood detectors for the Gaussian and Laplace noise distributions based on the mean and median, as well as Huber's conventional detector, have zero stability.

Table 11.2   Detection efficiency and stability for various detectors and noise distributions (the best detector performances for each noise distribution are boldfaced except the maximum likelihood and the minimum error sensitivity cases).

		Gaussian	Contaminated Gaussian	Laplace	Cauchy
ψ_{ML}	eff	1	1	1	1
	stb	0	0.38	0	0.50
ψ_{Mean}	eff	1	0	0.50	0
	stb	0	0	0	0
ψ_{Med}	eff	0.64	0.69	1	0.81
	stb	0	0	0	0
ψ_H	eff	**0.92**	0.88	0.67	0.75
	stb	0	0	0	0
ψ_{25A}	eff	0.91	**0.89**	0.64	0.81
	stb	0.69	0.62	0.15	0.49
ψ_{MES}	eff	0.65	0.71	0.75	0.80
	stb	1	1	1	1
ψ_{Rad}	eff	0.84	**0.89**	**0.89**	**0.92**
	stb	**0.84**	**0.89**	**0.89**	**0.92**

Table 11.3   The factor $\kappa(\psi, f) \times 10^3$ for various detectors and noise distributions (the best detector performances for each noise distribution are boldfaced except the maximum likelihood case).

	ψ_{ML}	ψ_{Mean}	ψ_{Med}	ψ_H	ψ_{25A}	ψ_{MES}	ψ_{Rad}
Gaussian	1000	1000	798	**962**	955	932	956
Contaminated Gaussian	953	0	782	922	927	928	**935**
Laplace	1414	1000	1414	1157	1135	1178	**1180**
Cauchy	707	0	637	613	655	**680**	665

Gaussian noise

From Tables 11.2 and 11.3 it can be seen that the redescenders are slightly inferior in efficiency and in the probability of missing to Huber's detector. The worst performance among the competitors is exhibited by the detector with the sign score ψ_{Med} asymptotically equivalent to the sign test with the efficiency $2/\pi \approx 0.637$.

Cauchy and Cauchy contaminated Gaussian noises

From Tables 11.2 and 11.3 it follows that all the redescenders outperform Huber's detector both in efficiency and in the probability of missing, evidently under the

Cauchy noise and slightly under the contaminated Gaussian noise. Since Huber's detector is designed for the neighborhoods of a Gaussian, this observation is quite natural. Apparently, the performance of the detector with the linear score ψ_{Mean} is disastrous under the Cauchy and Cauchy contaminated Gaussian noises. On the whole, the detector with the redescending score ψ_{Rad} (henceforth called *a radical detector*) outperforms the others on the chosen set of noise distributions.

Laplace noise

In this case of relatively heavy tails, Huber's detector performs better than Hampel's one in efficiency and in the probability of detection, but the proposed stable redescenders slightly outperform Huber's detector. Although the Laplace density has lighter tails than the Cauchy, all the competitors are considerably inferior to the maximum likelihood detector with the sign score. This observation can be explained by the fact that the Laplace distribution is the least favorable minimizing Fisher information over nondegenerate distributions due to the unique combination of its relatively high sharp central peak and relatively heavy tails. Finally, we note that in this case Hampel's detector has an unexpectedly low stability equal to 0.15.

Small sample performance

In real-life applications when only small or moderate numbers of observations are available, the specification of the area of applicability of the asymptotic approach is of great importance. To achieve this, we perform the Monte Carlo experiment on samples $N = 20, 40, 60, 80$ and 100 (the number of trials is taken equal to 40,000). We use the detection rule (11.15) with the thresholds defined by the asymptotic formula (11.17) together with the so-called one-step M-estimates

$$\theta_n = \theta_n^{(0)} + \widehat{\sigma}_n^{(0)} \times \sum_{i=1}^{n} s_i \, \psi \left(\frac{x_i - \theta_n^{(0)} \, s_i}{\widehat{\sigma}_n^{(0)}} \right) \Big/ \sum_{i=1}^{n} s_i \, \psi' \left(\frac{x_i - \theta_n^{(0)} \, s_i}{\widehat{\sigma}_n^{(0)}} \right), \quad (11.28)$$

where $\theta_n^{(0)}$ and $\widehat{\sigma}_n^{(0)}$ are initial estimates of location and scale. Being the first step of the Newton–Raphson iterative algorithm applied to the numerical solution of the estimating equation (11.16), the one-step M-estimates (11.28) fortunately have the same asymptotic behavior as their fully iterated versions (for details, see Hampel *et al.* 1986, p. 106). Here we take highly robust estimates of location and scale as the L_1-norm estimate $\theta_n^{(0)} = \arg \min_{\theta} \sum_{i=1}^{n} |x_i - \theta \, s_i|$ and the median absolute deviation $\widehat{\sigma}_n^{(0)} = \mathrm{MAD}_n = 1.483 \, \mathrm{med}\{|x_1 - \theta_n^{(0)} \, s_1|, \ldots, |x_n - \theta_n^{(0)} \, s_n|\}$.

In our experiment, we consider the particular case of the signal $\{s_i\}$ equal to a unit; hence the initial estimates of location and scale are given by the sample median $\theta_n^{(0)} = \mathrm{med} \, x$ and the median absolute deviation $\widehat{\sigma}_n^{(0)} = \mathrm{MAD} \, x = 1.483 \, \mathrm{med} \, |x - \mathrm{med} \, x|$.

For the numerical computation of the L_1-norm estimate and the MAD_n, we use the re-weighted least squares (RWLS) method (Green 1984): our experience shows that 3 or 4 iterations of the RWLS are sufficient to get the results practically coinciding with the results of application of precise sorting or linear programming-based methods.

The results of the Monte Carlo experiment for the probability of missing are displayed in Figs 11.4 to 11.6, from which it follows that, generally, the small sample results are qualitatively similar to the asymptotic ones.

- In the Gaussian noise, the best is the ψ_{Mean}-detector; Huber's and the redescenders except the ψ_{MES}-detector are slightly inferior to it; the worst are the ψ_{Med} and ψ_{MES}-detectors. In Figs 11.4 to 11.6, the signal-to-noise ratio SNR is defined as the ratio of the signal energy E and the noise variance σ_n^2 (in our study, $\sigma_n^2 = 1$ for the Gaussian and Laplace noises).

- In the Laplace noise, the best is the ψ_{Med}-detector; the proposed ψ_{MES} and radical detectors are slightly inferior to the optimal one whereas Huber's and Hampel's ψ_{25A} are next in performance; the worst is the ψ_{Mean}-detector.

- In the Cauchy noise, the radical detector dominates over the others.

From Fig. 11.6 it can be seen that the performance of the best in the Gaussian noise detectors stabilizes approximately beginning from the sample size $n = 60$.

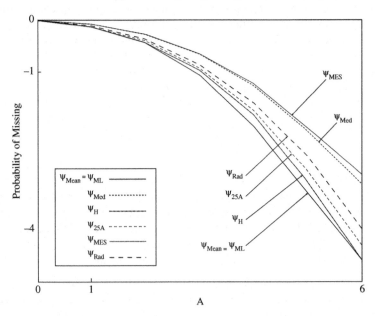

Figure 11.4   Probability of missing in the Gaussian noise: $n = 20$, $A = \sqrt{SNR}$, $\alpha = 0.1$.

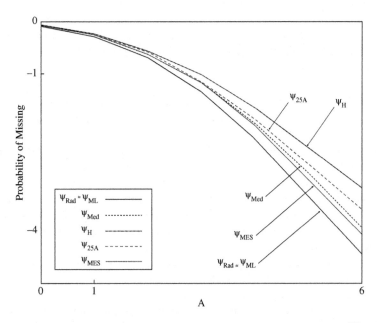

Figure 11.5   Probability of missing in the Cauchy noise: $n = 20$, $A = \sqrt{E}$, $\alpha = 0.1$ (E is the signal energy).

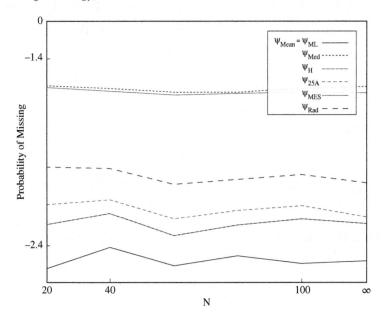

Figure 11.6   Probability of missing in the Gaussian noise: $SNR = 16$, $\alpha = 0.1$.

A similar effect is also observed in the case of the Laplace noise. On the contrary, in the Cauchy noise, the stabilization of detection performance is not observed even for the relatively large sample sizes about $n = 100$ – this can be explained as the effect of infinite noise variance.

Final remark

Our aim is to apply some new results in robust estimation to robust detection in the case of the Neyman–Pearson detection of a weak signal.

Since the considered detection problem allows for the direct use of M-estimates of location, the indicators of robustness of estimation, namely, the estimator variance sensitivity and estimate stability, are reformulated for the aims of detection.

The designed asymptotically stable detectors with the redescending minimum detection error sensitivity and radical score functions outperform conventional Huber's linear bounded detectors in heavy-tailed noises, being slightly inferior to the latter in the Gaussian noise.

Moreover, the proposed radical detector outperforms Hampel's detector based on the conventional three-part redescending score function.

Finally, we recommend to use the radical detector as the best among the other competitors under heavy-tailed noises and Huber's linear bounded detector under light-tailed noises.

11.4   A Unified Neyman–Pearson Detection of Weak Signals in a Fusion Model with Fading Channels and Non-Gaussian Noises

11.4.1   Introductory remarks

In this section we extend Huber's asymptotic minimax approach to fusion detection, an important area of real-life applications. Distributed detection and information fusion have received recent research interest due to the success of emerging wireless sensor network (WSN) technologies (Chamberland and Veeravalli 2007; Chen *et al.* 2004). For the problem of distributed detection in WSNs under energy constraints, a weak signal model in the canonical parallel fusion scheme with additive non-Gaussian noises and fading channels is considered. To apply a weak signal approach in this problem, the sample size is replaced by the number of sensor nodes. To solve this problem in the Neyman–Pearson setting, a unified asymptotic fusion rule generalizing the maximum ratio combiner (MRC) fusion rule is proposed. Explicit formulas for the threshold and detection probability applicable for wide classes of fading channels and noise distributions are written out. Both asymptotic analysis and Monte Carlo modeling are used to examine the performance of the proposed detection fusion rule. The results conveyed below are based on our recent works (Park *et al.* 2011a, 2011b, 2012).

The problem of distributed detection using multiple sensors has been extensively studied over the past few decades. The recent success of emerging wireless sensor network (WSN) technology has encouraged researchers to develop distributed algorithms in this field (Chamberland and Veeravalli 2007). Many applications in WSNs can be regarded as a two-hypotheses detection problem corresponding to the target-present or target-absent hypotheses. In practice, it is frequently observed that one hypothesis, target-absent, is more likely. The energy constraint in WSNs is typically connected with densely deployed sensor nodes having limited capabilities such as lifetime, processing power, etc., due to the cost. Thus, in order to save energy, each sensor node only transmits signals when its decision corresponds to the target-present hypothesis, and it is designed to use the minimum amount of transmitting power. A research effort to the energy efficient detection for the send/no-send transmission scenario can be found in Appadwedula *et al.* (2005).

With advances in wireless technologies, the wireless channel layer has become an important bottleneck in the design of the detection framework in WSNs (Chamberland and Veeravalli 2007). Recently, decision fusion problems have been studied based on the parallel fusion model incorporated with fading and Gaussian noise channels (Chen *et al.* 2004; Niu *et al.* 2006), where several suboptimal fusion rules are obtained by the signal-to-noise ratio (SNR) approximation method. In particular, the low SNR approximation is used in Chen *et al.* (2004), where $\lim_{\sigma^2 \to \infty} \log \Lambda$ is considered with the Gaussian noise variance σ^2 and the likelihood ratio Λ. In this section, we extend those results to non-Gaussian noise cases to consider more practical channel noises known to have impulsive characteristics (Bouvet and Schwartz 1989; Kassam 1988). Thus, our aim is to find a unified fusion rule that can be directly applicable for any noise probability density function (PDF).

In order to propose a unified decision fusion rule, we first concurrently integrate non-Gaussian noise channels into the fusion model using the asymptotic weak signal approach as in Kassam (1988) and Miller and Thomas (1972). Note that, for the Gaussian noise, both the low SNR approximation (Chen *et al.* 2004) and the weak signal approach give the maximum ratio combiner (MRC) fusion rule. In this section, by replacing the sample size with the number of sensor nodes, the weak signal is used to model each sensor output for decision of target-present. For detection of a weak signal, one necessarily uses a relatively large sample size to get a reasonable value for the probability of detection (Kassam 1988). This can be achieved by fusing decisions with the large number of participating sensor nodes. Under standard assumptions of regularity imposed on channel noise PDFs, we can thereby derive an asymptotic fusion rule of a general form that is directly applicable for wide classes of fading channels and non-Gaussian noise PDFs.

The remainder of this section is organized as follows. In subsection 11.4.2, we formulate the parallel fusion problem using the weak signal approach and propose a unified asymptotic fusion rule. In subsection 11.4.3, the asymptotic performance analysis is conducted for the Neyman–Pearson setting. In subsection 11.4.4, the asymptotic optimality of the proposed fusion rule in the case of a known channel noise PDF is numerically justified basing both on theoretical and experimental results. Finally, we conclude in subsection 11.4.5.

11.4.2   Problem setting—an asymptotic fusion rule

Figure 11.7 depicts a parallel fusion model used in Chen *et al.* (2004), where $u_k \in \{0, \theta\}$ is the kth sensor output having the false alarm and detection probabilities of $P_{fk} = P(u_k = \theta | H_0)$ and $P_{dk} = P(u_k = \theta | H_1)$, respectively. After passing through the communication channel, the received decision $y_k = h_k u_k + n_k$, where $h_k > 0$, is the attenuation of fading (assumed known) and n_k is the additive channel noise with a symmetric PDF f. A similar constant signal model can be found in Miller and Thomas (1972). Here, we assume the phase coherent reception, which can be accomplished by either training-based channel estimation for stationary channels or by employing different encoding for a fast fading with a small cost of SNR degradation, as mentioned in Chen *et al.* (2004). In this model, the sensor decision θ is a weak signal in the sense that its amplitude decreases with K as $\theta = \theta_K = v/\sqrt{K}$ with the finite constant $v > 0$. Although the weak signal model naturally arises within the analog signal sensor outputs $u_k \in \{0, \theta\}$, it is a useful mathematical tool allowing theoretical results to be obtained on the asymptotic behavior of detection procedures in scenarios with a low SNR approximation for non-weak decision signals, being, in our opinion, technically more convenient and powerful than the low SNR approximation method.

The decision fusion problem can be regarded as a two-hypothesis detection problem with individual sensor decisions; therefore, the optimal fusion rule is given by

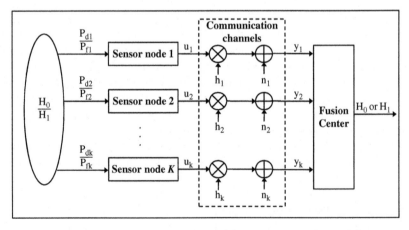

Figure 11.7   The parallel fusion model with the K sensor nodes and fusion center.
Source: *Shevlyakov (2012), Reproduced with permissions from IEEE.*

the likelihood ratio (*LR*). As shown in Chen *et al.* (2004), given the conditional independence assumption of local observations, the *LR* can be written as

$$\Lambda_K(\bar{y}) = \prod_{k=1}^{K} \frac{P_{dk}f(y_k - h_k\theta) + (1 - P_{dk})f(y_k)}{P_{fk}f(y_k - h_k\theta) + (1 - P_{fk})f(y_k)} ,$$ (11.29)

where $\bar{y} = [y_1, \ldots, y_K]^T$ is the observation vector. The fusion rule utilizing this *LR* statistic is optimal for any noise PDF f satisfying regularity conditions. From (11.29) it directly follows (Kassam 1988) that with the asymptotic weak signal approach ($\theta \to 0$ as $K \to \infty$) the locally optimum detector fusion statistic is given by

$$T_K^{ML}(\bar{y}) = \frac{1}{\sqrt{K}} \sum_{k=1}^{K} (P_{dk} - P_{fk}) h_k \psi_{ML}(y_k) ,$$ (11.30)

where $\psi_{ML}(x) = -f'(x)/f(x)$ is the maximum likelihood (*ML*) score function.

In the Gaussian noise case, the detector fusion statistic (11.30) is linear with respect to observations and equivalent to the maximum ratio combiner (*MRC*) statistic (Chen *et al.* 2004):

$$T_K^{ML} \propto \sum_{k=1}^{K} (P_{dk} - P_{fk}) h_k y_k.$$

Generalizing the *MRC* fusion rule, we consider the detector fusion statistic $T_K(\bar{y})$ based on an arbitrary score function ψ

$$T_K(\bar{y}) = \frac{1}{\sqrt{K}} \sum_{k=1}^{K} (P_{dk} - P_{fk}) h_k \psi(y_k),$$ (11.31)

which is compared with the threshold λ_α providing the required rate of the false alarm probability $P_F = \alpha$.

11.4.3   Asymptotic performance analysis

In what follows, we use the decision rule based on the detector fusion statistic (11.31) for which the related performance evaluation characteristics are computed. In the Neyman–Pearson setting, the false alarm probability is defined as

$$P_F = P\{T_K > \lambda_\alpha | H_0\} \leq \alpha.$$ (11.32)

All further results hold under general regularity conditions imposed on score functions ψ and noise PDFs f (see, for example, Hampel *et al.* 1986, pp. 125–127). To derive the asymptotic formulas for the false alarm and detection probabilities, we need to have the conditional means $E(T_K | H_i)$ and variances $var(T_K | H_i)$ ($i = 0, 1$) computed under the hypotheses H_0 and H_1. Those results are given by the following lemmas.

Lemma 11.4.1. *The conditional means of statistic T_K under hypotheses H_0 and H_1 are given by*

$$E(T_K|H_0) = v\overline{\psi'}H_{fK} \quad E(T_K|H_1) = v\overline{\psi'}H_{dK}, \tag{11.33}$$

where

$$\overline{\psi'} = \int_{-\infty}^{\infty} \psi'(x)f(x)\,dx, \quad H_{fK} = \frac{1}{K}\sum_{k=1}^{K} P_{fk}(P_{dk} - P_{fk})h_k^2 + o(1/K),$$

and

$$H_{dK} = \frac{1}{K}\sum_{k=1}^{K} P_{dk}(P_{dk} - P_{fk})h_k^2 + o(1/K).$$

Proof. Given h_k, from (11.30) it follows that

$$E(T_K|H_0) = \frac{1}{\sqrt{K}}\sum_{k=1}^{K} h_k(P_{dk} - P_{fk})E(\psi(y_k)|H_0).$$

Since $\theta = v/\sqrt{K} \to 0$ as $K \to \infty$, the following expansion holds with the remainder of order $o(1/\sqrt{K})$:

$$\psi(n_k + h_k\theta) = \psi(n_k) + h_k\theta\psi'(n_k) + o(1/\sqrt{K}).$$

Therefore, under H_0,

$$\psi(y_k) = \psi(n_k) \quad \text{with} \quad 1 - P_{fk}$$

and

$$\psi(y_k) = \psi(n_k) + h_k\theta\psi'(n_k) + o(1/\sqrt{K}) \quad \text{with} \quad P_{fk}.$$

Next,

$$E(\psi(y_k)|H_0) = \overline{\psi} + P_{fk}h_k\frac{v\overline{\psi'}}{\sqrt{K}} + o(1/\sqrt{K}),$$

where $\overline{\psi} = \int \psi(x)f(x)dx = 0$ and $\overline{\psi'} = \int \psi'(x)f(x)dx$. Hence, since n_k are i.i.d. random variables, the conditional mean takes the form

$$E(T_K|H_0) = v\overline{\psi'}H_{fK},$$

where $H_{fK} = \frac{1}{K}\sum_{k=1}^{K} P_{fk}(P_{dk} - P_{fk})h_k^2 + o(1/K)$. Similarly, under the alternative hypothesis H_1,

$$E(T_K|H_1) = v\overline{\psi'}H_{dK},$$

where $H_{dK} = \frac{1}{K}\sum_{k=1}^{K} P_{dk}(P_{dk} - P_{fk})h_k^2 + o(1/K).$

Lemma 11.4.2. *The conditional variances of statistic* T_K *under hypotheses* H_0 *and* H_1 *are given by*

$$var(T_K|H_0) = var(T_K|H_1) = \overline{\psi^2} H_K, \tag{11.34}$$

where

$$\overline{\psi^2} = \int \psi^2(x) f(x) \, dx, \quad H_K = \frac{1}{K} \sum_{k=1}^{K} (P_{dk} - P_{fk})^2 h_k^2 + o(1/K).$$

Proof. The conditional variance of statistic T_K under hypothesis H_0 is given by

$$var(T_K|H_0) = E(T_K^2|H_0) - (E(T_K|H_0))^2 .$$

Then, by using asymptotic expansions based on elementary but tedious transformations, it can be shown that

$$E(T_K^2|H_0) = \frac{1}{K} \sum_{k=1}^{K} (P_{dk} - P_{fk})^2 h_k^2 \overline{\psi^2} + v^2 (\overline{\psi'})^2 H_{fK}^2 ,$$

where $\overline{\psi^2} = \int \psi^2(x) f(x) dx$. Thus, the variance $var(T_K|H_0)$ is given by

$$var(T_K|H_0) = \overline{\psi^2} H_K , \tag{11.35}$$

where $H_K = \frac{1}{K} \sum_{k=1}^{K} (P_{dk} - P_{fk})^2 h_k^2 + o(1/K)$. By a similar procedure it can be shown that the conditional variance under hypothesis H_1 is the same as in (11.35):

$$var(T_K|H_1) = \overline{\psi^2} H_K.$$

Now assume that the attenuations h_k $(k = 1, \ldots, K)$ are random variables with a common density $p(h)$ such that the quantities H_{dK}, H_{fK}, and H_K defined by the above lemmas converge in probability to the finite values

$$H_d = \overline{h^2} \lim_{K \to \infty} \frac{1}{K} \sum_{k=1}^{K} P_{dk}(P_{dk} - P_{fk}),$$

$$H_f = \overline{h^2} \lim_{K \to \infty} \frac{1}{K} \sum_{k=1}^{K} P_{fk}(P_{dk} - P_{fk}),$$

and

$$H = \overline{h^2} \lim_{K \to \infty} \frac{1}{K} \sum_{k=1}^{K} (P_{dk} - P_{fk})^2,$$

where $\overline{h^2} = \int_{-\infty}^{\infty} h^2 p(h) \, dh < \infty$. Then the asymptotic expressions for the false alarm and detection probabilities are given by Theorem 11.4.1.

Theorem 11.4.1. *The false alarm probability has the following asymptotic form:*

$$P_F = \lim_{K \to \infty} P\{T_K > \lambda_\alpha | H_0\} = 1 - \Phi\left(\frac{\lambda_\alpha - v\overline{\psi'}H_f}{\sqrt{\overline{\psi^2}H}}\right), \qquad (11.36)$$

where $\Phi(z) = (2\pi)^{-1/2} \int_{-\infty}^{z} \exp(-t^2/2)dt$ *is the Gaussian cumulative. The threshold* λ_α *is obtained from (11.36) equating* P_F *to* α

$$\lambda_\alpha = v\overline{\psi'}H_f + \sqrt{\overline{\psi^2}H}\Phi^{-1}(1 - \alpha) . \qquad (11.37)$$

The asymptotic expression for the detection probability is given by

$$P_D = \lim_{K \to \infty} P\{T_K > \lambda_\alpha | H_1\}$$

$$= 1 - \Phi\left[\Phi^{-1}(1 - \alpha) - \frac{v\overline{\psi'}(H_d - H_f)}{\sqrt{\overline{\psi^2}H}}\right] . \qquad (11.38)$$

Proof. By the CLT, the ratio $(T_K - E(T_K))/\sqrt{var(T_K)}$ converges weakly to the standard normal random variable so that the false alarm probability can be written as

$$P_F = \lim_{K \to \infty} P\{T_K > \lambda_\alpha | H_0\}$$

$$= \lim_{K \to \infty} P\left\{\frac{T_K - E(T_K|H_0)}{\sqrt{var(T_K|H_0)}} > \frac{\lambda_\alpha - E(T_K|H_0)}{\sqrt{var(T_K|H_0}}\right\} .$$

Similarly to the aforementioned reasoning, the asymptotic expression for the detection probability is obtained by using the CLT with λ_α from (11.37), and the conditional mean and variance under hypothesis H_1 given by lemmas, respectively.

The obtained results are connected with the well-known Huber's formula for the asymptotic variance of M-estimates of location (Huber 1964; Kassam and Poor 1985): $V(\psi,f) = \overline{\psi^2}/(\overline{\psi'})^2$. Thus, the detection probability can be rewritten as

$$P_D = 1 - \Phi\left(\Phi^{-1}(1 - \alpha) - \frac{v(H_d - H_f)}{\sqrt{V(\psi,f)\,H}}\right) . \qquad (11.39)$$

In the particular case of the maximum likelihood score function $\psi = \psi_{ML} = -f'/f$, formulas (11.38) and (11.39) take the following form:

$$P_D = 1 - \Phi\left(\Phi^{-1}(1 - \alpha) - \frac{v\sqrt{I(f)}(H_d - H_f)}{\sqrt{H}}\right) , \qquad (11.40)$$

where $I(f) = \int (f'(x)/f(x))^2 f(x)\, dx$ is Fisher's information for location, hence providing the maximum value of detection probability when the channel noise PDF f is known.

11.4.4 Numerical results

In this subsection, we focus on the asymptotic optimality of the proposed fusion rule in case of a known channel noise PDF, and compare both the analytical and experimental results under the Gaussian, Laplace, and Cauchy noises. Throughout this subsection, we assume that the attenuations $h_k, k = 1, \dots, K$ have the Rayleigh distribution of unit power with the following PDF: $p_R(h) = h\, a^{-2} e^{-h^2/2a^2}$, where $a^2 = 2/(4 - \pi)$. Also we assume that all sensor nodes are identical: $P_{dk} = 0.5$ and $P_{fk} = 0.05$ for all k.

For the Gaussian noise with PDF $f_G(x) = (\sqrt{2\pi}\sigma)^{-1} \exp(-x^2/2\sigma^2)$ and with the maximum likelihood score function $\psi_{ML}(x) = x/\sigma^2$, the detection probability is given by

$$P_{DG} = 1 - \Phi(\Phi^{-1}(1 - \alpha) - (P_d - P_f)v/\sigma).$$

For the Laplace noise with PDF $f_L(x) = (2s)^{-1} \exp(-|x|/s)$ and $\psi_{ML}(x) = s^{-1}\operatorname{sign}(x)$, the detection probability is given by

$$P_{DL} = 1 - \Phi(\Phi^{-1}(1 - \alpha) - (P_d - P_f)v/s).$$

Finally, the detection probability for the Cauchy noise with PDF $f_C(x) = \gamma/(\pi(x^2 + \gamma^2))$ and $\psi_{ML}(x) = 2x/(x^2 + \gamma^2)$ is given by

$$P_{DC} = 1 - \Phi(\Phi^{-1}(1 - \alpha) - (P_d - P_f)v/(\gamma\sqrt{2})).$$

Next, we define SNR in a common way as the ratio of the network energy and the noise variance: $SNR = \mathcal{E}/\sigma^2(f)$. Therefore, we set the variances of the Gaussian and Laplace PDFs equal to unit, i.e., $\sigma^2 = 1$ and $s = 1/\sqrt{2}$. However, since the Cauchy distribution does not have moments, we use a geometric SNR (GSNR) (Gonzalez 1997): $GSNR = 2C_g A^2/S_0^2$ where $C_g = e^{C_e} \approx 1.78$ is the exponential of the Euler constant, A is the amplitude of a modulated signal, and S_0 is the geometric power given by $S_0 = S_0(X) = e^{E \log |X|}$ with a logarithmic-order random variable X. For the Cauchy density, geometric power is given by $S_0 = \gamma$. To obtain $GSNR = \mathcal{E}$, we set $A^2 = \mathcal{E}$ and $\gamma = \sqrt{2C_g}$.

Figure 11.8 shows the receiver operating characteristic (ROC) curves for the Gaussian noise case with $K = 100$; the results of the asymptotic analysis can be directly obtained using P_{DG}. Since we apply the weak signal approach, it is quite natural that the smaller the SNR, the better the match between theoretical and experimental results.

The numerical results for the Cauchy noise shown in Fig. 11.9 are similar to the Gaussian example from the viewpoint of asymptotic optimality. The ROC curves for

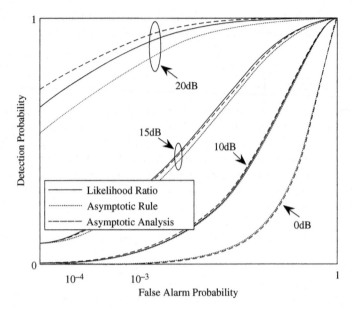

Figure 11.8   ROC curves for detection in the Gaussian noise at SNR = 0, 10, 15, and 20 dB with K=100. Source: *Shevlyakov (2012), Reproduced with permissions from IEEE.*

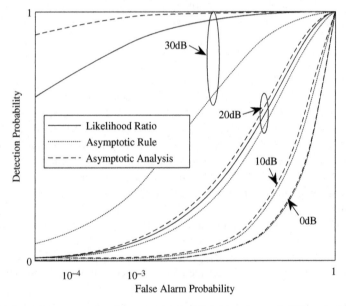

Figure 11.9   ROC curves for detection in the Cauchy noise at GSNR = 0, 10, 20, and 30 dB with K=100. Source: *Shevlyakov (2012), Reproduced with permissions from IEEE.*

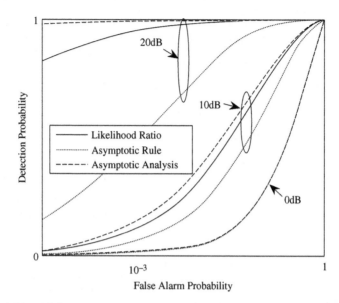

Figure 11.10   ROC curves for detection in the Laplace noise at SNR = 0, 10, and 20 dB with K=100. Source: Shevlyakov (2012), Reproduced with permissions from IEEE.

the Laplace noise are exhibited in Fig. 11.10, and the asymptotic optimality is also justified at 0 dB SNR. In this case, the discordance at 10 dB SNR is larger than in the other cases because of less accuracy of the CLT approximations for PDFs that are not bell-shaped (Zolotarev 1997).

On the whole, Monte Carlo modeling shows that analytical and experimental results match quite well, in particular, they practically coincide on samples $K > 60$.

11.4.5   Concluding remarks

The decision fusion problem in WSNs having constraints in energy efficiency is studied in this chapter. By using a weak signal model and non-Gaussian noise channels in a parallel fusion scheme, a unified asymptotic decision fusion rule based on an arbitrary score function was subsequently proposed. In the case of a known noise PDF, this rule is optimal with the maximum likelihood score function.

In the Neyman–Pearson setting, the performance characteristics of the proposed fusion rule such as the false alarm and detection probabilities together with the threshold value have been analytically obtained in the closed form. These results naturally incorporate the available information about channel noises and fading channel attenuation. The asymptotic optimality of the proposed rule is confirmed based on numerical, theoretical and experimental results.

The obtained results hold under the restrictive assumption of known attenuations in fading channels; the generalization for unknown attenuations is an open problem. The form of the obtained results allows for the robust choice of a score function in the conditions of the uncertainty of noise PDF models, prospectively based on Huber's minimax approach. However, these issues deserve a separate study.

11.5 Summary

In this chapter, first, we extend Huber's minimax approach to robustness on a number of settings in robust signal detection, namely, robust detection of a weak signal in a new class of the Weber–Hermite noise densities and in robust fusion detection; second, we apply robust stable methods of estimation of a location parameter to robust detection of a weak signal.

References

Andrews DF, Bickel PJ, Hampel FR, Huber PJ, Rogers WH, and Tukey JW 1972 *Robust Estimates of Location*, Princeton Univ. Press, Princeton.

Appadwedula S, Veeravalli VV, and Jones DL 2005 Energy-efficient detection in sensor networks. *IEEE J. Sel. Areas. Commun.* **23**, 639–702.

Bouvet M and Schwartz SC 1989 Comparison of adaptive and robust receivers for signal detection in ambient underwater noise. *IEEE Trans. Acoust. Speech, Signal Processing* **37**, 621–626.

Chamberland JF and Veeravalli VV 2007 Wireless sensors in distributed detection applications. *IEEE Signal Processing Mag.* **24**, 16–25.

Chelpanov IB and Shevlyakov GL 1983 Robust recognition algorithms based on approximation criteria. *Automation and Remote Control* **44**, 777–780.

Chen B, Jiang R, Kasetkasem T, and Varshney PK 2004 Channel aware decision fusion in wireless sensor networks. *IEEE Trans. Signal Process.* **52**, 3454–3458.

El-Sawy AH and Vandelinde DV 1977 Robust detection of known signals. *IEEE Trans. Inform. Theory* **23**, 722–727.

Gonzalez JG 1997 *Robust Techniques for Wireless Communications in Non-Gaussian Environments*. Ph.D. dissertation, Dept. Electr. Eng., Univ. Delaware, Newark.

Green PJ 1984 Weighted least squares procedures: a survey. *J. Roy. Statist. Soc.* **46**, 149–170.

Hampel FR, Ronchetti E, Rousseeuw PJ, and Stahel WA 1986 *Robust Statistics. The Approach Based on Influence Functions*, Wiley, New York.

Hossjer O and Mettiji M 1993 Robust multiple classification of known signals in additive noise—an asymptotic weak signal approach. *IEEE Trans. Inf. Theory* **39**, 594–608.

Huber PJ 1964 Robust estimation of a location parameter. *Ann. Math. Statist.* **35**, 73–101.

Huber PJ 1965 A robust version of the probability ratio test. *Ann. Math. Statist.* **36**, 1753–1758.

Huber PJ 1981 *Robust Statistics*, Wiley, New York.

Kassam SA 1988 *Signal Detection in Non-Gaussian Noise*, Springer, Berlin.

Kassam SA and Poor HV 1985 Robust techniques for signal processing: a survey. *Proc. IEEE* **73**, 433–481.

Kendall MG and Stuart A 1963 *The Advanced Theory of Statistics. Inference and Relationship*, Vol. 2, Griffin, London.

Kim K and Shevlyakov GL 2008 Why Gaussianity? *IEEE Signal Processing Magazine* **25**, 102–113.

Krishnaiah PR and Sen PK (eds) 1984 *Handbook of Statistics*, Vol. 4: *Nonparametric Methods*, Elsevier, Amsterdam.

Kuznetsov VP 1976 Robust detection of an approximately known signal. *Probl. Inform. Transmission* **12**, 47–61–(in Russian).

Lehmann EL 1959 *Testing Statistical Hypothesis*, Wiley, New York.

Miller JH and Thomas JB 1972 Detectors for discrete-time signals in non-Gaussian noise. *IEEE Trans. Inform. Theory* **18**, 241–250.

Niu R, Chen B, and Varshney PK 2006 Fusion of decisions transmitted over Rayleigh fading channels in wireless sensor networks. *IEEE Trans. Signal Process.* **54**, 1018–1027.

Noether GE 1967 *Elements of Nonparametric Statistics*, Wiley, New York.

Park J, Shevlyakov GL, and Kim K 2011a Maximin distributed detection in the presence of impulsive alpha-stable noise. *IEEE Trans. on Wireless Communications* **10**,1687–1691.

Park J, Shevlyakov GL, and Kim K 2011b Robust distributed detection with total power constraint in large wireless sensor networks. *IEEE Trans. on Wireless Communications* **10**, 2058–2062.

Park J, Shevlyakov GL, and Kim K 2012 Distributed detection and fusion of weak signals in fading channels with non-Gaussian noises. *IEEE Communications Letters* **16**, 220–223.

Shevlyakov GL 1976 *Robust Estimation and Detection of Signals Based on the Method of Least Absolute Deviations*. Ph.D. Thesis, Leningrad Polytechnic Institute (in Russian).

Shevlyakov GS and Kim K 2006 Robust minimax detection of a weak signal in noise with a bounded variance and density value at the center of symmetry. *IEEE Trans. Inform. Theory* **52**, 1206–1211.

Shevlyakov GL and Vilchevski NO 2002 *Robustness in Data Analysis: criteria and methods*, VSP, Utrecht.

Shevlyakov GL and Vilchevski NO 2011 *Robustness in Data Analysis*, De Gruyter, Boston.

Shevlyakov GL, Morgenthaler S, and Shurygin AM 2008 Redescending M-estimators. *J. Statist. Planning and Inference* **138**, 2906–2917.

Shevlyakov GL, Lee JW, Lee KM, Shin VI, and Kim K 2010 Robust detection of a weak signal with redescending M-estimators: a comparative study. *Int. J. Adapt. Control and Signal Proc.* **24**, 33–40.

Shurygin AM 1994a New approach to optimization of stable estimation. In *Proc. 1 US/Japan Conf. on Frontiers of Statist. Modeling*, Kluwer Academic Publishers, Netherlands, pp. 315–340.

Shurygin AM 1994b Variational optimization of the estimator stability. *Automation and Remote Control* **55**, 1611–1622.

Zolotarev VM 1997 *Modern Theory of Summation of Random Variables*, VSP, Utrecht.

12

Final Remarks

Although there exist many unsolved problems in statistics of univariate data, we may roughly state that, generally, there are two big parts of statistics of a principal interest and importance: multivariate statistics and time series analysis.

Finalizing the book, we briefly mark the "points of growth" in these two parts: robust multivariate statistics and robust time series analysis. In particular, we focus on these points in robust correlation analysis, as we see and/or try to foresee them. Wherever possible, we specify our vision of open problems.

12.1 Points of Growth: Open Problems in Multivariate Statistics

1. One of the most challenging problems in robust multivariate statistics as well as in multivariate statistics on the whole still remains the old problem of the curse of dimensionality. It is a well-known fact that the performance of almost all multivariate statistical procedures, including the procedures of parameter estimation, hypothesis testing, classification, and outlier detection, fastly and drastically deteriorates with increasing dimensionality of the data processed. There are many aspects of this problem–here we focus on an open problem of estimation of the critical values of dimensionality beyond that it becomes senseless to use classical methods of multivariate statistics. Here we observe the obvious demand in elaboration of novel statistical methods that can provide a reasonable rate of performance over wide ranges of data dimensionalities and data volumes–generally, similar problems arise in big data processing.

Robust Correlation: Theory and Applications, First Edition. Georgy L. Shevlyakov and Hannu Oja.
© 2016 John Wiley & Sons, Ltd. Published 2016 by John Wiley & Sons, Ltd.
Companion Website: www.wiley.com/go/Shevlyakov/Robust

2. There are huge challenges in analysis of structured multivariate or functional data. In graphical modeling, the inverse of covariance or correlation matrix is often hoped to be sparse, that is, to include a lot of zeros (meaning conditional independence). For matrix or tensor valued data, the covariance matrix is often assumed to have the corresponding Kronecker structure and that should also be taken into account in robust estimation. For functional data, robust competitors of the covariance operator are still needed.

3. The results on robust stable estimation of location for scalar data presented in Chapter 3 proved to be advatageous, outperforming the conventional Huber's and Hampel's methods in heavy-tailed distribution models. We expect that the solution of an open problem of design of robust stable methods for estimation of multivariate distribution parameters, namely, stable estimation of multivariate scale and correlation, especially of covariance and correlation matrices, may be prospective as compared to conventional robust methods of multivariate statistics.

4. The next open problem in robust multivariate statistics is the use of the independent component distribution (ICD) models as an alternative to the conventional class of multivariate elliptical distributions. Based on robust estimates of the correlation coefficient, in particular robust minimax bias and variance estimates, elaborated in Chapter 5 in the bivariate ICD model, we expect that those methods can be generalized in the multidimensional case. In this case, the primary goal is design of robust estimates of the covariance (correlation) matrices eigenvalues and eigenvectors, and the secondary–robust estimates of the covariance (correlation) matrices themselves.

5. Direct application of robust estimates of covariance (correlation) matrices to robust principal component and canonical correlation analysis is briefly discussed in Chapters 7 and 8. Detailed elaboration of these methods together with the related numerical algorithms of their implementation is an actual goal of research in multivariate statistics.

12.2   Points of Growth: Open Problems in Applications

In Chapter 10, we have only lightly touched on an important area of robust signal processing, robust estimation of a power spectrum. For parametric and nonparametric time series signal models and for two main classes of estimates of a power spectrum, the periodograms and Blackman–Tukey formula methods, we propose a number of robust estimates partially based on robust versions of the discrete Fourier transform and partially on robust estimates of the autocorrelation function. Since the obtained preliminary results are advantageous, we enlist the following open problems: (i) asymptotic analysis of the power spectrum estimate properties, (ii) reduction of the power spectrum estimate biases and variances on finite samples, (iii) examining the qualitative and quantative properties of robust versions of the Fourier transform, in particular, of their L_p-norm analogs and their inverse transforms.

Finally, we point to an open problem of robust detection of a weak multivariate signal in multivariate non-Gaussian noise, for the solution of which highly efficient and robust estimates of correlation may be helpful.

Index

Robust Correlation: Theory and Applications, First Edition. Georgy L. Shevlyakov and Hannu Oja.
© 2016 John Wiley & Sons, Ltd. Published 2016 by John Wiley & Sons, Ltd.
Companion Website: www.wiley.com/go/Shevlyakov/Robust

WILEY SERIES IN PROBABILITY AND STATISTICS
ESTABLISHED BY WALTER A. SHEWHART AND SAMUEL S. WILKS

Editors: *David J. Balding, Noel A. C. Cressie, Garrett M. Fitzmaurice, Geof H. Givens, Harvey Goldstein, Geert Molenberghs, David W. Scott, Adrian F. M. Smith, Ruey S. Tsay, Sanford Weisberg*
Editors Emeriti: *J. Stuart Hunter, Iain M. Johnstone, Joseph B. Kadane, Jozef L. Teugels*

The *Wiley Series in Probability and Statistics* is well established and authoritative. It covers many topics of current research interest in both pure and applied statistics and probability theory. Written by leading statisticians and institutions, the titles span both state-of-the-art developments in the field and classical methods.

Reflecting the wide range of current research in statistics, the series encompasses applied, methodological and theoretical statistics, ranging from applications and new techniques made possible by advances in computerized practice to rigorous treatment of theoretical approaches.

This series provides essential and invaluable reading for all statisticians, whether in academia, industry, government, or research.

† ABRAHAM and LEDOLTER · Statistical Methods for Forecasting
AGRESTI · Analysis of Ordinal Categorical Data, Second Edition
AGRESTI · An Introduction to Categorical Data Analysis, Second Edition
AGRESTI · Categorical Data Analysis, Third Edition
AGRESTI · Foundations of Linear and Generalized Linear Models
ALSTON, MENGERSEN and PETTITT (editors) · Case Studies in Bayesian Statistical Modelling and Analysis
ALTMAN, GILL, and McDONALD · Numerical Issues in Statistical Computing for the Social Scientist
AMARATUNGA and CABRERA · Exploration and Analysis of DNA Microarray and Protein Array Data
AMARATUNGA, CABRERA, and SHKEDY · Exploration and Analysis of DNA Microarray and Other High-Dimensional Data, Second Edition
ANDĚL · Mathematics of Chance
ANDERSON · An Introduction to Multivariate Statistical Analysis, Third Edition
* ANDERSON · The Statistical Analysis of Time Series
ANDERSON, AUQUIER, HAUCK, OAKES, VANDAELE, and WEISBERG · Statistical Methods for Comparative Studies
ANDERSON and LOYNES · The Teaching of Practical Statistics
ARMITAGE and DAVID (editors) · Advances in Biometry
ARNOLD, BALAKRISHNAN, and NAGARAJA · Records
* ARTHANARI and DODGE · Mathematical Programming in Statistics
AUGUSTIN, COOLEN, DE COOMAN and TROFFAES (editors) · Introduction to Imprecise Probabilities
* BAILEY · The Elements of Stochastic Processes with Applications to the Natural Sciences
BAJORSKI · Statistics for Imaging, Optics, and Photonics
BALAKRISHNAN and KOUTRAS · Runs and Scans with Applications

† Now available in a lower priced paperback edition in the Wiley Classics Library.
* Now available in a lower priced paperback edition in the Wiley-Interscience Paperback Series.

BALAKRISHNAN and NG · Precedence-Type Tests and Applications
BALI, ENGLE, and MURRAY · Empirical Asset Pricing: The Cross-Section of Stock
  Returns
BARNETT · Comparative Statistical Inference, Third Edition
BARNETT · Environmental Statistics
BARNETT and LEWIS · Outliers in Statistical Data, Third Edition
Bartholomew, Knott, and Moustaki · Latent Variable Models and Factor Analysis: A
  Unified Approach, Third Edition
BARTOSZYNSKI and NIEWIADOMSKA-BUGAJ · Probability and Statistical Inference,
  Second Edition
BASILEVSKY · Statistical Factor Analysis and Related Methods: Theory and Applications
BATES and WATTS · Nonlinear Regression Analysis and Its Applications
BECHHOFER, SANTNER, and GOLDSMAN · Design and Analysis of Experiments for
  Statistical Selection, Screening, and Multiple Comparisons
BEH and LOMBARDO · Correspondence Analysis: Theory, Practice and New Strategies
BEIRLANT, GOEGEBEUR, SEGERS, TEUGELS, and DE WAAL · Statistics of
  Extremes: Theory and Applications
BELSLEY · Conditioning Diagnostics: Collinearity and Weak Data in Regression
† BELSLEY, KUH, and WELSCH · Regression Diagnostics: Identifying Influential Data and
  Sources of Collinearity
BENDAT and PIERSOL · Random Data: Analysis and Measurement Procedures, Fourth
  Edition
BERNARDO and SMITH · Bayesian Theory
BHAT and MILLER · Elements of Applied Stochastic Processes, Third Edition
BHATTACHARYA and WAYMIRE · Stochastic Processes with Applications
BIEMER, GROVES, LYBERG, MATHIOWETZ, and SUDMAN · Measurement Errors in
  Surveys
BILLINGSLEY · Convergence of Probability Measures, Second Edition
BILLINGSLEY · Probability and Measure, Anniversary Edition
BIRKES and DODGE · Alternative Methods of Regression
Bisgaard and Kulahci · Time Series Analysis and Forecasting by Example
Biswas, Datta, Fine, and Segal · Statistical Advances in the Biomedical Sciences: Clinical
  Trials, Epidemiology, Survival Analysis, and Bioinformatics
BLISCHKE and MURTHY (editors) · Case Studies in Reliability and Maintenance
BLISCHKE and MURTHY · Reliability: Modeling, Prediction, and Optimization
BLOOMFIELD · Fourier Analysis of Time Series: An Introduction, Second Edition
BOLLEN · Structural Equations with Latent Variables
BOLLEN and CURRAN · Latent Curve Models: A Structural Equation Perspective
BONNINI, CORAIN, MAROZZI and SALMASO · Nonparametric Hypothesis Testing:
  Rank and Permutation Methods with Applications in R
BOROVKOV · Ergodicity and Stability of Stochastic Processes
BOSQ and BLANKE · Inference and Prediction in Large Dimensions
BOULEAU · Numerical Methods for Stochastic Processes
* BOX and TIAO · Bayesian Inference in Statistical Analysis
BOX · Improving Almost Anything, Revised Edition
* BOX and DRAPER · Evolutionary Operation: A Statistical Method for Process
  Improvement
BOX and DRAPER · Response Surfaces, Mixtures, and Ridge Analyses, Second Edition

---

† Now available in a lower priced paperback edition in the Wiley Classics Library.
* Now available in a lower priced paperback edition in the Wiley-Interscience Paperback Series.

BOX, HUNTER, and HUNTER · Statistics for Experimenters: Design, Innovation, and Discovery, Second Editon

BOX, JENKINS, REINSEL, and LJUNG · Time Series Analysis: Forecasting and Control, Fifth Edition

BOX, LUCEÑO, and Paniagua-QuiÑones · Statistical Control by Monitoring and Adjustment, Second Edition

* BROWN and HOLLANDER · Statistics: A Biomedical Introduction

CAIROLI and DALANG · Sequential Stochastic Optimization

CASTILLO, HADI, BALAKRISHNAN, and SARABIA · Extreme Value and Related Models with Applications in Engineering and Science

CHAN · Time Series: Applications to Finance with R and S-Plus®, Second Edition

CHARALAMBIDES · Combinatorial Methods in Discrete Distributions

CHATTERJEE and HADI · Regression Analysis by Example, Fourth Edition

CHATTERJEE and HADI · Sensitivity Analysis in Linear Regression

Chen · The Fitness of Information: Quantitative Assessments of Critical Evidence

CHERNICK · Bootstrap Methods: A Guide for Practitioners and Researchers, Second Edition

CHERNICK and FRIIS · Introductory Biostatistics for the Health Sciences

CHILÈS and DELFINER · Geostatistics: Modeling Spatial Uncertainty, Second Edition

CHIU, STOYAN, KENDALL and MECKE · Stochastic Geometry and Its Applications, Third Edition

CHOW and LIU · Design and Analysis of Clinical Trials: Concepts and Methodologies, Third Edition

CLARKE · Linear Models: The Theory and Application of Analysis of Variance

CLARKE and DISNEY · Probability and Random Processes: A First Course with Applications, Second Edition

* COCHRAN and COX · Experimental Designs, Second Edition

COLLINS and LANZA · Latent Class and Latent Transition Analysis: With Applications in the Social, Behavioral, and Health Sciences

CONGDON · Applied Bayesian Modelling, Second Edition

CONGDON · Bayesian Models for Categorical Data

CONGDON · Bayesian Statistical Modelling, Second Edition

CONOVER · Practical Nonparametric Statistics, Third Edition

COOK · Regression Graphics

COOK and WEISBERG · An Introduction to Regression Graphics

COOK and WEISBERG · Applied Regression Including Computing and Graphics

CORNELL · A Primer on Experiments with Mixtures

CORNELL · Experiments with Mixtures, Designs, Models, and the Analysis of Mixture Data, Third Edition

COX · A Handbook of Introductory Statistical Methods

CRESSIE · Statistics for Spatial Data, Revised Edition

CRESSIE and WIKLE · Statistics for Spatio-Temporal Data

CSÖRGÖ and HORVÁTH · Limit Theorems in Change Point Analysis

Dagpunar · Simulation and Monte Carlo: With Applications in Finance and MCMC

DANIEL · Applications of Statistics to Industrial Experimentation

DANIEL · Biostatistics: A Foundation for Analysis in the Health Sciences, Eighth Edition

* DANIEL · Fitting Equations to Data: Computer Analysis of Multifactor Data, Second Edition

DASU and JOHNSON · Exploratory Data Mining and Data Cleaning

* Now available in a lower priced paperback edition in the Wiley-Interscience Paperback Series.

DAVID and NAGARAJA · Order Statistics, Third Edition

DAVINO, FURNO and VISTOCCO · Quantile Regression: Theory and Applications

* DEGROOT, FIENBERG, and KADANE · Statistics and the Law

DEL CASTILLO · Statistical Process Adjustment for Quality Control

DeMaris · Regression with Social Data: Modeling Continuous and Limited Response Variables

DEMIDENKO · Mixed Models: Theory and Applications with R, Second Edition

Denison, Holmes, Mallick, and Smith · Bayesian Methods for Nonlinear Classification and Regression

DETTE and STUDDEN · The Theory of Canonical Moments with Applications in Statistics, Probability, and Analysis

DEY and MUKERJEE · Fractional Factorial Plans

DILLON and GOLDSTEIN · Multivariate Analysis: Methods and Applications

* DODGE and ROMIG · Sampling Inspection Tables, Second Edition

* DOOB · Stochastic Processes

DOWDY, WEARDEN, and CHILKO · Statistics for Research, Third Edition

DRAPER and SMITH · Applied Regression Analysis, Third Edition

DRYDEN and MARDIA · Statistical Shape Analysis

DUDEWICZ and MISHRA · Modern Mathematical Statistics

DUNN and CLARK · Basic Statistics: A Primer for the Biomedical Sciences, Fourth Edition

DUPUIS and ELLIS · A Weak Convergence Approach to the Theory of Large Deviations

EDLER and KITSOS · Recent Advances in Quantitative Methods in Cancer and Human Health Risk Assessment

* ELANDT-JOHNSON and JOHNSON · Survival Models and Data Analysis

ENDERS · Applied Econometric Time Series, Third Edition

† ETHIER and KURTZ · Markov Processes: Characterization and Convergence

EVANS, HASTINGS, and PEACOCK · Statistical Distributions, Third Edition

EVERITT, LANDAU, LEESE, and STAHL · Cluster Analysis, Fifth Edition

FEDERER and KING · Variations on Split Plot and Split Block Experiment Designs

FELLER · An Introduction to Probability Theory and Its Applications, Volume I, Third Edition, Revised; Volume II, Second Edition

FITZMAURICE, LAIRD, and WARE · Applied Longitudinal Analysis, Second Edition

* FLEISS · The Design and Analysis of Clinical Experiments

FLEISS · Statistical Methods for Rates and Proportions, Third Edition

† FLEMING and HARRINGTON · Counting Processes and Survival Analysis

FUJIKOSHI, ULYANOV, and SHIMIZU · Multivariate Statistics: High-Dimensional and Large-Sample Approximations

FULLER · Introduction to Statistical Time Series, Second Edition

† FULLER · Measurement Error Models

GALLANT · Nonlinear Statistical Models

GEISSER · Modes of Parametric Statistical Inference

GELMAN and MENG · Applied Bayesian Modeling and Causal Inference from ncomplete-Data Perspectives

GEWEKE · Contemporary Bayesian Econometrics and Statistics

GHOSH, MUKHOPADHYAY, and SEN · Sequential Estimation

GIESBRECHT and GUMPERTZ · Planning, Construction, and Statistical Analysis of Comparative Experiments

---

† Now available in a lower priced paperback edition in the Wiley Classics Library.

* Now available in a lower priced paperback edition in the Wiley-Interscience Paperback Series.

---

† Now available in a lower priced paperback edition in the Wiley Classics Library.

* Now available in a lower priced paperback edition in the Wiley-Interscience Paperback Series.

† HUBER and Ronchetti · Robust Statistics, Second Edition
HUBERTY · Applied Discriminant Analysis, Second Edition
HUBERTY and OLEJNIK · Applied MANOVA and Discriminant Analysis, Second
 Edition
HUITEMA · The Analysis of Covariance and Alternatives: Statistical Methods for
 Experiments, Quasi-Experiments, and Single-Case Studies, Second Edition
HUNT and KENNEDY · Financial Derivatives in Theory and Practice, Revised Edition
HURD and MIAMEE · Periodically Correlated Random Sequences: Spectral Theory and
 Practice
HUSKOVA, BERAN, and DUPAC · Collected Works of Jaroslav Hajek— with
 Commentary
HUZURBAZAR · Flowgraph Models for Multistate Time-to-Event Data
Jackman · Bayesian Analysis for the Social Sciences
† JACKSON · A User's Guide to Principle Components
JOHN · Statistical Methods in Engineering and Quality Assurance
JOHNSON · Multivariate Statistical Simulation
JOHNSON and BALAKRISHNAN · Advances in the Theory and Practice of Statistics: A
 Volume in Honor of Samuel Kotz
JOHNSON, KEMP, and KOTZ · Univariate Discrete Distributions, Third Edition
JOHNSON and KOTZ (editors) · Leading Personalities in Statistical Sciences: From the
 Seventeenth Century to the Present
JOHNSON, KOTZ, and BALAKRISHNAN · Continuous Univariate Distributions,
 Volume 1, Second Edition
JOHNSON, KOTZ, and BALAKRISHNAN · Continuous Univariate Distributions,
 Volume 2, Second Edition
JOHNSON, KOTZ, and BALAKRISHNAN · Discrete Multivariate Distributions
JUDGE, GRIFFITHS, HILL, LÜTKEPOHL, and LEE · The Theory and Practice of
 Econometrics, Second Edition
JUREK and MASON · Operator-Limit Distributions in Probability Theory
KADANE · Bayesian Methods and Ethics in a Clinical Trial Design
KADANE AND SCHUM · A Probabilistic Analysis of the Sacco and Vanzetti Evidence
KALBFLEISCH and PRENTICE · The Statistical Analysis of Failure Time Data, Second
 Edition
KARIYA and KURATA · Generalized Least Squares
KASS and VOS · Geometrical Foundations of Asymptotic Inference
† KAUFMAN and ROUSSEEUW · Finding Groups in Data: An Introduction to Cluster
 Analysis
KEDEM and FOKIANOS · Regression Models for Time Series Analysis
KENDALL, BARDEN, CARNE, and LE · Shape and Shape Theory
KHURI · Advanced Calculus with Applications in Statistics, Second Edition
KHURI, MATHEW, and SINHA · Statistical Tests for Mixed Linear Models
* KISH · Statistical Design for Research
KLEIBER and KOTZ · Statistical Size Distributions in Economics and Actuarial Sciences
Klemelä · Smoothing of Multivariate Data: Density Estimation and Visualization
KLUGMAN, PANJER, and WILLMOT · Loss Models: From Data to Decisions, Third
 Edition
KLUGMAN, PANJER, and WILLMOT · Loss Models: Further Topics

---

† Now available in a lower priced paperback edition in the Wiley Classics Library.
* Now available in a lower priced paperback edition in the Wiley-Interscience Paperback Series.

KLUGMAN, PANJER, and WILLMOT · Solutions Manual to Accompany Loss Models: From Data to Decisions, Third Edition

KOSKI and NOBLE · Bayesian Networks: An Introduction

KOTZ, BALAKRISHNAN, and JOHNSON · Continuous Multivariate Distributions, Volume 1, Second Edition

KOTZ and JOHNSON (editors) · Encyclopedia of Statistical Sciences: Volumes 1 to 9 with Index

KOTZ and JOHNSON (editors) · Encyclopedia of Statistical Sciences: Supplement Volume

KOTZ, READ, and BANKS (editors) · Encyclopedia of Statistical Sciences: Update Volume 1

KOTZ, READ, and BANKS (editors) · Encyclopedia of Statistical Sciences: Update Volume 2

KOWALSKI and TU · Modern Applied U-Statistics

Krishnamoorthy and Mathew · Statistical Tolerance Regions: Theory, Applications, and Computation

Kroese, Taimre, and Botev · Handbook of Monte Carlo Methods

KROONENBERG · Applied Multiway Data Analysis

KULINSKAYA, MORGENTHALER, and STAUDTE · Meta Analysis: A Guide to Calibrating and Combining Statistical Evidence

Kulkarni and Harman · An Elementary Introduction to Statistical Learning Theory

KUROWICKA and COOKE · Uncertainty Analysis with High Dimensional Dependence Modelling

KVAM and VIDAKOVIC · Nonparametric Statistics with Applications to Science and Engineering

LACHIN · Biostatistical Methods: The Assessment of Relative Risks, Second Edition

LAD · Operational Subjective Statistical Methods: A Mathematical, Philosophical, and Historical Introduction

LAMPERTI · Probability: A Survey of the Mathematical Theory, Second Edition

LAWLESS · Statistical Models and Methods for Lifetime Data, Second Edition

LAWSON · Statistical Methods in Spatial Epidemiology, Second Edition

LE · Applied Categorical Data Analysis, Second Edition

LE · Applied Survival Analysis

Lee · Structural Equation Modeling: A Bayesian Approach

LEE and WANG · Statistical Methods for Survival Data Analysis, Fourth Edition

LePAGE and BILLARD · Exploring the Limits of Bootstrap

LESSLER and KALSBEEK · Nonsampling Errors in Surveys

LEYLAND and GOLDSTEIN (editors) · Multilevel Modelling of Health Statistics

LIAO · Statistical Group Comparison

LIN · Introductory Stochastic Analysis for Finance and Insurance

LINDLEY · Understanding Uncertainty, Revised Edition

LITTLE and RUBIN · Statistical Analysis with Missing Data, Second Edition

Lloyd · The Statistical Analysis of Categorical Data

LOWEN and TEICH · Fractal-Based Point Processes

MAGNUS and NEUDECKER · Matrix Differential Calculus with Applications in Statistics and Econometrics, Revised Edition

MALLER and ZHOU · Survival Analysis with Long Term Survivors

MARCHETTE · Random Graphs for Statistical Pattern Recognition

MARDIA and JUPP · Directional Statistics

MARKOVICH · Nonparametric Analysis of Univariate Heavy-Tailed Data: Research and Practice

MARONNA, MARTIN and YOHAI · Robust Statistics: Theory and Methods

---

† Now available in a lower priced paperback edition in the Wiley Classics Library.
* Now available in a lower priced paperback edition in the Wiley-Interscience Paperback Series.

PIANTADOSI · Clinical Trials: A Methodologic Perspective, Second Edition
POURAHMADI · Foundations of Time Series Analysis and Prediction Theory
POURAHMADI · High-Dimensional Covariance Estimation
POWELL · Approximate Dynamic Programming: Solving the Curses of Dimensionality, Second Edition
POWELL and RYZHOV · Optimal Learning
PRESS · Subjective and Objective Bayesian Statistics, Second Edition
PRESS and TANUR · The Subjectivity of Scientists and the Bayesian Approach
PURI, VILAPLANA, and WERTZ · New Perspectives in Theoretical and Applied Statistics
† PUTERMAN · Markov Decision Processes: Discrete Stochastic Dynamic Programming
QIU · Image Processing and Jump Regression Analysis
* RAO · Linear Statistical Inference and Its Applications, Second Edition
RAO · Statistical Inference for Fractional Diffusion Processes
RAUSAND and HØYLAND · System Reliability Theory: Models, Statistical Methods, and Applications, Second Edition
Rayner, THAS, and BEST · Smooth Tests of Goodnes of Fit: Using R, Second Edition
RENCHER and SCHAALJE · Linear Models in Statistics, Second Edition
RENCHER and CHRISTENSEN · Methods of Multivariate Analysis, Third Edition
RENCHER · Multivariate Statistical Inference with Applications
RIGDON and BASU · Statistical Methods for the Reliability of Repairable Systems
* RIPLEY · Spatial Statistics
* RIPLEY · Stochastic Simulation
ROHATGI and SALEH · An Introduction to Probability and Statistics, Third Edition
ROLSKI, SCHMIDLI, SCHMIDT, and TEUGELS · Stochastic Processes for Insurance and Finance
ROSENBERGER and LACHIN · Randomization in Clinical Trials: Theory and Practice
ROSSI, ALLENBY, and McCULLOCH · Bayesian Statistics and Marketing
† ROUSSEEUW and LEROY · Robust Regression and Outlier Detection
Royston and Sauerbrei · Multivariate Model Building: A Pragmatic Approach to Regression Analysis Based on Fractional Polynomials for Modeling Continuous Variables
* RUBIN · Multiple Imputation for Nonresponse in Surveys
RUBINSTEIN and KROESE · Simulation and the Monte Carlo Method, Second Edition
RUBINSTEIN and MELAMED · Modern Simulation and Modeling
RUBINSTEIN, RIDDER, and VAISMAN · Fast Sequential Monte Carlo Methods for Counting and Optimization
RYAN · Modern Engineering Statistics
RYAN · Modern Experimental Design
RYAN · Modern Regression Methods, Second Edition
Ryan · Sample Size Determination and Power
RYAN · Statistical Methods for Quality Improvement, Third Edition
SALEH · Theory of Preliminary Test and Stein-Type Estimation with Applications
SALTELLI, CHAN, and SCOTT (editors) · Sensitivity Analysis
Scherer · Batch Effects and Noise in Microarray Experiments: Sources and Solutions
* SCHEFFE · The Analysis of Variance
SCHIMEK · Smoothing and Regression: Approaches, Computation, and Application
SCHOTT · Matrix Analysis for Statistics, Second Edition

† Now available in a lower priced paperback edition in the Wiley Classics Library.
* Now available in a lower priced paperback edition in the Wiley-Interscience Paperback Series.

---

† Now available in a lower priced paperback edition in the Wiley Classics Library.

* Now available in a lower priced paperback edition in the Wiley-Interscience Paperback Series.

---

† Now available in a lower priced paperback edition in the Wiley Classics Library.

* Now available in a lower priced paperback edition in the Wiley-Interscience Paperback Series.